Congruences for L-Functions

Mathematics and Its Applications

Managing Editor:

M. HAZEWINKEL

Centre for Mathematics and Computer Science, Amsterdam, The Netherlands

Volume 511

Congruences for L-Functions

by

Jerzy Urbanowicz

Institute of Mathematics,
Polish Academy of Sciences,
Warszawa, Poland

and

Kenneth S. Williams

Centre for Research in Algebra and Number Theory,
School of Mathematics and Statistics,
Carleton University,
Ottawa, Ontario, Canada

KLUWER ACADEMIC PUBLISHERS
DORDRECHT / BOSTON / LONDON

A C.I.P. Catalogue record for this book is available from the Library of Congress.

ISBN 978-90-481-5490-6

Published by Kluwer Academic Publishers,
P.O. Box 17, 3300 AA Dordrecht, The Netherlands.

Sold and distributed in North, Central and South America
by Kluwer Academic Publishers,
101 Philip Drive, Norwell, MA 02061, U.S.A.

In all other countries, sold and distributed
by Kluwer Academic Publishers,
P.O. Box 322, 3300 AH Dordrecht, The Netherlands.

Printed on acid-free paper

Contents

Preface   ix

I. SHORT CHARACTER SUMS   1
  1.   Introduction   1
  2.   Bernoulli Numbers and Polynomials   9
  3.   Generalized Bernoulli Numbers   12
  4.   Dirichlet L-functions   16
  5.   The values of $L(1, \chi)$ and $L(2, \chi)$ 18
  6.   The Dedekind Zeta Function   21
  7. K-theoretic Background   23
  8.   Quadratic Fields   27
  9.   Power Sums Involving Dirichlet Characters   30
  10.   Some Elementary Lemmas   45

II. CLASS NUMBER CONGRUENCES   51
  1.   Imaginary Quadratic Fields   51
  2.   Real Quadratic Fields   72

III. CONGRUENCES BETWEEN THE ORDERS OF K_2-GROUPS   77
  1.   Real Quadratic Fields   78
  2.   Congruences for Higher Bernoulli Numbers   96

IV. CONGRUENCES AMONG THE VALUES
    OF 2-ADIC L-FUNCTIONS   117
  1.   Notation   117
  2. p-adic L-functions   119
  3.   Coleman's Results   122
  4.   Some Auxiliary Lemmas   128
  5.   Linear Congruence Relations   160

V. APPLICATIONS OF ZAGIER'S FORMULA (I)   181
  1.   Introduction   181

2.   Dirichlet Characters with Certain Properties   184
3.   Character Sums in Terms of Bernoulli Numbers   192
4.   Applications   195
5.   Tables   198

VI. APPLICATIONS OF ZAGIER'S FORMULA (II)   203
1.   Preliminaries   203
2.   Gauss' Congruence from Dirichlet's Class Number Formula 208
3.   Character Power Sums in Terms of Bernoulli Numbers   210
4.   The Main Results   213

Bibliography   231

Author Index   247

Subject Index   249

List of symbols   253

Preface

In [Hardy and Williams, 1986] the authors exploited a very simple idea to obtain a linear congruence involving class numbers of imaginary quadratic fields modulo a certain power of 2. Their congruence provided a unified setting for many congruences proved previously by other authors using various means. The Hardy-Williams idea was as follows. Let d be the discriminant of a quadratic field. Suppose that d is odd and let $d = p_1 p_2 \cdots p_n$ be its unique decomposition into prime discriminants. Then, for any positive integer k coprime with d, the congruence

$$\prod_{i=1}^{n} \left(1 - \left(\frac{p_i}{k} \right) \right) \equiv 0 \, (\mathrm{mod}\, 2^n)$$

holds trivially as each Legendre-Jacobi-Kronecker symbol $\left(\dfrac{p_i}{k} \right)$ has the value $+1$ or -1. Expanding this product gives

$$\sum_{\substack{e|d \\ e \equiv 1 \,(\mathrm{mod}\, 4)}} (-1)^{\nu(e)} \left(\frac{e}{k} \right) \equiv 0 \, (\mathrm{mod}\, 2^n),$$

where e runs through the positive and negative divisors of d and $\nu(e)$ denotes the number of distinct prime factors of e. Summing this congruence for $0 < k < |d|/8$, $\gcd(k, d) = 1$, gives

$$\sum_{\substack{e|d \\ e \equiv 1 \,(\mathrm{mod}\, 4)}} (-1)^{\nu(e)} \sum_{\substack{0 < k < |d|/8 \\ \gcd(k,d)=1}} \left(\frac{e}{k} \right) \equiv 0 \, (\mathrm{mod}\, 2^n).$$

The term with $e = 1$ is just

$$\sum_{\substack{0 < k < |d|/8 \\ \gcd(k,d)=1}} 1,$$

ix

which by a simple arithmetic argument can be computed explicitly. The terms with $e \neq 1$ can be expressed in terms of the short character sums

$$\sum_{(r-1)|e|/8 < m < r|e|/8} \left(\frac{e}{m}\right), \quad r = 1, 2, \ldots, 8.$$

These character sums can be expressed by means of Dirichlet's class number formula as rational linear combinations of

$$h(-4e), \quad h(-8e),$$

if $e > 0$, and of

$$h(e), \quad h(8e),$$

if $e < 0$, where $h(\mathcal{D})$ denotes the class number of the quadratic field of discriminant $\mathcal{D}$. Putting all this information together yields a congruence of the type

$$\sum_{\substack{e|d, e>1 \\ e \equiv 1 \,(\mathrm{mod}\, 4)}} (c_1(d,e)h(-4e) + c_2(d,e)h(-8e))$$

$$+ \sum_{\substack{e|d, e<0 \\ e \equiv 1 \,(\mathrm{mod}\, 4)}} (c_3(d,e)h(e) + c_4(d,e)h(8e)) \equiv c(d) \,(\mathrm{mod}\, 2^{n+2})$$

for certain integers $c_1(d,e)$, $c_2(d,e)$, $c_3(d,e)$, $c_4(d,e)$, $c(d)$ which are given explicitly. Since 1986 this congruence has provided the impetus for a great deal of research which has led to extensions in many directions. For example in [Gras, 1989] the author found the analogous linear congruence relating the class numbers of both real and imaginary quadratic fields, and in [Urbanowicz, 1990b, 1991] the author found the analogous linear congruence relating the orders of K_2-groups of the integers of real quadratic fields. This book is the story of some of these extensions. At each stage the necessary background material is presented.

The title of the book was chosen for brevity and as such does not exactly indicate the contents of the book. The contents can perhaps be more accurately described as being concerned with congruences among special values of L-functions attached to quadratic characters.

The book is organized into six chapters. The first six sections and the final two sections of Chapter I contain basic definitions and facts about short character sums, Bernoulli numbers and polynomials, complex Dirichlet L-functions and the Dedekind zeta functions. More details concerning these topics can be found in the references given after the definitions. The seventh section gives the K-theory background for the quantities discussed in Chapter III. Section 8 illustrates in the case of quadratic fields the material introduced in the other sections of Chapter I. The key result of this chapter is Zagier's formula which is proved in section 9.

Chapter II contains a survey of congruences modulo powers of 2 between class numbers of quadratic fields. These congruences were obtained mostly by complex analytic methods. The first (resp. second) section of Chapter II is devoted to imaginary (resp. real) quadratic fields. The main element of Chapter II is the Hardy and Williams congruence contained in section 1.

Chapter III is devoted to congruences modulo powers of 2 between higher generalized Bernoulli numbers $B_{k,\chi}$, where k is an integer greater than 1 and χ is a quadratic character. The main result of this chapter is the complex analytic extension of the Hardy and Williams congruence proved in [Szmidt, Urbanowicz and Zagier, 1995]. This extension is the most general congruence of the Hardy and Williams type for the values of classical L-functions at nonpositive integers obtained by complex analytic methods until now.

Chapter IV presents Uehara's p-adic approach to congruences of the type discovered by Hardy-Williams and Gras. In the first three sections the basic material on p-adic L-functions and Coleman's formulae is presented. The final three sections contain recent results from [Urbanowicz and Wójcik, 1995/1996], where the authors extended the above mentioned congruences using Uehara's methods and Coleman's results.

Chapters V and VI contain some amazing applications of Zagier's formula, which was given in Chapter I. Chapter V contains recent results from [Schinzel, Urbanowicz and van Wamelen, 1999]. The first four sections of this chapter give the necessary material on Dirichlet characters to construct the tables given in the last section. These tables give all known formulae expressing class numbers as single sums of Kronecker symbols as well as all known cases of vanishing sums of Kronecker symbols.

Chapter VI presents recent results from [Fox, Urbanowicz and Williams, 1999]. In the first two sections of this chapter we make some elementary observations concerning Dirichlet's class number formula. We show in an elementary manner how Gauss' congruence $h(d) \equiv 0 \,(\bmod\, 2^{n-1})$ for imaginary quadratic fields of discriminant d can be deduced from Dirichlet's formula for $h(d)$. This is done by induction on the number n of distinct prime divisors of the fundamental discriminant d. In the final section of Chapter VI the Gauss congruence is generalized to 2-integral rational numbers $B_{k,\chi_d}/k$. We deduce from Zagier's formula, by induction on the pairs (k,n), that $(B_{k,\chi_d}/k) \equiv 0 \,(\bmod\, 2^{n-1})$.

Research of the first author was supported by K.B.N. (State Committee for Scientific Research of Poland) grant 2 P301 037 05. He would like to thank K.B.N. for financial support during the writing of this book and the Institute of Mathematics of the Polish Academy of Sciences for its encouragement. Both authors are grateful to their friends and colleagues, Professors Jerzy Browkin of Warsaw University, Jürgen Hurrelbrink of Louisiana State University, Pierre Kaplan of Université de Nancy I, Manfred Kolster of McMaster University and Andrzej Schinzel of the Institute of Mathematics of the Polish Academy of Sciences for reading the manuscript and for their valuable comments and suggestions. The authors would especially like to thank Professor Hurrelbrink for his many detailed improvements to Chapter III. We are also grateful to Dr. Jan Kowalski of the Institute of Mathematics of the Polish Academy of Sciences who provided technical assistance in the production of the final manuscript.

Note: References are given in the form [author(s), year]; in the case of multiple works by the same author(s) in the same year, we use a, b, ... after the date to indicate the order in which they are listed in the Bibliography.

JERZY URBANOWICZ
KENNETH S. WILLIAMS

July 1999

Chapter I

SHORT CHARACTER SUMS

In this chapter we give some basic definitions as well as some theorems (without proofs) which are needed in later chapters. Most of the notation used in the book is explained in this chapter. Notation related to p-adic numbers and functions, especially p-adic L-functions, is given in Chapter VI. The main result of this chapter is an identity given in [Szmidt, Urbanowicz and Zagier, 1995]. We also give some applications of this identity.

1.   INTRODUCTION

1.1   *Legendre and Jacobi Symbols*

By definition the Legendre symbol

$$\left(\frac{n}{p}\right),$$

where $p \geq 3$ is a prime number and $n \in \mathbb{Z}$, $n \not\equiv 0 \, (\mathrm{mod}\, p)$ is equal to 1, if n is a quadratic residue modulo p, and to -1 otherwise. In other words, this means that

$$\left(\frac{n}{p}\right) \equiv n^{(p-1)/2} \, (\mathrm{mod}\, p)$$

(Euler's criterion).

Let $m > 0$ be an odd integer relatively prime to n. The Jacobi symbol

$$\left(\frac{n}{m}\right)$$

is defined as the product of Legendre symbols

$$\left(\frac{n}{m}\right) = \left(\frac{n}{p_1}\right) \cdots \left(\frac{n}{p_r}\right),$$

where $m = p_1 \cdots p_r$ and each p_i is a prime.

1.2   *Quadratic Residues and Nonresidues*

In [Dirichlet, 1839] it is proved that for a prime number $p \equiv 3 \,(\mathrm{mod}\,4)$ the number of quadratic residues in the interval $(0, p/2)$ always exceeds the number of quadratic nonresidues in that interval. In other words, the Legendre symbol sum over intervals of length $p/2$ is positive:

$$\sum_{0 < n < (p/2)} \left(\frac{n}{p}\right) > 0 \,. \tag{1}$$

When $p \equiv 1 \,(\mathrm{mod}\,4)$ this sum is equal to 0.

1.3   *Class Number Formulae of Dirichlet Type*

Dirichlet deduced (1) from his class number formula for quadratic fields. Denoting by $h(d)$ the class number of the quadratic field $\mathbb{Q}(\sqrt{d})$ of discriminant d, we have

$$h(-p) = \left(2 - \left(\frac{2}{p}\right)\right)^{-1} \sum_{0 < \nu < (p/2)} \left(\frac{\nu}{p}\right),$$

where $p \,(> 3)$ is a prime with $p \equiv 3 \,(\mathrm{mod}\,4)$. Furthermore, Dirichlet showed if p is a prime with $p \equiv 1 \,(\mathrm{mod}\,4)$

$$h(-4p) = \frac{1}{2} \sum_{0 < \nu < (p/4)} \left(\frac{\nu}{p}\right).$$

1.4   *Kronecker Symbols*

Let d be the discriminant of a quadratic field, and denote by

$$\chi_d(n) = \left(\frac{d}{n}\right)$$

the associated quadratic character (Kronecker symbol). We also allow $d = 1$ for which χ_d is the trivial primitive character. These numbers d are the so-called fundamental discriminants. The set of all fundamental discriminants comprises the squarefree integers of the form $4n + 1$ and 4 times squarefree numbers not of this form. The corresponding characters give all primitive quadratic characters.

The Kronecker symbols are Dirichlet characters defined as products of the Jacobi symbols. The reader will find the basic properties of Dirichlet characters described for example in the books of Hasse [Hasse, 1964], Davenport [Davenport, 1980] or Washington [Washington, 1997]. For odd d we have

$$\left(\frac{d}{\cdot}\right) = \left(\frac{\cdot}{|d|}\right).$$

If $d = -4$ or $d = \pm 8$ then we have for a nonnegative integer n

$$\left(\frac{-4}{n}\right) = \left(\frac{-1}{n}\right), \quad \left(\frac{8}{n}\right) = \left(\frac{2}{n}\right), \quad \left(\frac{-8}{n}\right) = \left(\frac{-1}{n}\right) \cdot \left(\frac{2}{n}\right),$$

if n is odd, where the right hand side symbols are Jacobi symbols. The left hand side symbols are equal to 0 if n is even. For more details, see [Ireland and Rosen, 1990, Chapters 5 and 16], or [Borevich and Shafarevich, 1966, Chapter 3].

1.5   *Further Class Number Formulae of Dirichlet Type*

There are many known class number formulae of imaginary quadratic fields of the type given in section 1.3. Following [Hudson and Williams, 1982] we consider identities of the form

$$h(-sp) = c \sum_{(k-1)p/ls < \nu < kp/ls} \left(\frac{\nu}{p}\right), \quad c \in \mathbb{Q} \tag{2}$$

for $p (\geq 3)$ prime and $s, l, k, r, t \in \mathbb{N}$ such that $-sp$ is a fundamental discriminant and

$$p \equiv r \,(\mathrm{mod}\, t), \quad 1 \leq r \leq t, \quad t = 4 \prod_{\substack{q|ls \\ q\,\mathrm{prime}}} q,$$

$$1 \leq k \leq (ls+1)/2, \quad \gcd(r, 2sl) = 1,$$

where in the above product q runs through primes dividing ls.

Some formulae of this kind already appear in [Dirichlet, 1840], [Karpinski, 1904], [Lerch, 1905], [Holden, 1906b], [Johnson and Mitchell, 1977] and [Hudson and Williams, 1982]. In the last of these papers a computer search was designed to find all such formulae for which s and l satisfy $1 \leq s \leq 28$, $1 \leq ls \leq 30$. Within these limits the authors found 127 formulae of type (2). 84 of them were found in the earlier literature and 43 were new. We give below the tables presented there. In Table 1 the letters a, b, c, d, e, f refer to the indicated line and all the lines below which are not marked, and denote the references:

a) [Dirichlet, 1840],
b) [Lerch, 1905],
c) [Holden, 1906b],
d) [Karpinski, 1904],
e) [Hudson and Williams, 1982],
f) [Johnson and Mitchell, 1977].

TABLE 1

Dirichlet type class number formulae

l,k	c	r,t	
		$s=1$	
2,1	1/3	3,8	a)
2,1	1	7,8	
3,1	1/2	7,12	b)
3,1	1	11,12	
4,1	1	7,8	a)
4,2	1/3	3,8	
6,1	1	7,24	c)
6,1	1	11,24	
6,1	−1	19,24	
6,1	1	23,24	
6,2	1	7,24	
6,2	1/3	19,24	
6,3	−1	7,24	
6,3	1/2	11,24	
6,3	1	19,24	
10,1	1	11,40	d)
10,1	1	19,40	
12,3	−1	11,24	
12,3	1	19,24	
12,4	1	7,24	
12,4	1	11,24	
12,4	1/2	19,24	e)
14,3	−1	11,56	
14,3	−1	43,56	
14,3	−1	51,56	
14,6	1	11,56	
14,6	1	43,56	
14,6	1	51,56	d)
18,8	−1	19,24	

l,k	c	r,t	
20,6	1	11,40	e)
20,6	1	19,40	
30,10	1	19,120	
30,10	1/2	43,120	f)
30,10	1/2	67,120	f)
30,10	1	83,120	e)
30,10	1	91,120	
30,10	1	107,120	
		$s=3$	
1,1	2	1,12	b)
1,1	2	5,12	
1,2	−1	1,12	
1,2	−1	5,12	
2,1	1	1,24	
2,1	1	17,24	
2,2	−2	1,24	
2,2	−2	17,24	
2,2	2	5,24	
2,2	2	13,24	
2,3	−2	1,24	
2,3	−2	5,24	
2,3	−2	13,24	
2,3	−2	17,24	e)
4,1	2	17,24	d)
4,2	2	17,24	
4,5	−2	5,24	
4,5	−1	17,24	
4,6	2	17,24	
10,7	2	41,120	e)
10,7	2	89,120	

l,k	c	r,t	
		$s=4$	
1,1	2	1,8	a)
1,1	2	5,8	
1,2	−2	1,8	
1,2	−2	5,8	
3,1	2	13,24	d)
3,2	−2	13,24	
3,3	2	5,24	
3,3	2	13,24	
3,6	−2	13,24	
		$s=5$	
1,2	2	3,20	b)
1,2	2	7,20	
1,2	2	11,20	
1,2	2	19,20	
2,2	−4	11,40	d)
2,2	−4	19,40	
2,2	4	31,40	
2,2	4	39,40	
2,3	4	11,40	
2,3	4	19,40	
2,3	4/3	31,40	
2,3	4/3	39,40	
2,4	4	11,40	
2,4	4	19,40	
2,4	−4	31,40	
2,4	−4	39,40	
2,5	−4	31,40	
2,5	−4	39,40	
3,6	4	11,60	e)

l,k	c	r,t	l,k	c	r,t	l,k	c	r,t
3,6	4	59,60	6,10	−2	71,120	3,9	4	24,24
4,5	4	31,40	6,10	−2	119,120	$s=12$		
4,5	4	39,40	$s=7$			1,2	4	7,24
4,6	2	7,40	2,6	−2	5,56	1,2	4	11,24
4,6	2	23,40	2,6	−2	13,56	1,2	4	23,24
4,6	2	31,40	2,6	−2	29,56	1,5	4	11,24
4,6	2	39,40	2,6	−2	37,56	1,5	4	19,24
6,6	−4	11,120	2,6	−2	45,56	1,5	4	23,24
6,6	−4	31,120	2,6	−2	53,56	1,6	−4	7,24 e)
6,6	−4	59,120	1,4	−4	7,8	1,6	−4	23,24
6,6	−4	71,120	3,4	4	7,24 d)	$s=15$		
6,6	−4	79,120	3,4	4	23,24	2,12	−4	13,120 d)
6,6	−4	119,120	$s=8$			2,12	−4	37,120
6,10	−2	23,120	1,2	4	7,8 a)	2,12	−4	61,120 e)
6,10	−2	47,120	1,3	4	7,8	2,12	−4	109,120

In section 9.7 we give the class number formulae of Lerch and Mordell. Using these formulae and the elementary properties of short character sums established in [Johnson and Mitchell, 1977] the asserted formulae of Table 1 can easily be proved. This will be illustrated in section 9.7 in the case

$$h(-7p) = -2 \sum_{(5p/14)<\nu<(3p/7)} \left(\frac{\nu}{p}\right),$$

where $p \equiv 5 \pmod 8$.

Recently in [Schinzel, Urbanowicz and van Wamelen, 1999] the authors considered class number formulae of a similar type to (2) but of a more general form, namely,

$$h(-e|\mathcal{D}|) = c \sum_{q_1|\mathcal{D}|<\nu<q_2|\mathcal{D}|} \left(\frac{\mathcal{D}}{\nu}\right),$$

where q_1 and q_2 are rational numbers satisfying $0 \le q_1 < q_2$, $q_1 + q_2 \le 1$, $q_2 \le \frac{1}{2}$ if $q_1 = 0$ or $D < 0$. Here $\mathcal{D}(\neq 1)$ denotes a fundamental discriminant, which is taken to be in certain arithmetical progressions and coprime with the least common multiple of the denominators of q_1 and q_2, e is a positive integer satisfying $e|\mathcal{D}| > 4$ and c is a rational number. They found 222 such formulae and wondered if these are all the formulae of this type. When $q_2 - q_1 = 1/m$, where m is the least common multiple of the denominators of q_1 and q_2 and $|\mathcal{D}| = p$ is a prime, these formulae comprise the 127 formulae listed in Table 1 together with the new formula

$$h(-12p) = 4 \sum_{\frac{5}{60}p<\nu<\frac{6}{60}p} \left(\frac{\nu}{p}\right) = -4 \sum_{\frac{54}{60}p<\nu<\frac{55}{60}p} \left(\frac{\nu}{p}\right),$$

where $p \equiv 11$ or $59 \,(\bmod\, 60)$. The parameters for this formula are outside the range considered by Hudson and Williams [Hudson and Williams, 1982].

The authors in [Schinzel, Urbanowicz and van Wamelen, 1999] proved their 222 formulae using an identity given in [Szmidt, Urbanowicz and Zagier, 1995], which is proved in section 9.2 of Chapter 1, Theorem 2. These formulae are presented in Table 4 in Chapter 5.

1.6   *Short Character Sums*

Let χ be a Dirichlet character and let F be a continuous function on the interval $[a, b]$. This chapter is devoted to the study of the so-called short character sums:

$$\sum_{a\leq n\leq b} \chi(n)F(n). \tag{3}$$

These sums were systematically studied by many authors beginning with Gauss and Dirichlet. The sums for quadratic characters χ and $F(x) = 1$ were considered in [Lerch, 1905], [Holden, 1906a-c, 1907a-c, 1908], [Rédei, 1949/1950], [Mordell, 1963] and [Bölling, 1979]. In [Lerch, 1905] the case $F(x) = x$ was also considered and in [Berndt, 1975a,b, 1976] the case of any Dirichlet character χ and any continuous function F (see also [Berndt and Schoenfeld, 1975]). The case of $F(x) = x$ and any primitive quadratic character χ was investigated in [Urbanowicz, 1990a] and the case of $F(x) = x^m$ $(m \geq 0)$ for any Dirichlet character χ in [Szmidt, Urbanowicz and Zagier, 1995]. Moreover short character sums were examined in [Urbanowicz, 1983, 1990c] and in [Szmidt and Urbanowicz, 1994].

The sums were applied in [Hardy and Williams, 1986], [Urbanowicz, 1990b], [Szmidt and Urbanowicz, 1994] and in [Szmidt, Urbanowicz and Zagier, 1995] to obtain a new type of congruence between special values of L-functions, the so-called linear congruences proved in [Hardy and Williams, 1986].

1.7   *Short Character Sums which Vanish*

The short character sums listed in Table 1 are all equal to class numbers of imaginary quadratic fields multiplied by rational constants. We are going to deal with the cases when the short character sums are rational linear combinations of more than one class number as well as of the orders of other well known groups in algebraic number theory, the so-called K_2-groups. These identities will give us many new congruences between class numbers and the orders of K_2-groups modulo powers of 2.

Each short character sum listed in Table 1 is either positive or negative. In [Wolke, 1973], [Berndt and Chowla, 1974] and [Johnson and Mitchell, 1977] cases when these short character sums are zero are given. These results were obtained by purely elementary techniques.

Denote by $s(n, k)$ with $n = ls$ the sum on the right hand side of (2). In view of the relation

$$s(n, k) = (-1)^{(p-1)/2} s(n, n - k + 1)$$

it suffices to evaluate $s(n, k)$ only for those k satisfying $1 \le k \le [(n + 1)/2]$.

When n is odd and $p \equiv 3 \pmod 4$ it is easy to check that

$$s(n, (n + 1)/2) = 0.$$

Following [Johnson and Mitchell, 1977] in Table 2 we list cases when n is even and

$$\sum_{k=k_1}^{k_2} s(n, k) = 0,$$

where $k_2 \ge k_1$.

In [Schinzel, Urbanowicz and van Wamelen, 1999], the authors gave, in addition to the 222 class number formulae described in section 1.5, 55 formulae for which the short sum of the Kronecker symbols

$$\sum_{q_1|\mathcal{D}|<\nu<q_2|\mathcal{D}|} \left(\frac{\mathcal{D}}{\nu}\right)$$

vanishes, where the meanings of q_1, q_2 and $\mathcal{D}$ are as in section 1.5. When $q_1 - q_2 = 1/m$, where m is the least common multiple of the denominators of q_1 and q_2 and $|\mathcal{D}| = p$ is a prime, these formulae reduce to the 25 formulae given in Table 2 with $k_1 = k_2$. The 55 formulae are presented in Table 3 in Chapter 5.

In Table 2 the letters a, b, c, d, e, f refer to the indicated line and all the lines below which are not marked, and denote the references:

a) [Dirichlet, 1839],

b) [Karpinski, 1904],

c) [Holden, 1906b],

d) [Berndt, 1976],

e) [Holden, 1907c],

f) [Johnson and Mitchell, 1977].

TABLE 2

$$\text{Cases where } \sum_{k=k_1}^{k_2} s(n,k) = 0, p \equiv r \,(\text{mod}\, t)$$

n	k_1, k_2	r	t		n	k_1, k_2	r	t	
2	1,1	1	4	a)	12	4,4	23	24	
4	1,1	3	8	b)	12	4,5	13	24	e)
4	2,2	7	8	c)	12	6,6	5	24	d)
6	1,1	5	8	b)	14	2,4	3,19,27	56	f)
6	2,2	11	12		14	3,3	3,19,27	56	
6	2,3	5	8		14	6,6	3,19,27	56	
6	3,3	23	24		18	2,4	11	24	
10	1,1	3,27	40		18	8,8	11	24	
10	2,3	3,11,19,27	40		20	6,6	3,27	40	
12	1,1	5	24	d)	24	6,7	5	24	b)
12	2,2	5	24		30	4,5	11,59	120	f)
12	2,3	13	24	e)	30	7,7	17,113	120	
12	2,4	17	24		30	10,10	11,59	120	
12	3,3	7	8	d)					

1.8   *Character Analogue of the Poisson Summation Formula*

The identities listed in Tables 1 and 2 are special cases of some more complicated linear relations between short character sums (see for example [Currie and Williams, 1982] or [Berndt and Chowla, 1974]). All these relations are implied by Berndt's character analogue of the Poisson summation formula (see [Berndt, 1975a,b] or [Berndt and Schoenfeld, 1975]). This very pretty formula can undoubtedly be applied to many situations.

Let F be a continuous function on the interval $[a,b]$. The famous Poisson summation formula states

$$\sum_{a\leq n\leq b}^{*} F(n) = \int_a^b F(x)dx + 2\sum_{n=1}^{\infty}\int_a^b F(x)\cos(2\pi nx)dx ,$$

where the asterisk * on the summation sign on the left hand side indicates that if a or b is an integer, then the associated summands must be halved (see [Apostol, 1976]).

For a Dirichlet character χ modulo M, the Gauss sum attached to χ is defined as usual by

$$\tau(\chi,\zeta) = \sum_{a=1}^{M} \chi(a)\zeta^a ,$$

where ζ is a primitive Mth root of unity. If $\zeta = \exp(2\pi i/M)$ write $\tau(\chi) = \tau(\chi,\zeta)$.

THEOREM 1 ([Berndt, 1975a,b], [Berndt and Schoenfeld, 1975]) *If F is a continuous function on the interval $[a,b]$ and χ is a primitive Dirichlet character modulo M then*

$$\sideset{}{^*}\sum_{a\leq l\leq b} \chi(l)F(l)$$

$$= \delta_{\chi,\chi_1} \int_a^b F(x)dx + \frac{1}{\tau(\overline{\chi})} \sum_{\substack{n=-\infty \\ n\neq 0}}^{\infty} \overline{\chi}(n) \int_a^b F(x)\exp(2\pi inx/M)dx ,$$

where $\delta_{\chi,\chi_1} = 1$ if $\chi = \chi_1$ and is zero otherwise.

PROOF. To prove the assertion of Theorem 1 one can write $\chi(l)$ as

$$\tau(\overline{\chi})^{-1} \sum_{k=1}^{M} \overline{\chi}(k)\exp(2\pi kl/M)$$

and apply the usual Poisson summation formula to $F(x)\exp(2\pi kx/M)$. ∎

The above (unpublished) proof is due to Zagier. Our purpose in §§2–8 is to set up the notation and explain the terminology which will be used throughout the book. We review some of the standard facts on Dirichlet characters and L-series attached to them, as well as the Dedekind zeta function and Bernoulli numbers. Most of these facts are presented without proof, however, the reader will find appropriate references listed there. All of the facts presented in this introductory chapter will be used as we develop congruences between class numbers, the orders of K-groups of integers of number fields, and classical and generalized Bernoulli numbers, in subsequent chapters.

2.   BERNOULLI NUMBERS AND POLYNOMIALS

2.1   *Bernoulli Numbers and Polynomials*

The (ordinary) Bernoulli numbers B_n ($n = 0, 1, 2, \ldots$) are rational numbers defined formally (with no regard to convergence) by

$$\frac{t}{e^t - 1} = \sum_{n=0}^{\infty} B_n \frac{t^n}{n!} .$$

By definition $B_0 = 1$, $B_1 = -\frac{1}{2}$, $B_{2k+1} = 0$, if $k > 1$, and $B_2 = \frac{1}{6}$, $B_4 = -\frac{1}{30}$, $B_6 = \frac{1}{42}$, $B_8 = -\frac{1}{30}$, $B_{10} = \frac{5}{66}$, etc.

These numbers were discovered by Jakob Bernoulli (for the first time discussed in his posthumous work *Ars Conjectandi* in 1713) and play a role in many different areas of mathematics. Bernoulli was led to define the numbers B_n when seeking the sum of the nth powers of consecutive natural numbers. More precisely, for x a positive integer, we set

$$S_m(x) = \sum_{a=1}^{x-1} a^m .$$

Then for $k \geq 1$ we have

$$S_k(x) = \frac{1}{k+1} \left(B_{k+1}(x) - B_{k+1} \right) ,$$

where the $B_n(x)$ $(n = 0, 1, 2, \ldots)$ are the Bernoulli polynomials defined by the generating function

$$\frac{te^{Xt}}{e^t - 1} = \sum_{n=0}^{\infty} B_n(X) \frac{t^n}{n!} .$$

It follows easily that

$$B_n(X) = \sum_{i=0}^{n} \binom{n}{i} B_i X^{n-i} .$$

We have $B_n(0) = B_n$, and $B_0(X) = 1$, $B_1(X) = X - \frac{1}{2}$, $B_2(X) = X^2 - X + \frac{1}{6}$, etc. For a thorough treatment of Bernoulli numbers and polynomials we refer the reader to [Borevich and Shafarevich, 1966, Chapter 5], [Ireland and Rosen, 1990, Chapter 15] or [Washington, 1997]. See also [Dilcher, Skula and Slavutskiĭ, 1991, 1994].

2.2   Kummer's Congruence

Let p be a prime number and let m be an even natural number. It is well known that the Bernoulli number B_m contains p in its denominator if and only if m is divisible by $p - 1$. Moreover for these p the congruence

$$pB_m \equiv -1 \,(\bmod\, p)$$

holds. These two facts comprise the von Staudt and Clausen theorem (see [Borevich and Shafarevich, 1966, Chapter 5, §8, Theorem 4]). This theorem was discovered independently by both the authors in 1840. Furthermore, Carlitz in [Carlitz, 1953] generalized this congruence by proving that if $p^{\nu}(p-1)|m$, $\nu \geq 0$ then we have

$$pB_m \equiv p - 1 \,(\bmod\, p^{\nu+1}) .$$

It is not too difficult to prove that there are infinitely many Bernoulli numbers with the same denominators, and even more that for any even integer $n \geq 2$ there exist infinitely many integers $m \geq n$ such that $B_m - B_n \in \mathbb{Z}$.

If the even integer $m \geq 2$ is not divisible by $p - 1$ then the numbers B_m are p-integral. In fact we have more, the quotients B_m/m are p-integral and the congruence

$$\frac{B_{m+p-1}}{m+p-1} \equiv \frac{B_m}{m} \, (\mathrm{mod}\, p)$$

holds.

The above congruence is due to Kummer. It is the simplest case of a more general congruence referred to as the Kummer congruence (see [Borevich and Shafarevich, 1966, Chapter 5, §8, Theorem 5]). Let n and m be even natural numbers. If $n \equiv m \, (\mathrm{mod}\, \phi(p^e))$ (throughout the book ϕ denotes the Euler phi function) and $n \not\equiv 0 \, (\mathrm{mod}\, p - 1)$ then for $e \geq 1$ we have

$$\left(1 - p^{n-1}\right) \frac{B_n}{n} \equiv \left(1 - p^{m-1}\right) \frac{B_m}{m} \, (\mathrm{mod}\, p^e) .$$

There are several proofs of the above congruence. The most common is a proof using the Voronoï congruence (see [Ireland and Rosen, 1990, Chapter 15, §2]). The Kummer congruence has a very beautiful interpretation and a natural proof in the language of p-adic numbers (see [Koblitz, 1984] or [Washington, 1997]).

2.3   Riemann Zeta Function

The Riemann zeta function is defined as

$$\zeta(s) = \sum_{n=1}^{\infty} \frac{1}{n^s} = \prod_{p} \left(1 - p^{-s}\right)^{-1}$$

for $\mathrm{Re}(s) > 1$, where the infinite product runs over all rational primes. This function has an analytic continuation to the whole complex plane, except for a simple pole at $s = 1$ with residue 1. Moreover putting

$$\Phi(s) = \pi^{-s/2} \Gamma\left(\frac{s}{2}\right) \zeta(s)$$

we have Riemann's functional equation for $\zeta(s)$

$$\Phi(s) = \Phi(1 - s)$$

(for more details, see [Apostol, 1976]). Here Γ denotes the Gamma function defined by

$$\Gamma(s) = \int\limits_0^{+\infty} e^{-t} t^{s-1} dt$$

for $\mathrm{Re}(s) > 0$. The Gamma function admits a meromorphic continuation to the complex plane with only (simple) poles at $s = 0, -1, -2, \ldots$. It has a functional equation, namely

$$\Gamma(s+1) = s\Gamma(s),$$

and it is nowhere zero.

2.4   *Euler's Formula for B_{2k}*

The (ordinary) Bernoulli numbers are related to values of the Riemann zeta function at even integers. In 1736 Euler found the following famous formula:

$$B_{2k} = (-1)^{k-1} \frac{2(2k)!}{(2\pi)^{2k}} \zeta(2k)$$

(see [Euler, 1740, 1924] or [Borevich and Shafarevich, 1966, Chapter 5, §8], [Ireland and Rosen, 1990, Chapter 15, §1]). Euler's formula shows that the sign of the Bernoulli number B_{2k} is $(-1)^{k-1}$ and implies the inequality

$$|B_{2k}| > 2\left(\frac{k}{\pi e}\right)^{2k}.$$

Hence we see that $|B_{2k}|/(2k)$ (and even $\log |B_{2k}|/(2k)$) tends to ∞ as k tends to ∞.

2.5   *Values of the Riemann Zeta Function at Negative Integers*

By means of the functional equation for the Riemann zeta function we obtain from the Euler formula

$$\zeta(1 - m) = -\frac{B_m}{m},$$

which is valid for $m > 1$. If $m = 1$ we have $\zeta(1 - m) = \frac{B_m}{m} = -\frac{1}{2}$.

3.   GENERALIZED BERNOULLI NUMBERS

3.1   *Generalized Bernoulli Numbers*

Following [Leopoldt, 1958] (see also [Iwasawa, 1972, §2], or [Washington, 1997, Chapter 4]) we make the following definition. Let M be a natural number. For a Dirichlet character χ modulo M, the generalized Bernoulli numbers $B_{k,\chi}$ are defined by the formal identity

$$\sum_{a=1}^{M} \frac{\chi(a) t e^{at}}{e^{Mt} - 1} = \sum_{n=0}^{\infty} B_{n,\chi} \frac{t^n}{n!}.$$

Note that if $\chi = \chi_1$ is the trivial primitive character then we have $B_{n,\chi} = B_n$, except for $n = 1$, when $B_{1,\chi_1} = \frac{1}{2} = -B_1$. The generalized Bernoulli numbers belonging to quadratic Dirichlet characters appeared in [Hurwitz, 1882], [Berger, 1890/1891] and in [Ankeny, Artin and Chowla, 1952]. Leopoldt in [Leopoldt, 1958] used them extensively. By definition, the generalized Bernoulli numbers belong to the field $\mathbb{Q}(\chi(1), \chi(2), \ldots)$. For example, for quadratic characters they are rational. If $\chi = \chi_{-4}$ and $n \geq 1$ then we have $B_{n,\chi} = -\frac{1}{2}nE_{n-1}$, where the E_n $(n = 0, 1, 2, \ldots)$ are the usual Euler numbers defined formally by

$$\frac{1}{\cosh t} = \sum_{n=0}^{\infty} E_n \frac{t^n}{n!},$$

where

$$\cosh t = \frac{e^{-t} + e^t}{2}$$

denotes the hyperbolic cosine. We have $E_{2k-1} = 0$, if $k \geq 1$, $E_0 = 1$, $E_2 = -1$, $E_4 = 5$, $E_6 = -61$, $E_8 = 1385$, $E_{10} = -50521$, etc. If $\chi = \chi_{-3}$ and $n \geq 1$ then we have $B_{n,\chi} = -\frac{1}{3}nD_{n-1}$, where the D_n $(n = 0, 1, 2, \ldots)$ are the so-called D-numbers introduced in [Kleboth, 1955] and investigated in [Ernvall, 1979]. These numbers are defined formally by

$$\frac{3}{2\cosh t + 1} = \sum_{n=0}^{\infty} D_n \frac{t^n}{n!},$$

and we have $D_{2k-1} = 0$ if $k \geq 1$, $D_0 = 1$, $D_2 = 1$, $D_4 = -\frac{2}{3}$, $D_6 = 2$, $D_8 = -14$, $D_{10} = \frac{1618}{9}$, etc.

By definition the number $B_{0,\chi}$ equals $\phi(M)/M$ if χ is the principal character and 0 otherwise. If $m \geq 1$, then $B_{m,\chi} = 0$ if $\chi(-1) = (-1)^{m-1}$ (unless $M = m = 1$). In fact, for $m > 1$, $B_{m,\chi} = 0$ if and only if $\chi(-1) = (-1)^{m-1}$.

3.2   *Generalized Bernoulli Polynomials*

By definition, for $n \geq 0$ and N a positive integral multiple of M we have

$$B_{n,\chi} = N^{n-1} \sum_{a=1}^{N} \chi(a) B_n \left(\frac{a}{N}\right), \qquad (4)$$

see [Washington, 1997, Proposition 4.1]. For $m \geq 1$ write

$$S_m(N, \chi) = \sum_{u=0}^{N-1} \chi(a) a^m.$$

We have the following formula

$$S_m(N, \chi) = \frac{1}{m+1} \left(B_{m+1,\chi}(N) - B_{m+1,\chi}\right), \qquad (5)$$

where the generalized Bernoulli polynomials $B_{m,\chi}(X)$ are defined by

$$\sum_{a=1}^{M} \frac{\chi(a)te^{(a+X)t}}{e^{Mt}-1} = \sum_{n=0}^{\infty} B_{n,\chi}(X)\frac{t^n}{n!}$$

so that

$$B_{n,\chi}(X) = \sum_{i=0}^{n} \binom{n}{i} B_{i,\chi} X^{n-i}.$$

We have

$$B_{n,\chi}(-X) = (-1)^n \chi(-1)B_{n,\chi}(X) + n(-1)^n X^{n-1}\delta_{\chi,\chi_1}. \tag{6}$$

Let χ be a Dirichlet character modulo $M > 1$ and let $m \geq 1$ be a natural number. It follows easily from (4) that

$$\frac{B_{m,\chi}}{m} = \frac{S_m(M,\chi)}{mM} + \sum_{r=1}^{m} \binom{m-1}{r-1} \frac{B_r}{r} M^{r-1} S_{m-r}(M,\chi). \tag{7}$$

3.3   *Frobenius Polynomials*

We are going to study later in the book the values of L-functions attached to quadratic characters χ_d, where d is the discriminant of a quadratic field. These values at nonpositive integers $1 - m$ are the numbers $B_{m,\chi_d}/m$ which are integers unless $d = -4$ or $d = \pm p$, where p is an odd prime number such that $2m/(p-1)$ is an odd integer, in which cases they have denominators 2 or p respectively. For these primes p, we have

$$pB_{m,\chi_d} \equiv p - 1 \pmod{p^{\nu+1}}, \tag{8}$$

where $p^{\nu} \parallel m$. If $d = -4$ and m is odd, the number $2B_{m,\chi_{-4}}$ is an odd integer. Carlitz ([Carlitz, 1959]) used formula (7) and the so-called Frobenius polyniomials $R_n(u) \in \mathbb{Z}[u]$ $(n \geq 0)$ defined in [Frobenius, 1910] by

$$\frac{1-u}{e^t - u} = \sum_{n=0}^{\infty} \frac{R_n(u)}{(1-u)^n} \frac{t^n}{n!}$$

to prove the formula

$$\frac{B_{m,\chi}}{m} = \frac{\tau(\chi,\zeta_M)}{M} \sum_{a=1}^{M-1} \bar{\chi}(a)\zeta_M^a \frac{R_{m-1}(\zeta_M^a)}{(1-\zeta_M^a)^m},$$

where $M > 1$, $m \geq 1$, ζ_M is a primitive Mth root of unity in $\mathbb{C}$ and χ is a primitive Dirichlet character modulo M. From this formula, the congruence (8) and other congruences follow easily. For more details, see [Leopoldt, 1958] or [Carlitz, 1959].

Washington (see [Washington, 1997, Lemma 5.20]) gave a more explicit formula for the quotient $B_{m,\chi}/m$, namely,

$$\frac{B_{m,\chi}}{m} = -\frac{\tau(\chi, \zeta_M)}{M} \sum_{a=1}^{M-1} \sum_{i=1}^{m} \frac{\overline{\chi}(a)}{i(\zeta_M^a - 1)^i} \sum_{j=1}^{i} \binom{i}{j} (-1)^{i-j} j^m,$$

but we will not use this.

3.4 Note on Kummer Congruences for Generalized Bernoulli Numbers

Ernvall ([Ernvall, 1983]) generalized Kummer's congruence cited in section 2.2 by proving its analogue for generalized Bernoulli numbers. Let p be a prime number and let χ be a primitive Dirichlet character with conductor not being a power of p. Suppose that m and n are natural numbers such that $m \equiv n$ (mod $\phi(p^e)$), where $e \geq 1$, $p^e > 2$. Then the congruence

$$\left(1 - \chi(p)p^{n-1}\right) \frac{B_{n,\chi}}{n} \equiv \left(1 - \chi(p)p^{m-1}\right) \frac{B_{m,\chi}}{m} \ (\bmod \, p^e)$$

holds in the sense of section 1.2, Chapter IV of this book. For quadratic characters χ the congruence is the usual congruence for p-integral rational numbers. For any Dirichlet character χ the congruence is a special case of a more general congruence between generalized Bernoulli numbers belonging to the same character also known as Kummer's congruence. For details see [Ernvall, 1979, Theorem 1.7, p. 17]. For another generalization in the sense of section 1.2, Chapter IV see also [Washington, 1997, Exercise 7.5, p. 141].

3.5 Lower Bound for $|B_{m,\chi}|$

It is known that

$$\frac{|B_{m,\chi}|}{m} \to \infty \text{ as } m \to \infty$$

for m satisfying $m \equiv \delta \, (\bmod \, 2)$, where throughout the book $\delta \in \{0, 1\}$ is defined by $\chi(-1) = (-1)^\delta$. In fact for these m we have the inequality

$$|B_{m,\chi}| > \frac{2c}{\sqrt{e}} \left(\frac{mM}{2\pi e}\right)^{m-\frac{1}{2}},$$

where $c = \exp\left(-\pi^2/6\right)$ (see [Ernvall, 1979, Theorem 1.3] or [Ernvall, 1983]).

4.   DIRICHLET L-FUNCTIONS

4.1   *Dirichlet L-functions*

In order to express the short character sums attached to a Dirichlet character χ in terms of class numbers or other special values of L-functions, we will make use of the famous Dirichlet class number formula.

For a Dirichlet character χ modulo M, the L-function $L(s, \chi)$ attached to χ is defined by

$$L(s, \chi) = \sum_{n=1}^{\infty} \frac{\chi(n)}{n^s} = \prod_p \left(1 - \chi(p)p^{-s}\right)^{-1}$$

for $\mathrm{Re}(s) > 1$. The above infinite product is taken over all primes and is called an Euler product. Its factors are called Euler factors. If χ is the trivial primitive character then $L(s, \chi) = \zeta(s)$. For more details see [Lang, S., 1986, Chapter 8, §2], [Ireland and Rosen, 1990, Chapter 16], [Lang, S., 1990, Chapter 5], [Narkiewicz, 1990, Chapter 7, §1], or for a short discussion [Washington, 1997, Chapter 4]. For any Dirichlet character, $L(s, \chi)$ can be extended to a meromorphic function in the whole complex plane being holomorphic, except for a simple pole at $s = 1$ if χ is trivial. Moreover putting

$$\Psi(s, \chi) = \left(\frac{M}{\pi}\right)^{s/2} \Gamma\left(\frac{s + \delta}{2}\right) L(s, \chi)$$

we have

$$\Psi(s, \chi) = \frac{\tau(\chi)}{i^\delta \sqrt{M}} \Psi(1 - s, \overline{\chi}). \tag{9}$$

4.2   *Dirichlet L-functions for Nonprimitive Characters*

If the character χ modulo M is not primitive but induced from a character ψ modulo some divisor of M, then we have the evident identity

$$L(s, \chi) = L(s, \psi) \prod_{\substack{p \mid M \\ p \text{ prime}}} \left(1 - \psi(p)p^{-s}\right). \tag{10}$$

4.3   *Values of the Dirichlet L-function at Nonpositive Integers*

From the Euler product expansion we see that $L(s, \chi) \neq 0$, if $\mathrm{Re}(s) > 1$. Consequently, by the functional equation for primitive Dirichlet characters χ we obtain

$$L(1 - m, \chi) \neq 0$$

if $m \geq 1$ and $m \equiv \delta \, (\mathrm{mod} \, 2)$, and

$$L(1 - m, \chi) = 0$$

if $m \not\equiv \delta \, (\mathrm{mod} \, 2)$ unless $\chi = \chi_1$ and $m = 1$, in which case $L(0, \chi) = \zeta(0) = -\frac{1}{2}$. All the above zeros $1 - m$ are simple. This will be seen at once if we rewrite (9) in the form

$$\Gamma(s) \cos\left(\frac{\pi(s - \delta)}{2}\right) L(s, \chi) = \frac{\tau(\chi)}{2i^\delta} \left(\frac{2\pi}{M}\right)^s L\left(1 - s, \overline{\chi}\right).$$

4.4 Gauss Sums

The Gauss sum $\tau(\chi)$ attached to a Dirichlet character χ modulo M was defined in section 1.8. If χ is a primitive Dirichlet character then it is easy to check that $\tau(\overline{\chi})\tau(\chi) = \chi(-1)M$ and $|\tau(\chi)| = \sqrt{M}$ (see [Washington, 1997, Lemmas 4.7 and 4.8]). It is much more difficult to prove that for an abelian field F with discriminant d we have

$$\prod_{\chi \in X} \tau(\chi) = i^{r_2}\sqrt{|d|}, \tag{11}$$

where X denotes the group of primitive characters of F and r_2 denotes the number of nonconjugate complex embeddings of F into $\mathbb{C}$ (see for example [Lang, S., 1990, Theorem 3.1 in Chapter 3]). By the Kronecker-Weber theorem (see for example [Narkiewicz, 1990, Theorem 6.5], [Washington, 1997, Chapter 14]), every finite abelian field F is a subfield of a cyclotomic field. Let n denote the smallest natural number for which $F \subseteq \mathbb{Q}(\zeta_n)$. Then the Galois group $G := \mathrm{Gal}(F/\mathbb{Q})$ may be regarded as a quotient of $(\mathbb{Z}/n\mathbb{Z})^*$ and X, and consists of primitive characters of the group $(\mathbb{Z}/n\mathbb{Z})^*$ vanishing on the kernel of the natural homomorphism $(\mathbb{Z}/n\mathbb{Z})^* \to G$.

For quadratic characters χ_d formula (11) gives Gauss' famous result

$$\tau(\chi_d) = \begin{cases} \sqrt{d}, & \text{if } d > 0, \\ i\sqrt{|d|}, & \text{if } d < 0. \end{cases}$$

For a natural number M denote by ζ_M a primitive complex Mth root of unity. Let χ, χ_1 and χ_2 be primitive Dirichlet characters modulo M, M_1 and M_2 respectively. It is well known that

$$\tau(\chi, \zeta_M^a) = \chi^{-1}(a)\tau(\chi, \zeta_M)$$

if $a \in \mathbb{Z}$, and

$$\tau(\chi, \zeta_{M_1}\zeta_{M_2}) = \tau(\chi_1, \zeta_{M_1})\tau(\chi_2, \zeta_{M_2})$$

if $\chi = \chi_1\chi_2$, $M = M_1M_2$, $\gcd(M_1, M_2) = 1$. Moreover, we have

$$\tau(\chi) = \chi_1(M_2)\chi_2(M_1)\tau(\chi_1)\tau(\chi_2).$$

4.5   Connection between Dirichlet L-series and Generalized Bernoulli Numbers

One of the most important properties of the generalized Bernoulli numbers is that they give the values of Dirichlet L-functions at nonpositive integers. Namely, we have

$$L(1 - m, \chi) = -\frac{B_{m,\chi}}{m},$$

where $m \geq 1$ (see [Washington, 1997, Theorem 4.2]). By the functional equation for L-series we can rewrite the above equation in the form

$$B_{m,\chi} = (-1)^{(m-\delta+2)/2} \frac{2i^{\delta} m! M^m}{(2\pi)^m \tau(\overline{\chi})} L(m, \overline{\chi}). \tag{12}$$

(Hence we see that the sign of the number $B_{m,\chi}$, for a quadratic character χ is $(-1)^{(m-\delta+2)/2}$.) Moreover, we can express the values $L(k, \chi)$ for $k \leq 0$ in terms of Frobenius polynomials (see section 3.3)

$$L(k, \chi) = -\frac{\tau(\chi, \zeta_M)}{M} \sum_{a=1}^{M-1} \overline{\chi}(a) \zeta_M^a \frac{R_{-k}(\zeta_M^a)}{(1 - \zeta_M^a)^{1-k}}.$$

5.   THE VALUES OF $L(1,\chi)$ AND $L(2,\chi)$

5.1   Formulae for $L(1, \chi)$

Let ζ_M be a primitive Mth root of unity in $\mathbb{C}$. Applying the well known formula

$$\sum_{r=1}^{M} \zeta_M^{rt} = \begin{cases} M, & \text{if } t \equiv 0 \,(\text{mod } M), \\ 0, & \text{otherwise} \end{cases}$$

one obtains easily that

$$L(s, \chi) = \frac{\tau(\chi, \zeta_M)}{M} \sum_{a=1}^{M} \overline{\chi}(a) \sum_{n=1}^{\infty} \frac{\zeta_M^{-na}}{n^s} \tag{13}$$

for $\text{Re}(s) > 1$ (see [Lang, S., 1990, Chapter 3, §2, Theorem 2.1]). Thus, using the well known expansion

$$-\log(1 - z) = \sum_{n=1}^{\infty} \frac{z^n}{n}$$

(which is valid for $|z| \leq 1$, $z \neq 1$), we obtain for nontrivial primitive Dirichlet characters χ

$$L(1, \chi) = -\frac{\tau(\chi, \zeta_M)}{M} \sum_{a=1}^{M} \overline{\chi}(a) \log(1 - \zeta_M^{-a}). \tag{14}$$

The above formula is a special case, for $k = 1$, of a more general formula expressing $L(k, \chi)$ in terms of complex multilogarithms. See the second formula of section 3.1, Chapter IV.

Formula (14) can be transformed as follows. If χ is even, that is $\chi(-1) = \chi(1)$, we have

$$L(1, \chi) = -\frac{\tau(\chi, \zeta_M)}{M} \sum_{a=1}^{M} \overline{\chi}(a) \log|1 - \zeta_M^a|,$$

or

$$L(1, \chi) = -\frac{\tau(\chi)}{M} \sum_{a=1}^{M} \overline{\chi}(a) \log \sin\left(\frac{\pi a}{M}\right).$$

If χ is odd, that is $\chi(-1) = -\chi(1)$, then

$$L(1, \chi) = \frac{\pi i \tau(\chi)}{M} \sum_{a=1}^{M} \overline{\chi}(a) \left(\frac{a}{M} - \frac{1}{2}\right).$$

Thus, using generalized Bernoulli numbers, we obtain by (4)

$$L(1, \chi) = \frac{\pi i \tau(\chi)}{M} B_{1,\overline{\chi}}.$$

For details see [Lang, S., 1990, Theorem 2.2 and p. 74], [Washington, 1997, Theorem 4.9], or [Borevich and Shafarevich, 1966, §2].

5.2   *Formula for $L(2, \chi)$ for Even χ*

If χ is even and $m = 2$ then by (12) we have the formula:

$$L(2, \chi) = \frac{\pi^2 \tau(\chi)}{M^2} B_{2,\overline{\chi}}.$$

5.3   *Dilogarithms*

For $|z| \leq 1$, we define Euler's dilogarithm $\mathrm{Li}_2(z)$ by the formula:

$$\mathrm{Li}_2(z) = \sum_{n=1}^{\infty} \frac{z^n}{n^2}.$$

Next we fix branches of the multivalued functions $\arg z$ and $\log z$. For $z \in \mathbb{C} \setminus (-\infty, 0)$, $z = x + iy$, we define as usual

$$\arg z = \mathrm{sgn}\, y \cdot \arccos \frac{x}{|z|}$$

and

$$\log z = \log|z| + i \arg z.$$

We can extend the function $\mathrm{Li}_2(z)$ to $\mathbb{C} \setminus (-\infty, 0)$ by the integral

$$\mathrm{Li}_2(z) = \frac{\pi^2}{6} - \int_1^z \frac{\log(1-t)}{t} dt,$$

$z \in \mathbb{C} \setminus (-\infty, 0)$.

Rogers' dilogarithm $\mathrm{L}(z)$ is defined by the formula:

$$\mathrm{L}(z) = \mathrm{Li}_2(z) + \frac{1}{2} \log z \cdot \log(1-z)$$

for $z \in \mathbb{C} \setminus ((-\infty, 0) \cup (1, \infty))$.

The Clausen function $\mathrm{Cl}_2(t)$ is defined for $t \in \mathbb{R}$ as the imaginary part of $\mathrm{Li}_2(\exp(it))$:

$$\mathrm{Cl}_2(t) = \mathrm{Im}(\mathrm{Li}_2(\exp(it))) = \sum_{n=1}^{\infty} \frac{\sin nt}{n^2}.$$

The dilogarithm of Wigner and Bloch $D(z)$ is defined for $z \in \mathbb{C} \setminus \mathbb{R}$ by the formula:

$$D(z) = -\mathrm{Im} \int_1^z \frac{\log(1-t)}{t} dt + \arg(1-z) \cdot \log|z|$$

and $D(z) = 0$ for $z \in \mathbb{R} \cup \{\infty\}$. The dilogarithm $D : \mathbb{C} \cup \{\infty\} \to \mathbb{R}$ is a real-analytic function on $\mathbb{C} \setminus \{0, 1\}$, continuous on $\mathbb{C} \cup \{\infty\}$. For $z \in \mathbb{C} \setminus \mathbb{R}$, we have

$$D(z) = \frac{1}{2} \left(\mathrm{Cl}_2(2\theta) + \mathrm{Cl}_2(2\omega) - \mathrm{Cl}_2(2\theta + 2\omega) \right),$$

where $\theta = \arg z$, $\omega = \arg(1 - \bar{z})$. In particular, for $t \in \mathbb{R}$ we observe that

$$D(\exp(it)) = \mathrm{Cl}_2(t).$$

For a fuller treatment we refer the reader to [Browkin, 1991].

5.4   *Formula for $L(2, \chi)$ for Odd χ*

If χ is odd then (13) implies

$$L(s, \chi) = -\frac{i\tau(\chi)}{M} \sum_{a=1}^{M} \bar{\chi}(a) \sum_{n=1}^{\infty} \frac{\sin(2\pi na/M)}{n^s}.$$

Hence we have

$$L(2, \chi) = -\frac{i\tau(\chi)}{M} \sum_{a=1}^{M} \bar{\chi}(a) \mathrm{Cl}_2(2\pi a/M)$$

and

$$L(2,\chi) = -\frac{i\tau(\chi,\zeta_M)}{M}\sum_{a=1}^{M}\overline{\chi}(a)D(\zeta_M^a)\,.$$

6.   THE DEDEKIND ZETA FUNCTION

6.1   *Dedekind Zeta Function*

Let O_F be the ring of integers of a number field F. The Dedekind zeta function is defined as

$$\zeta_F(s) := \sum_{\mathfrak{a}}(N\mathfrak{a})^{-s} = \prod_{\mathfrak{p}}\left(1-N(\mathfrak{p})^{-s}\right)^{-1}$$

for $\mathrm{Re}(s) > 1$, where the sum runs over all nonzero ideals $\mathfrak{a}$ of O_F and the infinite product over all nonzero prime ideals $\mathfrak{p}$ of O_F. Here $N(\mathfrak{a}) :=$ card$(O_F/\mathfrak{a})$. If $F = \mathbb{Q}$ then $\zeta_F(s) = \zeta(s)$. For more details see [Borevich and Shafarevich, 1966, Chapter 5], [Narkiewicz, 1990, Chapter 7].

6.2   *Dirichlet Class Number Formula*

It is well known that $\zeta_F(s)$ can be analytically continued to the whole complex plane, except for a pole at $s = 1$ and we have the classical regulator formula for units:

$$\lim_{s\to 1+0}(s-1)\zeta_F(s) = \frac{2^{r_1+r_2}\pi^{r_2}Rh}{w\sqrt{|d|}}, \tag{15}$$

where $h := h(F)$ denotes the class number of F, $R := R(F)$ stands for the Dirichlet regulator of F and $w := w(F)$ is the number of roots of unity in F. Appropriate references are [Borevich and Shafarevich, 1966, Chapter 15], [Lang, S., 1986, Chapter 8, §2 and Chapter 13], [Narkiewicz, 1990, Chapter 7], or [Washington, 1997, Chapter 4]. Here, as usual, r_1 (resp. r_2) denotes the number of real (resp. nonconjugate complex) embeddings of F into $\mathbb{C}$, so that $[F : \mathbb{Q}] = r_1 + 2r_2$. If $F/\mathbb{Q}$ is a Galois extension then either $r_2 = 0$ or $r_1 = 0$. In the former case all characters of F are even, and in the latter one half of them (r_2) are even and half (also r_2) are odd. Generally, $r_1 + r_2$ of the characters are even and r_2 of them are odd.

6.3   *Functional Equation*

Setting

$$\Phi(s) = C_F^s\Gamma\left(\frac{s}{2}\right)^{r_1}\Gamma(s)^{r_2}\zeta_F(s)\,,$$

where

$$C_F = 2^{-r_2}\pi^{-[F:\mathbb{Q}]/2}\sqrt{|d|}\,,$$

a famous theorem of Hecke asserts that

$$\Phi_F(s) = \Phi_F(1 - s).$$

For details, see e.g. [Narkiewicz, 1990, Chapter 7].

6.4   *Dirichlet Class Number Formula in the Abelian Case*

If F is an abelian number field then we have

$$\zeta_F(s) = \prod_{\chi \in X} L(s, \chi),$$

where $X := X(F)$ denotes the group of characters of F (for details see [Narkiewicz, 1990, Theorem 8.2], [Washington, 1997, Theorem 4.3], or [Lang, S., 1990, p. 75]). Therefore by (15) we obtain

$$\frac{2^{r_1 + r_2} \pi^{r_2} Rh}{w \sqrt{|d|}} = \prod_{\substack{\chi \in X \\ \chi \neq 1}} L(1, \chi).$$

6.5   *Multiplicities of the Zeros of the Dedekind Zeta Function*

By the functional equation for $\zeta_F(s)$, the multiplicities of the zeros of the Dedekind zeta function at $s = 1 - m$, $m > 1$ are easily seen to be

$$\operatorname{ord}_{s=1-m}(\zeta_F(s)) = \begin{cases} r_1 + r_2, & \text{if } m \text{ is odd,} \\ r_2, & \text{if } m \text{ is even.} \end{cases}$$

In the abelian case the above also follows from the first formula in section 6.4 because all the zeros of Dirichlet L-functions are simple.

For $m = 1$, formula (15) and the functional equation give immediately

$$\operatorname{ord}_{s=0}(\zeta_F(s)) = r_1 + r_2 - 1.$$

Then we can reformulate (15) in the form

$$\lim_{s \to 0} s^{1-r_1-r_2} \zeta_F(s) = -\frac{Rh}{w}.$$

6.6   *Values of the Dedekind Zeta Function at Nonpositive Integers*

For abelian fields F and integers $m \geq 1$, the first formula of section 4.5 and the first formula of section 6.4 give at once

$$\zeta_F(1 - m) = \prod_{\chi \in X} \left(-\frac{B_{m,\chi}}{m} \right).$$

6.7   *Evaluation of the Dedekind Zeta Function at* -1

For any arbitrary number field F, it is known that $\zeta_F(s)$ has a zero of multiplicity r_2 at $s = -1$ and so $\lim_{s\to-1}(s+1)^{-r_2}\zeta_F(s)$ exists and is nonzero. Moreover by the functional equation for $\zeta_F(s)$ we have:

$$\lim_{s\to-1}(s+1)^{-r_2}\zeta_F(s) = (-1)^{r_1+r_2}\frac{|d|^{3/2}}{2^{r_1+3r_2}\pi^{2r_1+3r_2}}\zeta_F(2),$$

where as usual d denotes the discriminant of F.

6.8   *Remark on the Relative Class Number of CM-fields*

Let F be a CM-field, i.e., a totally imaginary quadratic extension of a totally real number field. Denote by F^+ its maximal totally real subfield. Let $h := h(F)$ (resp. $h^+ := h(F^+)$) denote the class number of F (resp. F^+). Set $h^- = h/h^+$. It follows from class field theory that $h^+|h$ (see [Washington, 1997, Theorem 4.10]), so that h^- is a natural number. Let E and E^+ be the groups of units of F and F^+ respectively, and let W denote the group of roots of unity in F ($w = \operatorname{card} W$). Let $R(F)$ and $R(F^+)$ be the Dirichlet regulators of F and F^+ respectively.

It is not difficult to prove that

$$\frac{R(F)}{R(F^+)} = \frac{2^{([F:\mathbb{Q}]/2)-1}}{Q},$$

where $Q := \operatorname{card}(E/WE^+) = 1$ or 2 (see [Washington, 1997, Proposition 4.16 and Theorem 4.12] and [Hasse, 1952]). Combining the above with the second formula of section 6.4 (for F and F^+, respectively) and the last formula of section 5.1, we obtain:

$$h^-(F) = Qw \prod_{\substack{\chi\in X \\ \chi \text{ odd}}} \left(-\frac{1}{2}B_{1,\chi}\right)$$

(see [Washington, 1997, Theorem 4.17]). Cyclotomic fields and imaginary quadratic fields are CM-fields. We have $\mathbb{Q}(\zeta_n)^+ = \mathbb{Q}(\zeta_n + \zeta_n^{-1})$ ($Q = 1$ if n is a prime power and $Q = 2$ otherwise) and $\mathbb{Q}(\sqrt{d})^+ = \mathbb{Q}$, $d < 0$.

7.   K-THEORETIC BACKGROUND

7.1   *Quillen's K-functors*

Let $K_n(n \geq 0)$ denote Quillen's K-functors which are from the category of associative rings with 1 to the category of abelian groups. Let O_F denote the ring of integers of a number field F. The groups

$$K_n(O_F) = \pi_n(BGL^+O_F)$$

(here BGL^+ denotes the plus space of Quillen obtained by adding some cells to BGL, for details see [Quillen, 1973]) are finitely generated with rank

$$\operatorname{rank} K_n(O_F) = \begin{cases} 0, & \text{if } n \text{ is even}, n > 0, \\ r_2, & \text{if } n \equiv 3 \,(\operatorname{mod} 4), \\ r_1 + r_2, & \text{if } n \equiv 1 \,(\operatorname{mod} 4), \ n > 1, \\ r_1 + r_2 - 1, & \text{if } n = 1, \\ 1, & \text{if } n = 0 \end{cases}$$

(see [Borel, 1972]). Moreover rank $K_n(F) = \operatorname{rank} K_n(O_F)$ if $n > 1$ and the functor K_2 coincides with Milnor's functor (see [Milnor, 1971]). Garland ([Garland, 1971]) proved that the group $K_2(O_F)$ is finite.

7.2   *Some Results of Browkin and Schinzel on $K_2 O_F$*

Let $r_2(A)$ denote the 2-rank of a finite abelian group A, i.e., the number of cyclic direct summands of its Sylow 2-subgroup. Tate ([Tate, 1976, Theorem 6.2]) proved that

$$r_2(K_2 O_F) = r_1 + g(2) - 1 + r,$$

where r_1 is the number of real embeddings of F, $g(2)$ is the number of distinct prime ideals of O_F dividing the ideal $2O_F$, $r = r_2(Cl(F)/Cl_2(F))$, where as usual $Cl(F)$ is the group of ideal classes of F and $Cl_2(F)$ is its subgroup generated by classes containing prime ideals dividing the ideal $2O_F$. See also [Browkin, 1982] and [Keune, 1989, Corollary (3.9)].

If $F = \mathbb{Q}(\sqrt{D})$, where D squarefree has t odd prime factors and as usual $N = N_{F/\mathbb{Q}}$ denotes the norm, then Browkin and Schinzel ([Browkin and Schinzel, 1982]) proved that

$$r_2(K_2 O_F) = \begin{cases} t + s, & \text{if } D \geq 2, \\ t + s - 1, & \text{if } D \leq -1, \end{cases}$$

where 2^s is the number of elements of the set $\{\pm 1, \pm 2\}$ that are norms of an element of F. They derived from these formulae many results on the 2-primary subgroup of $K_2 O_F$. We state a few of their results (p, q denote distinct odd primes).

(i) If $D \not\equiv \pm 1 \,(\operatorname{mod} 8)$, $D > 2$ then $\operatorname{card}(K_2 O_F)_2 \geq 8$ unless $D = 2, p$ or $2p$, $p \equiv \pm 3 \,(\operatorname{mod} 8)$, in which case $(K_2 O_F)_2 = \mathbb{Z}/2\mathbb{Z} \oplus \mathbb{Z}/2\mathbb{Z}$.

(ii) If $D \equiv \pm 1 \,(\operatorname{mod} 8)$, $D > 2$ then $\operatorname{card}(K_2 O_F)_2 \geq 16$ unless $D = pq$, $p \equiv q \equiv 3 \,(\operatorname{mod} 8)$ or $d = p = u^2 - 2w^2$, $u > 0$, $u \equiv 3 \,(\operatorname{mod} 4)$, $w \equiv 0 \,(\operatorname{mod} 4)$, in which case $(K_2 O_F)_2 = \mathbb{Z}/2\mathbb{Z} \oplus \mathbb{Z}/2\mathbb{Z} \oplus \mathbb{Z}/2\mathbb{Z}$.

(iii) If $D < 0$ then card $(K_2 O_F)_2 \geq 2$ unless $D = -1, -2, -p$ or $-2p$, $p \equiv \pm 3 \,(\bmod\, 8)$, in which case card $(K_2 O_F)$ is odd.

(iv) If $D < 0$ and $D \equiv 1 \,(\bmod\, 8)$ then card $(K_2 O_F)_2 \geq 4$ unless $D = -p \equiv 1 \,(\bmod\, 8)$ or $D = -pq$, $p \equiv -q \equiv 3 \,(\bmod\, 8)$, in which case $(K_2 O_F)_2 = \mathbb{Z}/2\mathbb{Z}$.

The proofs of Browkin and Schinzel are algebraic in nature. For $D = 2$ we have $K_2 O_F = \mathbb{Z}/2\mathbb{Z} \oplus \mathbb{Z}/2\mathbb{Z}$ (see [Gebhardt, 1977]). For $D = -1, -2, -3, -5, -6, -11, -19$ the group $K_2 O_F$ is trivial and for $D = -7, -15, -35$ we have $K_2 O_F = \mathbb{Z}/2\mathbb{Z}$ (see [Tate, 1973], [Skałba, 1987, 1994] and [Qin, 1994b, 1996]). Recently, Browkin ([Browkin, 1999]) has proved that $K_2 O_F = \mathbb{Z}/2\mathbb{Z}$ for $D = -23, -31$. Belabas and Gangl ([Belabas and Gangl, 1999]) have determined the group $K_2 O_F$ for all negative discriminants down to -151.

For real quadratic fields F much more is known. The groups $K_2 O_F$ are determined up to several thousand. This is due to the fact that the orders of the groups $K_2 O_F$ can be expressed by means of corresponding Bernoulli numbers and so are easily computed. See also [Conner and Hurrelbrink, 1986], [Qin, 1994a, 1995a,b, 1998], [Vazzana, 1997a,b, 1999] and [Browkin and Gangl, 1999]. Browkin and Schinzel ([Browkin and Schinzel, 1982]) found a sequence of real quadratic fields $F = \mathbb{Q}(\sqrt{4^a - 1})$, where $a = 2^{r-1} - 1$, $r \geq 4$ such that the group $K_2 O_F$ contains an element of order 2^{r+1}. For other results on the group $(K_2 O_F)_2$ see [Kolster, 1986, 1987, 1992], [Conner and Hurrelbrink, 1988, Section 25, 1989a,b, 1995], [Candiotti and Kramer, 1989], [Keune, 1989], [Berger, 1990], [Boldy, 1991], [Brauckman, 1991], and [Hurrelbrink and Kolster, 1998].

7.3 *Remarks on* $R_2(F)$

For any field F by Matsumoto's theorem (see e.g. [Milnor, 1971, §11]) we have

$$K_2 F = (F^* \otimes F^*) / I,$$

where $F^* := F - \{0\}$ and I is the subgroup of $F^* \otimes F^*$ generated by elements $a \otimes (1 - a)$, $a \in F^*$, $a \neq 1$. The symbol $\{a, b\}$ is defined by $\{a, b\} = a \otimes b \,(\bmod\, I)$. For $a \in F^*$, we have $\{a, -a\} = 1$.

Denote by $F^* \wedge F^*$ a modified external product (with the usual relation $a \wedge a = 0$ replaced by $a \wedge (-a) = 0$, $a \in F^*$) and by $A(F)$ the free abelian group generated by the elements $[a]$, where $a \in F^*$, $a \neq 1$.

We have the following exact sequence

$$0 \to C(F) \to A(F) \xrightarrow{\lambda} F^* \wedge F^* \xrightarrow{\varphi} K_2 F \to 0,$$

where $\varphi(a \wedge b) = \{a, b\}$ and $\lambda([a]) = a \wedge (1 - a)$, $a \in F^*$, $a \neq 1$. Let $C(F) = \ker \lambda$.

Now let F be a number field. Consider the map

$$\mathbb{D} : C(F) \to \mathbb{R}^{r_2}$$

defined by the formula

$$\mathbb{D}(b) \to (D(\sigma_1(b)), \ldots, D(\sigma_{r_2}(b))),$$

where D is the dilogarithm of Wigner and Bloch extended by linearity to $C(F)$ and $\sigma_1, \ldots, \sigma_{r_2}$ are all the complex places of F (for details, see [Zagier, 1991] and [Browkin, 1991]).

$\mathbb{D}(C(F))$ is a lattice of rank r_2 in $\mathbb{R}^{r_2}$ and so the volume of $\mathbb{R}^{r_2}/\mathbb{D}(C(F))$ is finite. It is well known that this volume equals the second Borel regulator $R_2(F)$ of the field F defined in [Borel, 1977] (see also [Ramakrishnan, 1989]). For details, see [Bloch, 1977], [Borel, 1977, 1995] and [Suslin, 1987].

7.4   *Definition of $w_r(F)$*

For a number field F and a natural number r, let $w_r := w_r(F)$ denote the largest integer s such that the Galois group $G(F(\zeta_s)/F)$ is annihilated by r. We have $w_1(F) = w(F)$. It is easy to see that for $r = 2$ and a totally real number field F

$$w_2(F) = 2 \prod_{l\,\text{prime}} l^{n(l)},$$

where for any prime l, $n(l)$ is the largest integer n such that $\zeta_{l^n} + \zeta_{l^n}^{-1} \in F$ (see [Serre, 1971]). For example $w_2(F) = 24$, if $F = \mathbb{Q}$ or F is a quadratic field over $\mathbb{Q}$, with two exceptions, namely, $w_2(\mathbb{Q}(\sqrt{d})) = 24d$ for $d = 2$ or 5.

7.5   *Conjectures of Lichtenbaum, and Birch and Tate*

We state the Lichtenbaum conjecture (at $s = -1$) as formulated in [Lichtenbaum, 1973] and modified in [Borel, 1977] (see also [Ramakrishnan, 1989]).

CONJECTURE ([Lichtenbaum, 1973], [Borel, 1977]). For every number field F

$$\lim_{s \to -1} (s + 1)^{-r_2} \zeta_F(s) = \pm \frac{R_2(F) \cdot \text{card}(K_2 O_F)}{w_2(F)}.$$

This conjecture is still unproven. Up to powers of 2 it was proven in [Kolster, Nguyen Quang Do and Fleckinger, 1996] with a different regulator, the Beilinson regulator. The special case of the conjecture when F is a totally real number field was formulated by Birch [Birch, 1969] and Tate [Tate, 1971]. It asserts that

$$\text{card}(K_2 O_F) = w_2(F)|\zeta_F(-1)|$$

for a totally real number field F. In this case we have $r_2 = 0$, $R_2(F) = 1$ and

$$\zeta_F(-1) = \lim_{s \to -1} (s + 1)^{-r_2} \zeta_F(s).$$

From work on the main conjecture of Iwasawa theory [Mazur and Wiles, 1984] the odd part of the Birch-Tate conjecture was confirmed for abelian extensions F of $\mathbb{Q}$. Subsequently the odd part of the Birch-Tate conjecture was confirmed by Wiles ([Wiles, 1990]) for arbitrary totally real number fields F. Kolster [Kolster, 1989] proved that the even part of the Birch-Tate conjecture holds assuming that Federer's analogue of the Main Conjecture at $p = 2$ (see [Federer, 1982]) is true. Wiles [Wiles, 1990, the footnote on p. 499] has proved Federer's conjecture completing the proof of the Birch-Tate conjecture for abelian extensions F of $\mathbb{Q}$.

Kolster ([Kolster, 1986]) proved that the Birch-Tate conjecture holds provided the 2-primary subgroup $K_2 O_F$ is elementary abelian. It seems that the proof of the Birch-Tate conjecture has not been completed yet for nonabelian extensions F of $\mathbb{Q}$ for which the 2-primary subgroup $K_2 O_F$ is not elementary abelian. For other results on the nonabelian case see [Conner and Hurrelbrink, 1988] and [Hurrelbrink, 1989]. For more details on the subject see also [Coates, 1973, 1977], [Greither, 1992], [Washington, 1997, Chapter 13] and the Appendix by Rubin in [Lang, S., 1990]. For generalizations of the Birch-Tate conjecture to values of zeta functions at other odd negative integers and totally real number fields see [Kolster, 2000].

If F is a totally real number field of degree N over $\mathbb{Q}$ with discriminant d then for an even natural number m

$$\pi^{-Nm} |d|^{\frac{1}{2}} \zeta_F(m)$$

is a rational number. From the functional equation for $\zeta_F(s)$ we deduce that $\zeta_F(-m)$ is a nonzero rational number. For more details on the subject see [Hecke, 1924, 1959], [Klingen, 1962], [Siegel, 1937, 1966, 1968]), [Meyer, 1967], [Lang, H., 1968], and [Barner, 1969]. Furthermore Serre ([Serre, 1971]) noticed that the number $w_2(F)$ is divisible by the denominator of $\zeta_F(-1)$, i.e., $w_2(F)\zeta_F(-1) \in \mathbb{Z}$, and moreover that

$$2^N \mid w_2(F)\zeta_F(-1).$$

8. QUADRATIC FIELDS

8.1 *Class Number Formulae for Imaginary Quadratic Fields*

An important aspect of the numbers $B_{1,\chi}$ (χ odd) and $B_{2,\chi}$ (χ even) is that they occur, respectively, in the formulae for $h^-(F)$ when F is a CM-field and in the formulae for the orders of the groups $K_2 O_F$ when F is a totally real abelian field. See sections 6.7 and 6.6.

Let d be the discriminant of a quadratic field F. Let O_F denote the ring of integers of F and let χ_d denote its character. For negative d we have $h(d) = h^-(F/\mathbb{Q})$. Thus from the last equation in section 6.8, we have

$$h(d) = -\frac{w}{2}B_{1,\chi_d}, \tag{16}$$

where

$$w = w(d) = \begin{cases} 2, & \text{if } d < -4, \\ 4, & \text{if } d = -4, \\ 6, & \text{if } d = -3. \end{cases}$$

We also have the formula

$$h(d) = \frac{w\sqrt{|d|}}{2\pi}L(1,\chi_d) = \frac{w}{2}L(0,\chi_d).$$

These equalities are implied by the last equation in section 6.4 and the functional equation for Dirichlet L-functions given in section 4.1.

By the definition of generalized Bernoulli numbers, we have for odd characters χ modulo M

$$B_{1,\chi} = M^{-1}\sum_{a=1}^{M} \chi(a)a.$$

After some manipulation this formula can be rewritten as

$$B_{1,\chi} = \frac{1}{\overline{\chi}(2) - 2} \sum_{1 \leq a \leq M/2} \chi(a).$$

This reformulation also follows from Berndt's character analogue of the Poisson summation formula (see Theorem 1, section 1.8) and formula (12). We remark that the formula is valid for any odd character χ. In [Borevich and Shafarevich, 1966] the formula is only proved for quadratic characters χ.

The above formula together with formula (16) implies

$$h(d) = \frac{w}{2(2 - \chi_d(2))} \sum_{1 \leq a \leq |d|/2} \chi_d(a). \tag{17}$$

8.2   Class Number Formulae for Real Quadratic Fields

Similarly, the Dirichlet class number formula gives for a real quadratic field F of discriminant d

$$h(d) = \frac{\sqrt{d}}{2\log\varepsilon}L(1,\chi_d), \tag{18}$$

or by Gauss' evaluation of $\tau(\chi_d)$

$$h(d) = -\frac{1}{2\log\varepsilon}\sum_{a=1}^{d}\chi_d(a)\log(1-\zeta^{-a}).$$

Here $\zeta = \exp(2\pi i/d)$ and ε denotes the fundamental unit of F.

8.3 Order of K_2O_F for Real Quadratic Fields

For a real quadratic field F of discriminant d we have

$$w_2 = w_2(d) = \begin{cases} 24, & \text{if } d > 8, \\ 48, & \text{if } d = 8, \\ 120, & \text{if } d = 5. \end{cases}$$

By Wiles' Theorem (the Birch-Tate conjecture) and the formula for $\zeta_F(-m)$ (see section 6.6), we have

$$k_2(d) := \text{card}(K_2O_F) = \frac{w_2}{24}B_{2,\chi_d}. \tag{19}$$

Formula (19) follows from

$$k_2(d) = \frac{w_2 d^{3/2}}{24\pi^2}L(2,\chi_d) = -\frac{w_2}{12}L(-1,\chi_d),$$

which is a consequence of the Birch-Tate conjecture and the functional equation for Dirichlet L-functions.

By the definition of generalized Bernoulli numbers given in section 3.1, for any even character χ modulo M we have

$$B_{2,\chi} = M^{-1}\sum_{a=1}^{M}\chi(a)a^2$$

(recall the evident identity $\sum_{a=1}^{M}\chi(a)a = 0$).

It is not trivial to deduce in an elementary way from the above formula for $B_{2,\chi}$ that for even primitive Dirichlet characters χ modulo $M > 1$

$$B_{2,\chi} = \frac{4}{\overline{\chi}(2)-4}\sum_{1\le a\le M/2}\chi(a)a.$$

However it follows easily from Berndt's character analogue of the Poisson summation formula and formula (12). For the character χ_d the above formula implies:

$$k_2(d) = \frac{w_2}{6(\chi_d(2)-4)}\sum_{1\le a\le d/2}\chi_d(a)a. \tag{20}$$

8.4   *Lichtenbaum Conjecture for Imaginary Quadratic Fields*

The Lichtenbaum conjecture for an imaginary quadratic field F of discriminant d takes the form

$$k_2(d) = \frac{|d|^{3/2}}{2R_2(d)} L(2, \chi_d), \tag{21}$$

where $R_2(d) = R_2(F)$ (see sections 7.5, 6.4 and 6.7). Therefore by means of the formulae of section 5.4 and Gauss' evaluation of $\tau(\chi_d)$ we obtain

$$k_2(d) = \frac{|d|}{2R_2(d)} \sum_{a=1}^{|d|} \chi_d(a) D(\zeta^a),$$

where $\zeta = \exp(2\pi i/|d|)$.

8.5   *2-rank of the Narrow Class Group*

Let $h_0(d)$ denote the order of the group of narrow classes of ideals in a quadratic field F with discriminant d and let ε be the fundamental unit of this field if $d > 0$. It is known that $h_0(d) = h(d)$, if $d < 0$, or $d > 0$ and $N(\varepsilon) = -1$, and $h_0(d) = 2h(d)$, otherwise. Recall that as usual $N = N_{F/\mathbb{Q}}$ denotes the norm. Moreover, if ν denotes the number of distinct prime factors of d, Gauss' theory of ambiguous classes gives the 2-rank of the narrow class group as $\nu - 1$. This implies that

$$2^{\nu-1} \mid h(d), \text{ if } d < 0, \text{ or } d > 0, \ N(\varepsilon) = -1,$$

$$2^{\nu-2} \mid h(d), \text{ if } d > 0, \ N(\varepsilon) = 1.$$

For details, see [Borevich and Shafarevich, 1966, p. 247], [Narkiewicz, 1990, Chapter 8, §3] or [Conner and Hurrelbrink, 1988, p. 133].

9.   POWER SUMS INVOLVING DIRICHLET CHARACTERS

9.1   *Short Character Power Sums*

Let χ be a Dirichlet character. In this section we consider short character sums of the form (3), in section 1.6, with $F(x) = x^r$, where $r \geq 0$ is an integer. In section 1.6 we mentioned papers related to these sums. Now we will briefly review the methods used by the authors of these papers. Generally speaking, the authors expressed the sums as rational linear combinations of special values of L-functions with explicitly determined coefficients. In the case of quadratic characters χ the short character sums were expressed in terms of the class numbers of appropriate imaginary quadratic fields if $F(x) = 1$, or in terms of the class numbers of imaginary quadratic fields and the orders of K_2-groups related to real quadratic fields if $F(x) = x$. In view of (16) (resp. (16) and (19)) the sums were expressed in terms of the first generalized Bernoulli numbers if $F(x) = 1$ (resp. the first and second generalized Bernoulli numbers if $F(x) = x$). If $F(x) = x^r$, $r > 1$ one has to use higher generalized Bernoulli numbers. All such results can also be deduced from Berndt's analogue of the Poisson summation formula (Theorem 1, section 1.8). Historically, the authors used the following methods: elementary transformations of the formulae for $L(1, \chi)$ or $L(2, \chi)$ (see section 5), transformations of these formulae by means of Fourier series or contour integration, or Berndt's character analogue of the Poisson summation formula.

The most convenient way of producing new relations for short character sums is either by Berndt's formula or by Zagier's identity proved in [Szmidt, Urbanowicz and Zagier, 1995]. The latter formula is the key point of this chapter.

9.2   *Zagier's Identity*

In this section we express short character power sums of any length in terms of generalized Bernoulli numbers.

THEOREM 2 ([Szmidt, Urbanowicz and Zagier, 1995]) *Let χ be a Dirichlet character* modulo M, N *a positive integral multiple of M and r (> 1) a positive integer prime to N. Then for any integer $m \geq 0$ we have*

$$(m+1)r^m \sum_{0 < n < N/r} \chi(n)n^m = -B_{m+1,\chi}r^m + \frac{\overline{\chi}(r)}{\phi(r)} \sum_{\psi} \overline{\psi}(-N)B_{m+1,\chi\psi}(N),$$

where the sum on the right hand side of the above equation is taken over all Dirichlet characters ψ modulo r.

PROOF. The proof of the theorem falls naturally into two parts.

(i) If χ is a Dirichlet character modulo M, we define

$$\mathcal{L}_\chi(s) = \sum_{n=1}^{\infty} \chi(n) \exp(ns).$$

The series converges absolutely for $Re(s) < 0$. From the obvious identity

$$\sum_{n=1}^{M} \chi(n) \exp(ns) = (1 - \exp(Ms))\mathcal{L}_\chi(s) \tag{22}$$

and by the definition of $B_{n,\chi}$ we obtain the Laurent expansion

$$\mathcal{L}_\chi(s) = -\sum_{n=0}^{\infty} B_{n,\chi}\frac{s^{n-1}}{n!} \quad (s \to 0). \tag{23}$$

Comparing coefficients of $s^{m-1}/(m-1)!$ on both sides of (22) gives the identity

$$\sum_{n=1}^{M} \chi(n) n^{m-1} = \frac{1}{m} \sum_{k=1}^{m} \binom{m}{k} B_{m-k,\chi} M^k \quad (m \geq 1)$$

(cf. formula (5)) which can be used to compute the generalized Bernoulli numbers $B_{m,\chi}$ inductively. A generalization of the above argument is the basis of the proof of the formula of Theorem 2.

(ii) Now let N be a positive multiple of M and r a positive integer prime to N. Then

$$\sum_{0<n<N/r} \chi(n) \exp(rns)$$

$$= \sum_{n>0} \chi(n) \exp(rns) - \sum_{n>0,\, r|n+N} \overline{\chi}(r)\chi(n) \exp\left((n+N)s\right)$$

$$= \sum_{n=1}^{\infty} \chi(n) \exp(rns) - \exp(Ns) \sum_{n=1}^{\infty} \left(\frac{\overline{\chi}(r)}{\phi(r)} \sum_{\psi} \psi(n)\overline{\psi}(-N)\right) \chi(n) \exp(ns)$$

$$= \mathcal{L}_{\chi}(rs) - \frac{\overline{\chi}(r)}{\phi(r)} \exp(Ns) \sum_{\psi} \overline{\psi}(-N) \mathcal{L}_{\chi\psi}(s)$$

in virtue of the well known formula

$$\sum_{\psi} \psi(a)\overline{\psi}(b) = \begin{cases} \phi(r), & \text{if } a \equiv b \,(\operatorname{mod} r), \\ 0, & \text{otherwise}, \end{cases}$$

where the sum is over all Dirichlet characters ψ modulo r.

Comparing coefficients of $s^{m-1}/m!$ $(m \geq 0)$ on both sides and using (23) we obtain the theorem. ∎

REMARKS 1. The corresponding identities for a quadratic character χ and $m = 0$ were conjectured in [Akiyama, 1994].
2. Zagier has noticed that the first formula of section 4.5 giving the values of the Dirichlet series $L(s,\chi)$ at negative integers follows formally from (23), since if we ignore all questions of convergence then the "coefficient" of $s^r/r!$ in $\mathcal{L}_{\chi}(s)$ is $\sum_{n\geq 1} \chi(n)n^r = L(-r,\chi)$. To prove the formula rigorously one also uses equation (23): write $\Gamma(s)L(s,\chi)$ as a Mellin transform integral $\int_{0}^{\infty} \mathcal{L}_{\chi}(-t)t^{s-1}dt$, split up the integral into $\int_{0}^{1} + \int_{1}^{\infty}$, expand the first term, and compare residues at $s = 1 - m$.

9.3 The Case when r Is a Divisor of 24

When r is a divisor of 24, the group $(\mathbb{Z}/r\mathbb{Z})^*$ has exponent 2, so all the characters ψ are quadratic. Thus if we also restrict χ to be a quadratic character then all the generalized Bernoulli numbers occurring in the equation of Theorem 2 are attached to quadratic characters. We return to this case in Chapters 2 and 3.

9.4 Lerch's Formulae

For a nontrivial Dirichlet character χ modulo M, a natural number r, k an integer satisfying $1 \le k \le r$, and a nonnegative integer m, we set

$$s_m(r, k, \chi) = \sum_{(k-1)M/r < l < kM/r} \chi(l) l^m.$$

We remark that if s is a positive integer coprime with M then

$$s_m(r, k, \chi) = \sum_{j=(k-1)s+1}^{ks} s_m(rs, j, \chi).$$

The sums $s(n, k)$ considered in section 1.7 are special cases of the above sums. Namely, we have

$$s(n, k) = s_0(n, k, \chi_d),$$

where $d = \pm p$.

In the special case of Theorem 2 when $m = 0$ and $r = 8$ we obtain at once the famous formulae of Lerch proved in his Paris Academy prize memoir for 1900. See [Lerch, 1905].

THEOREM 3 ([Lerch, 1905], [Berndt, 1976]) *Let d be an odd discriminant of a quadratic field. Then we have*

$$4 \sum_{0 < l < |d|/8} \chi_d(l) = \begin{cases} \chi_d(2)h(-4d) + h(-8d), & \text{if } d > 0, \\ (5 - \chi_d(2))h(d) - h(8d) - \lambda(d), & \text{if } d < 0, \end{cases}$$

$$4 \sum_{|d|/8 < l < |d|/4} \chi_d(l) = \begin{cases} (2 - \chi_d(2))h(-4d) - h(-8d), & \text{if } d > 0, \\ 3(-1 + \chi_d(2))h(d) + h(8d) + \lambda(d), & \text{if } d < 0, \end{cases}$$

$$4 \sum_{|d|/4 < l < 3|d|/8} \chi_d(l) = \begin{cases} (-2 - \chi_d(2))h(-4d) + h(-8d), & \text{if } d > 0, \\ 3(1 - \chi_d(2))h(d) + h(8d) - \lambda(d), & \text{if } d < 0, \end{cases}$$

$$4 \sum_{3|d|/8 < l < |d|/2} \chi_d(l) = \begin{cases} \chi_d(2)h(-4d) - h(-8d), & \text{if } d > 0, \\ 3(1 - \chi_d(2))h(d) - h(8d) - \lambda(d), & \text{if } d < 0, \end{cases}$$

where $\lambda(-3) = 1$, *and* $\lambda(d) = 0$ *otherwise. Moreover for* $k = 5, 6, 7, 8$ *we have*

$$s_0(8, k, \chi_d) = \chi_d(-1)s_0(8, 9 - k, \chi_d).$$

PROOF. This theorem is an immediate consequence of Theorem 2 for $m = 0$, $\chi = \chi_d$, $N = M = |d|$, $r = 8$ and the formulae of section 8.   ∎

REMARK The corresponding formulae for an arbitrary Dirichlet character χ were given in [Berndt, 1976].

Let $E^*(x) = [x]$ if $x \notin \mathbb{Z}$, and $E^*(x) = [x] - \frac{1}{2}$ otherwise. Write $R(x) = x - [x - \frac{1}{2}]$, and $R^*(x) = x - E^*(x + \frac{1}{2})$. Here $[x]$ as usual denotes the integral part of x. Lerch [Lerch, 1905] investigated the Fourier series of the functions related to $E^*(x)$, $R^*(x)$, $\mathrm{sgn}\, R^*(x)$ and $|R(x)|$. Among others he proved the formulae

$$\sum_{l=1}^{[dx]} \chi_d(l) = d \sum_{n=1}^{\infty} \chi_d(n) \frac{\sin 2\pi n x}{\pi n}$$

for positive d, and

$$\sum_{l=1}^{[|d|x]} \chi_d(l) = h(d) - \sqrt{|d|} \sum_{n=1}^{\infty} \chi_d(n) \frac{\cos 2\pi n x}{\pi n}$$

for negative d, where in both cases $x + (l/|d|) \notin \mathbb{Z}$, $1 \le l \le |d| - 1$. One can obtain the equations of Theorem 3 by putting $x = 1/8, 1/4, 3/8$ and $1/2$ in the above formulae and using the obvious fact that $\sin(\pi n/2) = \chi_{-4}(n)$, if $n \in \mathbb{Z}$, and $\cos(\pi n/4) = (\sqrt{2})^{-1}\chi_8(n)$, $\sin(\pi n/4) = (\sqrt{2})^{-1}\chi_{-8}(n)$, if $n \in \mathbb{Z}$ is odd. Berndt ([Berndt, 1976]) noticed that similar arguments can be applied to the case when χ is any primitive Dirichlet character (not only quadratic). Using Fourier series, contour integration and his character analogue of the Poisson summation formula, Berndt expressed the short character sums $s_0(r, k, \chi)$ for various r and k as linear combinations of the values at $s = 1$ of L-functions attached to Dirichlet characters of the form $\chi_d\overline{\chi}$ with $d|r$. He investigated the sums $s_0(r, k, \chi)$ for $r = 2, 3, 4, 5, 6, 8, 10, 12, 16$ and 24 and the sums $s_1(r, k, \chi)$ for $r = 2$ and 4. The short character sums with $m = 1$ and $r = 2$ were also investigated in [Lerch, 1905] but only for quadratic characters χ.

When $m = 1$ and $r = 8$, the following theorem is deduced from Theorem 2 using similar arguments.

THEOREM 4 ([Urbanowicz, 1990a]) *Let d be the odd discriminant of a quadratic field. Then we have*

$$64 \sum_{0 < l < |d|/8} \chi_d(l)l = \begin{cases} k_2(8d) - (34 - \chi_d(2))k_2(d) \\ \quad + 2d(\chi_d(2)h(-4d) + h(-8d)) + 7\omega(d), \\ \hfill \text{if } d > 0, \\ k_2(-8d) + \chi_d(2)k_2(-4d) \\ \quad - 2d((1 - \chi_d(2))h(d) - h(8d)) - \nu(d), \\ \hfill \text{if } d < 0, \end{cases}$$

$$64 \sum_{|d|/8 < l < |d|/4} \chi_d(l)l = \begin{cases} -k_2(8d) + 3(2 - 3\chi_d(2))k_2(d) \\ \quad + 2d((4 - \chi_d(2))h(-4d) - h(-8d)) \\ \hfill - 3\omega(d), \text{ if } d > 0, \\ -k_2(-8d) + (4 - \chi_d(2))k_2(-4d) \\ \quad - 2d(5(-1 + \chi_d(2))h(d) + h(8d)) + 5\nu(d), \\ \hfill \text{if } d < 0, \end{cases}$$

$$64 \sum_{|d|/4 < l < 3|d|/8} \chi_d(l)l = \begin{cases} -k_2(8d) - 3(2 - 3\chi_d(2))k_2(d) \\ \quad + 2d((-4 - 3\chi_d(2))h(-4d) + 3h(-8d)) \\ \hfill + 3\omega(d), \text{ if } d > 0, \\ k_2(-8d) - (4 + \chi_d(2))k_2(-4d) \\ \quad - 2d(7(1 - \chi_d(2))h(d) + 3h(8d)) - 7\nu(d), \\ \hfill \text{if } d < 0, \end{cases}$$

$$64 \sum_{3|d|/8 < l < |d|/2} \chi_d(l)l = \begin{cases} k_2(8d) - 15(2 - \chi_d(2))k_2(d) \\ \quad + 2d(3\chi_d(2)h(-4d) - 3h(-8d)) + 9\omega(d), \\ \hfill \text{if } d > 0, \\ -k_2(-8d) + \chi_d(2)k_2(-4d) \\ \quad - 2d(13(1 - \chi_d(2))h(d) - 3h(8d)) - 13\nu(d), \\ \hfill \text{if } d < 0, \end{cases}$$

where $\omega(5) = \frac{1}{4}$, $\omega(d) = 0$ otherwise, and $\nu(-3) = \frac{1}{8}$, $\nu(d) = 0$ otherwise. Moreover for $k = 5, 6, 7, 8$ we have

$$s_1(8, k, \chi_d) = ds_0(8, 9 - k, \chi_d) - \chi_d(-1)s_1(8, 9 - k, \chi_d).$$

PROOF. This theorem is an immediate corollary to Theorem 2. It suffices to put $m = 1$, $\chi = \chi_d$, $N = M = |d|$, $r = 8$ in the equation of Theorem 2 and to use the formulae of section 8. ∎

In [Urbanowicz, 1990a] Theorem 4 is proved using the Fourier series of the function $|R(x)|$, namely the series

$$|R(x)| = \frac{1}{4} - 2 \sum_{2 \nmid n} \frac{\cos 2\pi nx}{\pi^2 n^2}.$$

9.5   *Formulae of Szmidt and Urbanowicz*

For a nonnegative integer m, x a positive real number, and χ a Dirichlet character, we set

$$S_m(x, \chi) = \sum_{0 \leq a < x} \chi(a) a^m .$$

Theorems 3 and 4 are implied by the following more general result.

THEOREM 5 ([Szmidt and Urbanowicz, 1994]) *Let χ be a primitive Dirichlet character with odd conductor M and let $N \geq 1$ be an odd multiple of M. Then for $m \geq 0$, we have*

$$2^{3m+2}(m+1)S_m(N/8, \chi)$$

$$= -\chi_{-8}(N)\overline{\chi}(8)B_{m+1,\chi_{-8}\chi}(N) - \chi_{-4}(N)\overline{\chi}(8)B_{m+1,\chi_{-4}\chi}(N)$$

$$+ \chi_8(N)\overline{\chi}(8)B_{m+1,\chi_8\chi}(N) + \overline{\chi}(8)B_{m+1,\chi}(N)$$

$$- 2^m\overline{\chi}(4)B_{m+1,\chi}(N/2) - 2^{3m+2}B_{m+1,\chi} .$$

PROOF. Theorem 5 follows from either Theorem 1 or Theorem 2. In [Szmidt and Urbanowicz, 1994] the authors used the formula of Theorem 1 to prove Theorem 5. Their proof is considerably longer than the proof using Theorem 2, which we leave to the reader. Here we demonstrate the standard method of attack.

Write

$$S_m^*(N/8, \chi) = \sum_{0 \leq l < N/8}^* \chi(l) l^m ,$$

where the asterisk on the summation sign has the same meaning as in Theorem 1, i.e., $S_m^*(N/8, \chi) = S_m(N/8, \chi) + (1/2)\varepsilon_{m,\chi}$, where $\varepsilon_{m,\chi} = 1$ if $m = 0$ and $\chi = \chi_1$, and $\varepsilon_{m,\chi} = 0$ otherwise.

Theorem 1 applied to the function $F(x) = x^m$ with $a = 0$, $b = N/8$ gives

$$S_m^*(N/8, \chi) = \delta_{\chi,\chi_1} \int_0^{N/8} x^m dx + \frac{1}{\tau(\overline{\chi})} \sum_{\substack{n=-\infty \\ n \neq 0}}^{\infty} \overline{\chi}(n) \int_0^{N/8} x^m \exp(2\pi inx/M) dx.$$

The latter integral on the right hand side of the above equation can be evaluated using the formula

$$\int x^m \exp(iax) dx = -i \exp(iax) \sum_{k=0}^{m} (-1)^{k/2} k! \binom{m}{k} \frac{x^{m-k}}{a^{k+1}}$$

(see for example [Gradshteyn and Ryzhik, 1994, formula 2.513]).

We have

$$S_m^*(N/8, \chi) = \frac{N^{m+1}\delta_{\chi,\chi_1}}{2^{3m+3}(m+1)}$$

$$- \frac{i}{\tau(\overline{\chi})} \sum_{\substack{n=-\infty \\ n\neq 0}}^{\infty} \overline{\chi}(n) \exp\left(\frac{\pi i n N}{4M}\right) \sum_{k=0}^{m} (-1)^{k/2} k! \binom{m}{k} \left(\frac{\acute{N}}{8}\right)^{m-k} \left(\frac{M}{2\pi n}\right)^{k+1}$$

$$+ \frac{im!}{\tau(\overline{\chi})} \left(\frac{M}{2\pi}\right)^{m+1} (-1)^{m/2} \sum_{\substack{n=-\infty \\ n\neq 0}}^{\infty} \frac{\overline{\chi}(n)}{n^{m+1}},$$

and in consequence we obtain

$$S_m^*(N/8, \chi) = \frac{N^{m+1}\delta_{\chi,\chi_1}}{2^{3m+3}(m+1)}$$

$$- \frac{i}{\tau(\overline{\chi})} \sum_{k=0}^{m} k! \binom{m}{k} \left(\frac{N}{8}\right)^{m-k} \left(\frac{M}{2\pi}\right)^{k+1} (-1)^{k/2} A_{k,\chi}(N/M) \quad (24)$$

$$+ \frac{im!}{\tau(\overline{\chi})} \left(\frac{M}{2\pi}\right)^{m+1} (-1)^{m/2} A_{m,\chi}(0),$$

where $A_{k,\chi}(R)$, for $R = 0$ or $R = N/M$, is defined by

$$A_{k,\chi}(R) = \sum_{\substack{n=-\infty \\ n\neq 0}}^{\infty} \overline{\chi}(n) \frac{\exp(\pi i n R/4)}{n^{k+1}}.$$

On the other hand, using the obvious identity

$$A_{k,\chi}(R) = \left(1 + (-1)^{\delta+k+1}\right) \sum_{n=1}^{\infty} \overline{\chi}(n) \frac{\cos(\pi n R/4)}{n^{k+1}}$$

$$+ i \left(1 - (-1)^{\delta+k+1}\right) \sum_{n=1}^{\infty} \overline{\chi}(n) \frac{\sin(\pi n R/4)}{n^{k+1}},$$

where $\chi(-1) = (-1)^\delta$, $\delta \in \{0,1\}$, for odd R we obtain

$$A_{k,\chi}(R) = \left(1 + (-1)^{\delta+k+1}\right) \left((\sqrt{2})^{-1}\chi_8(R)L(k+1, \chi_8\overline{\chi})\right.$$

$$+ \overline{\chi}(4)2^{-2(k+1)} \left(\overline{\chi}(2)2^{-k} - 1\right) L(k+1, \overline{\chi})\right)$$

$$+ i \left(1 - (-1)^{\delta+k+1}\right) \left((\sqrt{2})^{-1}\chi_{-8}(R)L(k+1, \chi_{-8}\overline{\chi})\right.$$

$$+ \overline{\chi}(2)2^{-(k+1)}\chi_{-4}(R)L(k+1, \chi_{-4}\overline{\chi})\right)$$

by virtue of the obvious facts that

$$\cos(\pi n R/4) = (\sqrt{2})^{-1}\chi_8(nR), \quad \sin(\pi n R/4) = (\sqrt{2})^{-1}\chi_{-8}(nR),$$

if n is odd, and

$$\cos(\pi n R/4) = 0, \quad \sin(\pi n R/4) = \chi_{-4}(nR/2),$$

if $n \equiv 2 \,(\bmod\, 4)$, and

$$\cos(\pi n R/r) = (-1)^{n/4}, \quad \sin(\pi n R/4) = 0,$$

if $n \equiv 0 \,(\bmod\, 4)$.

Therefore we have

$$A_{k,\chi}(R) = -2(-1)^{\frac{k+\delta}{2}+1}i^{\delta-1}\frac{\tau(\overline{\chi})\overline{\chi}(8)}{(k+1)!}\left(\frac{2\pi}{8M}\right)^{k+1}$$

$$\times \left(\chi_{-8}(RM)B_{k+1,\chi_{-8}\chi} + \chi_{-4}(RM)B_{k+1,\chi_{-4}\chi}\right),$$

if $k \equiv \delta \,(\bmod\, 2)$, and

$$A_{k,\chi}(R) = 2(-1)^{\frac{k+1-\delta}{2}+1}i^{-\delta}\frac{\tau(\overline{\chi})}{(k+1)!}\left(\frac{2\pi}{8M}\right)^{k+1}$$

$$\times \left(\chi_8(RM)\overline{\chi}(8)B_{k+1,\chi_8\chi} + \overline{\chi}(4)(\overline{\chi}(2) - 2^k)B_{k+1,\chi}\right),$$

if $k \not\equiv \delta \,(\bmod\, 2)$.

Consequently, we find that

$$A_{k,\chi}(R) = 2(-1)^{\delta+\frac{k-1}{2}}\left(\frac{2\pi}{8M}\right)^{k+1}\frac{\tau(\overline{\chi})}{(k+1)!}$$

$$\times \left(\chi_{-8}(RM)\overline{\chi}(8)B_{k+1,\chi_{-8}\chi} + \chi_{-4}(RM)\overline{\chi}(8)B_{k+1,\chi_{-4}\chi}\right.$$

$$\left. + \chi_8(RM)\overline{\chi}(8)B_{k+1,\chi_8\chi} + \overline{\chi}(4)(\overline{\chi}(2) - 2^k)B_{k+1,\chi}\right).$$

Similarly, we see that

$$A_{m,\chi}(0) = \left(1 + (-1)^{\delta+m+1}\right)\sum_{n=1}^{\infty}\frac{\overline{\chi}(n)}{n^{m+1}},$$

and in consequence

$$A_{m,\chi}(0) = 0,$$

if $m \equiv \delta \,(\bmod\, 2)$, and

$$A_{m,\chi}(0) = 2L(m+1,\overline{\chi}) = (-1)^{\frac{m+1-\delta}{2}+1}i^{-\delta}\frac{\tau(\overline{\chi})}{(m+1)!}\left(\frac{2\pi}{M}\right)^{m+1}B_{m+1,\chi},$$

if $m \not\equiv \delta \,(\bmod\, 2)$.

Thus we have

$$\mathcal{A}_{m,\chi}(0) = (-1)^{\delta + \frac{m-1}{2}} \frac{\tau(\overline{\chi})}{(m+1)!} \left(\frac{2\pi}{M}\right)^{m+1} (1 - \varepsilon_{m,\chi}) B_{m+1,\chi}.$$

Finally to obtain the theorem it suffices to substitute the obtained formulae for $\mathcal{A}_{k,\chi}$ into (24) and to make use of formula (6). We have

$$2^{3m+2}(m+1)(-1)^{m+1} S_m^*(N/8, \chi)$$
$$= \chi_{-8}(N)\overline{\chi}(-8) B_{m+1,\chi_{-8}\chi}(-N) + \chi_{-4}(N)\overline{\chi}(-8) B_{m+1,\chi_{-4}\chi}(-N)$$
$$+ \chi_8(N)\overline{\chi}(-8) B_{m+1,\chi_8\chi}(-N) + \overline{\chi}(-8) B_{m+1,\chi}(-N)$$
$$- \overline{\chi}(-4) 2^m B_{m+1,\chi}(-N/2) - 2^{3m+2}(-1)^{m+1}(1 - \varepsilon_{m,\chi}) B_{m+1,\chi}. \quad\blacksquare$$

A similar argument applies to the case $r = 24$. If χ is a quadratic character then all characters occuring in the equation of Theorem 2 are quadratic. In this case the formulae for the short character sums over intervals of length $M/24$ are so complicated that we omit them because of limitation of space, see [Berndt, 1976]. In this case the short character sums for $m = 0$ are linear combinations of

$$h(d), \ h(8d), \ h(12d), \ h(24d), \ \text{if } d < 0,$$
$$h(-3d), \ h(-4d), \ h(-8d), \ h(-24d), \ \text{if } d > 0,$$

and if $m = 1$ of

$$h(d), \ h(8d), \ h(12d), \ h(24d), \ k_2(-3d), \ k_2(-4d), \ k_2(-8d), \ k_2(-24d),$$
$$\text{if } d < 0,$$
$$h(-3d), \ h(-4d), \ h(-8d), \ h(-24d), \ k_2(d), \ k_2(8d), \ k_2(12d), \ k_2(24d),$$
$$\text{if } d > 0.$$

9.6 *Formulae for Short Character Sums*

All the classical formulae, as well as some new formulae, for short character sums as linear combinations of class numbers can be deduced easily from Theorem 2. For $r = 2$, $m = 0, 1$, and $\chi = \chi_d$ Theorem 2 gives the classical formulae (17) and (20). Formula (17) for $h(d)$ is one of Dirichlet's famous class numbers formulae (see [Dirichlet, 1839]) and formula (19) for $k_2(d)$ was already proved in [Lerch, 1905, p. 395]. See also [Cauchy, 1882], [Pepin, 1874, p. 205] or [Ayoub, Chowla and Walum, 1967].

For $r = 4$, $m = 0$ and odd d we obtain the other famous Dirichlet formulae

$$h(-4d) = 2 \sum_{0<a<d/4} \chi_d(a), \tag{25}$$

$$h(-8d) = \sum_{0<a<2d} \chi_{8d}(a),$$

if $d > 0$, and

$$h(d) = \frac{1}{2} \sum_{0<a<|d|} \chi_{-4d}(a),$$

$$h(8d) = \sum_{0<a<2|d|} \chi_{-8d}(a),$$

if $d < 0$ (see [Dirichlet, 1840]).

The above four formulae follow from the formula

$$s_0(4, 1, \chi) = \frac{\tau(\chi)}{\pi} L(1, \chi_{-4}\overline{\chi}),$$

which holds for any even Dirichlet character χ, and the formulae of section 8. See also [Berndt, 1976, Theorem 3.7] and [Lerch, 1905, pp. 402, 408].

Berndt [Berndt, 1976, Corollary 4.3] found the formulae for short character sums over intervals of length $|d|/3$

$$h(-3d) = 2 \sum_{0<a<d/3} \chi_d(a), \tag{26}$$

if $d > 0$, $3 \nmid d$, and

$$h(d) = \frac{2}{3 - \chi_d(3)} \sum_{0<a<|d|/3} \chi_d(a),$$

if $d < 0$ (see also [Holden, 1906b], [Yamamoto, Y., 1977]).

Both the above formulae follow from the formulae

$$s_0(3, 1, \chi) = \begin{cases} \dfrac{\sqrt{3}\tau(\chi)}{2\pi} L(1, \chi_{-3}\overline{\chi}), & \text{if } \chi \text{ is even,} \\[2ex] \dfrac{\tau(\chi)}{2\pi i} (3 - \overline{\chi}(3)) L(1, \overline{\chi}), & \text{if } \chi \text{ is odd.} \end{cases}$$

For more details see [Berndt, 1976, Theorem 4.1].

Lerch ([Lerch, 1905, pp. 407, 404]), Berndt ([Berndt, 1976, Corollary 5.3, p. 276]) and Yamamoto ([Yamamoto, Y., 1977]) proved that for $d < 0$, $5 \nmid d$ we have

$$h(5d) = 2 \sum_{|d|/5<a<2|d|/5} \chi_d(a), \tag{27}$$

$$h(d) = \frac{2}{5 - \chi_d(5)}\Big(2 \sum_{0<a<|d|/5} \chi_d(a) + \sum_{|d|/5<a<2|d|/5} \chi_d(a)\Big).$$

The above two formulae follow from

$$s_0(5,1,\chi) = \frac{\tau(\chi)}{4\pi i}\Big((5 - \overline{\chi}(5))L(1,\overline{\chi}) - \sqrt{5}L(1,\chi_5\overline{\chi})\Big),$$

$$s_0(5,2,\chi) = \frac{\tau(\chi)\sqrt{5}}{2\pi i}L(1,\chi_5\overline{\chi})$$

(see for example [Berndt, 1976, Theorem 5.1]).

The case of Theorem 3 when $r = 8$ and χ is a quadratic character was studied in an elementary way in [Osborn, 1896] and [Glaisher, 1901, 1903a,b]. Some of the formulae for $r = 8$ and χ quadratic can be traced back to Gauss [Gauss, 1828, 1897] with the proofs given by Dedekind [Dedekind, 1863]. Dirichlet in [Dirichlet, 1839] proved that for odd d

$$h(-8d) = 2\Big(\sum_{0<a<d/8} \chi_d(a) - \sum_{3d/8<a<d/2} \chi_d(a)\Big),$$

if $d > 0$, and

$$h(8d) = 2\Big(\sum_{|d|/8<a<|d|/4} \chi_d(a) + \sum_{|d|/4<a<3|d|/8} \chi_d(a)\Big),$$

if $d < 0$. These two formulae were also proved in [Lerch, 1905, pp. 407, 409]. See also [Pepin, 1874], [Hurwitz, 1895], [Glaisher, 1903b], [Holden, 1907c], [Karpinski, 1904] and [Rédei, 1949/1950].

The sums $s_0(10, k, \chi)$ for $1 \le k \le 5$ and quadratic characters χ were discussed in [Karpinski, 1904], [Rédei, 1949/1950] and [Berndt, 1976]. Berndt ([Berndt, 1976, Theorem 8.1]) expressed these sums in terms of $L(1,\overline{\chi})$ and $L(1,\chi_5\overline{\chi})$. In [Berndt, 1976, Theorem 9.1], [Lerch, 1905, pp. 407, 408, 414], [Holden, 1906b, 1907b,c], [Karpinski, 1904] and [Rédei, 1949/1950] the authors investigated $s_0(12, k, \chi)$ for $1 \le k \le 6$ and quadratic characters χ.

Some of these formulae had already been stated by Gauss [Gauss, 1828, 1863] with the proofs given by Dedekind [Dedekind, 1863]. In the above mentioned papers, the authors expressed the short character sums as linear combinations of $L(1,\overline{\chi})$, $L(1,\chi_{-3}\overline{\chi})$ and $L(1,\chi_{-4}\overline{\chi})$. Berndt considered arbitrary Dirichlet characters χ while the other authors considered only quadratic characters. In the case $r = 12$ the obtained expressions for the character sums were linear combinations of the class numbers $h(d)$, $h(-3d)$, $h(-4d)$ and $h(12d)$. A similar situation occurs with short character sums over intervals of length $|d|/24$. This case was considered in the papers [Lerch, 1905, pp. 409, 410], [Karpinski, 1904] and [Rédei, 1949/1950] for quadratic characters χ_d and in [Berndt, 1976, Theorem 11.1] for arbitrary characters χ. These formulae for quadratic characters imply 24 new and different class number formulae involving the sums $s_0(24, k, \chi_d)$. In [Berndt, 1976, Corollary 11.3] just two of the formulae were given because of limitation of space. If $d > 0$, $2 \nmid d$, $3 \nmid d$ then, for instance, we have

$$8 \sum_{0<a<|d|/24} \chi_d(a) = \chi_d(2)(1 + \chi_d(3))h(-4d) + (1 + \chi_d(2))h(-3d)$$

$$+ (\chi_d(3) - 1)h(-8d) + h(-24d),$$

and if $d < 0$, $2 \nmid d$, $3 \nmid d$ then we have

$$8 \sum_{|d|/24<a<|d|/12} \chi_d(a) = (2\chi_d(2) - 1)(\chi_d(2) - 1)(1 - \chi_d(3))h(d)$$

$$+ (1 + \chi_d(3))h(8d) + (\chi_d(2) - 2)h(12d) + h(24d).$$

We remark that the sums $s_0(24, k, \chi_d)$ serve as "building bricks" from which all sums $s_0(m, k, \chi_d)$, where m divides 24, can be obtained in terms of class numbers of imaginary quadratic fields (see section 9.4). When m is not a divisor of 24, not all the characters occuring in Zagier's formula for a short character sum are quadratic, and this fact suggests that such sums cannot be given in terms of class numbers of imaginary quadratic fields.

If $r = 16$ the situation is quite different. Berndt (in [Berndt, 1976], Theorem 10.1) could only express the sums $s_0(16, k, \chi_d) + \chi_d(-1)s_0(16, 9 - k, \chi_d)$, $1 \leq k \leq 4$ in terms of class numbers of imaginary quadratic fields. In view of Zagier's identity we have some understanding of why there are no formulae of this type for the single sums $s_0(16, k, \chi_d)$.

When $m = 1$ or 2, Berndt ([Berndt, 1976]) gave formulae for $s_1(2, 1, \chi)$ and $s_1(4, 2, \chi)$, as well as the formulae

$$s_2(1, 1, \chi) = \frac{\tau(\chi)M^2}{\pi^2}L(2, \overline{\chi}),$$

$$s_2(2, 1, \chi) = \frac{\tau(\chi)M^2}{4\pi^2}(\overline{\chi}(2) - 2)L(2, \overline{\chi}),$$

in case of even χ, and formulae for $s_1(4, 1, \chi)$ and $s_1(4, 3, \chi)$, as well as the formulae

$$s_1(2, 1, \chi) = \frac{i\tau(\chi)M}{2\pi}(\overline{\chi}(2) - 1)L(1, \overline{\chi}),$$

$$s_1(3, k, \chi) = \frac{i\tau(\chi)M}{2\pi}\left(\frac{1}{3}(\overline{\chi}(3) - 1)L(1, \overline{\chi}) - \frac{k\sqrt{3}}{2\pi}L(2, \chi_{-3}\overline{\chi})\right),$$

in case of odd χ and $k = 1, 2$. If $\chi = \chi_d$, $d < 0$ then the above formulae together with the evaluation of Gauss sums (see section 4.4) give at once

$$6s_1(3, k, \chi_d) - (\chi_d(3) - 1)dh(d) = \begin{cases} -\dfrac{k}{2}k_2(-3d), & \text{if } 3 \nmid d, \\ -\dfrac{27k}{2}k_2(-d/3), & \text{if } 3 \mid d. \end{cases}$$

9.7   The Class Number Formulae of Lerch and Mordell

We conclude section 9 by using Zagier's identity to prove a result (Theorem 7) which contains the class number formulae of Lerch ([Lerch, 1905]) and Mordell ([Mordell, 1963]) (Theorem 6) as special cases.

THEOREM 6 ([Lerch, 1905], [Mordell, 1963]) *Let* $d_1 < 0$ *and* $d_2 > 1$ *be coprime fundamental discriminants. Then*

$$h(d_1 d_2) = 2 \sum_{1 \le a \le |d_1|/2} \chi_{d_1}(a) \sum_{1 \le n \le a d_2/|d_1|} \chi_{d_2}(n)$$

$$= -2 \sum_{1 \le a \le d_2/2} \chi_{d_2}(a) \sum_{1 \le n \le a|d_1|/d_2} \chi_{d_1}(n).$$

PROOF. The theorem follows immediately from the following more general result and equation (16). ∎

THEOREM 7 *Let* χ_1 *and* χ_2 *be nontrivial Dirichlet characters* modulo M_1 *and* M_2 *respectively. Assume that* χ_1 *and* χ_2 *are of opposite parity and that* M_1 *and* M_2 *are relatively prime. Then we have*

$$B_{1,\chi_1\chi_2} = 2\chi_1(-M_2)\chi_2(M_1) \sum_{1 \le a \le M_1/2} \chi_1(a) \sum_{1 \le n \le aM_2/M_1} \chi_2(n).$$

PROOF. It follows from Theorem 2 with $m = 0$ that for $\gcd(a, M_1) = 1$

$$\sum_{1 \le n \le aM_2/M_1} \chi_2(n) = -B_{1,\chi_2} + \frac{\overline{\chi}_2(M_1)}{\phi(M_1)} \sum_{\psi \bmod M_1} \overline{\psi}(-aM_2)B_{1,\chi_2\psi}.$$

Therefore we have

$$\sum_{1 \le a \le M_1/2} \chi_1(a) \sum_{1 \le n \le aM_2/M_1} \chi_2(n) = -B_{1,\chi_2} \sum_{1 \le a \le M_1/2} \chi_1(a)$$

$$+ \frac{\overline{\chi}_2(M_1)}{\phi(M_1)} \sum_{\psi \bmod M_1} \psi(-M_2)B_{1,\chi_2\psi} \sum_{1 \le a \le M_1/2} \overline{\psi}(a)\chi_1(a).$$

As χ_1 and χ_2 are of opposite parity either $B_{1,\chi_2} = 0$ or

$$\sum_{1 \leq a \leq M_1/2} \chi_1(a) = 0$$

so that

$$B_{1,\chi_2} \sum_{1 \leq a \leq M_1/2} \chi_1(a) = 0.$$

If ψ has opposite parity to χ_1 then ψ has the same parity as χ_2 and $B_{1,\chi_2\psi} = 0$. If ψ has the same parity as χ_1 then

$$\sum_{1 \leq a \leq M_1/2} \overline{\psi}\chi_1(a) = \begin{cases} \frac{1}{2}\phi(M_1), & \text{if } \psi = \chi_1, \\ 0, & \text{if } \psi \neq \chi_1, \end{cases}$$

which follows easily from the well known formula

$$\sum_{1 \leq a \leq M_1} \overline{\psi}\chi_1(a) = \begin{cases} \phi(M_1), & \text{if } \psi = \chi_1, \\ 0, & \text{if } \psi \neq \chi_1. \end{cases} \qquad \blacksquare$$

REMARK Let p be a prime with $p \equiv 5 \,(\mathrm{mod}\,8)$. Taking $d_1 = -7$ and $d_2 = p$ in Theorem 6, we obtain

$$h(-7p) = 2 \sum_{1 \leq a \leq 3} \sum_{1 \leq n \leq ap/7} \left(\frac{p}{n}\right)$$

(see also [Hudson and Williams, 1982]).

Using the notation of section 1.7, namely,

$$s(n, k) = \sum_{(k-1)p/n < \nu < kp/n} \left(\frac{\nu}{p}\right),$$

we deduce that

$$h(-7p) = 2(s(7, 1) - s(7, 3))$$
$$= 2(s(14, 1) + s(14, 2) - s(14, 5) - s(14, 6)).$$

The formula

$$h(-7p) = -2s(14, 6)$$

now follows from the elementary result

$$s(14, 1) + s(14, 2) = s(14, 5) \quad (p \equiv 5 \,(\mathrm{mod}\,8))$$

proved in [Johnson and Mitchell, 1977, p. 122]. This result also follows from Zagier's identity (Theorem 2).

Both Lerch and Mordell's proofs of Theorem 6 are rather complicated and much longer than the above. Our proof of the theorem shows how useful Zagier's identity can be. For other formulae of this type, see [Berndt, 1976, Theorems 12.1-7], [Currie and Williams, 1982, Lemmas 1, 2, 3, 5 and 10], [Holden, 1906b,c, 1907c], [Johnson and Mitchell, 1977] and [Urbanowicz, 1990c].

10. SOME ELEMENTARY LEMMAS

10.1 *Notation*

In this section we shall prove some auxiliary lemmas concerning character sums of the form

$$S_m(x, r, s, \chi) := \sum_{\substack{0 < a \leq x \\ a \equiv r \,(\mathrm{mod}\, s)}} \chi(a)a^m \,,$$

where χ is a nontrivial Dirichlet character modulo M, $x \geq 0$ is a positive real number, m is a nonnegative integer and r, s are integers with $s \geq 1$.

10.2 *Two Lemmas*

We shall express the sums $S_m(M, r, 8, \chi)$ ($r = 0$, 2, 4, 6) in terms of the short character sums $S_m(x, \chi)$ ($x = M/8$, $M/4$, $M/2$) defined in section 9.5.

By the formulae of section 8 in Chapter I for odd Dirichlet characters χ modulo M, we have

$$S_1(M, \chi) = MB_{1,\chi} \text{ and } S_0(M/2, \chi) = (\overline{\chi}(2) - 2)B_{1,\chi} \,.$$

For even Dirichlet characters χ modulo M, we have

$$S_2(M, \chi) = MB_{2,\chi} \text{ and } S_1(M/2, \chi) = \frac{1}{4}M(\overline{\chi}(2) - 2)B_{2,\chi} \,.$$

LEMMA 1 ([Urbanowicz, 1990/1991a]) *Let χ be a Dirichlet primitive quadratic character modulo M, where M is odd.*

(i) *If χ is even then we have:*

$$S_0(M, M, 8, \chi) = \chi(2)S_0(M/8, \chi) \,,$$

$$S_0(M, M + 2, 8, \chi) = -(1 + \chi(2))S_0(M/4, \chi) + \chi(2)S_0(M/8, \chi) \,,$$

$$S_0(M, M + 4, 8, \chi) = S_0(M/4, \chi) - \chi(2)S_0(M/8, \chi) \,,$$

$$S_0(M, M + 6, 8, \chi) = \chi(2)S_0(M/4, \chi) - \chi(2)S_0(M/8, \chi) \,.$$

(ii) *If χ is odd then we have:*

$$S_0(M, M, 8, \chi) = -\chi(2)S_0(M/8, \chi),$$

$$S_0(M, M+2, 8, \chi) = \chi(2)S_0(M/4, \chi) - \chi(2)S_0(M/8, \chi),$$

$$S_0(M, M+4, 8, \chi) = -S_0(M/4, \chi) + \chi(2)S_0(M/8, \chi),$$

$$S_0(M, M+6, 8, \chi) = S_0(M/2, \chi) - (1 + \chi(2))S_0(M/4, \chi)$$
$$+ \chi(2)S_0(M/8, \chi).$$

PROOF. Let us note that for any Dirichlet character χ we have

$$S_0(M/2, M, 4, \chi)$$

$$= \chi(-1) \sum_{\substack{0 \le a \le M/2 \\ a \equiv M \,(\mathrm{mod}\,4)}} \chi(M-a) = \chi(-1) \sum_{\substack{M/2 \le a \le M \\ 4|a}} \chi(a)$$

$$= \chi(-1)(S_0(M, 0, 4, \chi) - S_0(M/2, 0, 4, \chi)) \tag{28}$$

$$= \chi(-1)(S_0(M/4, \chi) - S_0(M/8, \chi))$$

and

$$S_0(M/2, M+2, 4, \chi)$$

$$= \chi(-1) \sum_{\substack{0 \le a \le M/2 \\ a \equiv M+2 \,(\mathrm{mod}\,4)}} \chi(M-a) = \chi(-1) \sum_{\substack{M/2 \le a \le M \\ a \equiv 2 \,(\mathrm{mod}\,4)}} \chi(a)$$

$$= \chi(-1)\left(\sum_{\substack{M/2 \le a \le M \\ 2|a}} \chi(a) - \sum_{\substack{M/2 \le a \le M \\ 4|a}} \chi(a) \right) \tag{29}$$

$$= \chi(-1)(S_0(M, 0, 2, \chi) - S_0(M/2, 0, 2, \chi)$$
$$- S_0(M, 0, 4, \chi) + S_0(M/2, 0, 4, \chi))$$

$$= \chi(-1)(\chi(2)S_0(M/2, \chi) - (1 + \chi(2))S_0(M/4, \chi) + S_0(M/8, \chi)).$$

The lemma now follows as

$$S_0(M, M, 8, \chi) = \chi(-1)S_0(M, 0, 8, \chi) = \chi(-1)\chi(2)S_0(M/8, \chi),$$

$$S_0(M, M+2, 8, \chi) = \chi(-1)S_0(M, -2, 8, \chi)$$
$$= \chi(-1)\chi(2)S_0(M/2, 3, 4, \chi),$$

$$S_0(M, M+4, 8, \chi) = \chi(-1)(S_0(M, 0, 4, \chi) - S_0(M, 0, 8, \chi))$$
$$= \chi(-1)(S_0(M/4, \chi) - \chi(2)S_0(M/8, \chi)),$$

$$S_0(M, M+6, 8, \chi) = \chi(-1)S_0(M, 2, 8, \chi)$$
$$= \chi(-1)\chi(2)S_0(M/2, 1, 4, \chi). \qquad \blacksquare$$

LEMMA 2 ([Urbanowicz, 1990/1991b]) *Let χ be a Dirichlet primitive quadratic character modulo M, where M is odd.*

(i) *If χ is even then we have*

$$S_1\,(M,M,8,\chi) = \chi(2)(-8S_1(M/8,\chi) + MS_0(M/8,\chi)),$$
$$\begin{aligned}S_1\,(M,M+2,8,\chi) = {} &4(S_1(M/2,\chi) - (2\chi(2)+1)S_1(M/4,\chi)\\ &+ 2\chi(2)S_1(M/8,\chi))\\ &- M(-(1+\chi(2))S_0(M/4,\chi) + \chi(2)S_0(M/8,\chi)),\end{aligned}$$
$$\begin{aligned}S_1\,(M,M+4,8,\chi) = {} &-4(S_1(M/4,\chi) - 2\chi(2)S_1(M/4,\chi))\\ &+ M(S_0(M/4,\chi) - \chi(2)S_0(M/8,\chi)),\end{aligned}$$
$$\begin{aligned}S_1\,(M,M+6,8,\chi) = {} &8\chi(2)(S_1(M/4,\chi) - S_1(M/8,\chi))\\ &- \chi(2)M(S_0(M/4,\chi) - S_0(M/8,\chi)).\end{aligned}$$

(ii) *If χ is odd then we have*

$$S_1\,(M,M,8,\chi) = \chi(2)(8S_1(M/8,\chi) - MS_0(M/8,\chi)),$$
$$\begin{aligned}S_1\,(M,M+2,8,\chi) = {} &8\chi(2)(S_1(M/4,\chi) - S_1(M/8,\chi))\\ &- \chi(2)M(S_0(M/4,\chi) - S_0(M/8,\chi)),\end{aligned}$$
$$\begin{aligned}S_1\,(M,M+4,8,\chi) = {} &4(S_1(M/4,\chi) - 2\chi(2)S_1(M/8,\chi))\\ &+ M(-S_0(M/4,\chi) + \chi(2)S_0(M/8,\chi)),\end{aligned}$$
$$\begin{aligned}S_1\,(M,M+6,8,\chi) = {} &4(S_1(M/2,\chi) - (2\chi(2)+1)S_1(M/4,\chi)\\ &+ 2\chi(2)S_1(M/8,\chi))\\ &- M(S_0(M/2,\chi) - (1+\chi(2))S_0(M/4,\chi) + \chi(2)S_0(M/8,\chi)).\end{aligned}$$

PROOF. First, note that in view of (28) and (29) we have

$$S_1(M/2,M,4,\chi)$$

$$= -\chi(-1)\sum_{\substack{0\le a\le M/2\\ a\equiv M\,(\mathrm{mod}\,4)}} \chi(M-a)(M-a) + MS_0(M/2,M,4,\chi)$$

$$= -\chi(-1)\sum_{\substack{M/2\le a\le M\\ 4\mid a}} \chi(a)a + MS_0(M/2,M,4,\chi)$$

$$= -\chi(-1)(S_1(M,0,4,\chi) - S_1(M/2,0,4,\chi)) + MS_0(M/2,M,4,\chi)$$

$$= -4\chi(-1)(S_1(M/4,\chi) - S_1(M/8,\chi))$$

$$\quad + M\chi(-1)(S_0(M/4,\chi) - S_0(M/8,\chi)),$$

and similarly

$$S_1(M/2, M+2, 4, \chi)$$

$$= -\chi(-1) \sum_{\substack{M/2 \leq a \leq M \\ a \equiv 2 \,(\mathrm{mod}\, 4)}} \chi(a)a + MS_0(M/2, M+2, 4, \chi)$$

$$= -\chi(-1)(S_1(M, 0, 2, \chi) - S_1(M, 0, 4, \chi) - S_1(M/2, 0, 2, \chi)$$

$$\quad + S_1(M/2, 0, 4, \chi)) + MS_0(M/2, M+2, 4, \chi)$$

$$= -2\chi(-1)(\chi(2)S_1(M/2, \chi) - (2+\chi(2))S_1(M/4, \chi) + 2S_1(M/8, \chi))$$

$$\quad + M\chi(-1)(\chi(2)S_0(M/2, \chi) - (1+\chi(2))S_0(M/4, \chi) + S_0(M/8, \chi)).$$

Applying the above formulae and Lemma 1 we obtain Lemma 2 as

$$S_1(M, M, 8, \chi) = -\chi(-1)S_1(M, 0, 8, \chi) + MS_0(M, M, 8, \chi)$$

$$\quad = -8\chi(-1)\chi(2)S_1(M/8, \chi) + MS_0(M, M, 8, \chi),$$

$$S_1(M, M+2, 8, \chi) = -\chi(-1)S_1(M, -2, 8, \chi) + MS_0(M, M+2, 8, \chi)$$

$$\quad = -2\chi(-1)\chi(2)S_1(M/2, 3, 4, \chi) + MS_0(M, M+2, 8, \chi),$$

$$S_1(M, M+4, 8, \chi) = -\chi(-1)S_1(M, 4, 8, \chi) + MS_0(M, M+4, 8, \chi)$$

$$\quad = -\chi(-1)(S_1(M, 0, 4, \chi) - S_1(M, 0, 8, \chi)) + MS_0(M, M+4, 8, \chi)$$

$$\quad = -4\chi(-1)(S_1(M/4, \chi) - 2\chi(2)S_1(M/8, \chi)) + MS_0(M, M+4, 8, \chi),$$

$$S_1(M, M+6, 8, \chi) = -\chi(-1)S_1(M, 2, 8, \chi) + MS_0(M, M+6, 8, \chi)$$

$$\quad = 2\chi(-1)\chi(2)S_1(M/2, 1, 4, \chi) + MS_0(M, M+6, 8, \chi). \qquad \blacksquare$$

Lemmas 1 and 2 can be extended to any Dirichlet character χ modulo M (M odd) by slight modifications of the above proofs, since the only facts that we used for quadratic characters were $\chi(-1) = (-1)^{(M-1)/2}$ and $\chi(4) = 1$.

10.3   *Nagell's Formulae*

For a positive real number t and a natural number n let $A(t, n)$ denote the number of positive integers $\leq t$ that are prime to n.

It is easy to prove that

$$A\left(\frac{mn}{2}, n\right) = \frac{m}{2}\phi(n),$$

where $m = 1$ or 2, and

$$A\left(\frac{mn}{4}, n\right) = \begin{cases} \dfrac{m}{4}\phi(n), & \text{if } x > 0, \\[2mm] \dfrac{m}{4}\phi(n) + (-1)^{(n-m)/2}2^{y-2}, & \text{if } x = 0, \end{cases} \tag{30}$$

where n is odd, $m = 1$ or 3, and x (resp. y) is the number of prime factors of n of the form $4t + 1$ (resp. $4t + 3$).

It is much more difficult to prove that

$$A\left(\frac{mn}{8}, n\right) = \frac{m}{8}\phi(n) \tag{31}$$

$$+ \begin{cases} 0, & \text{if } x > 0, \text{ or } x = 0, y \geq 0, z > 0, u > 0, \\ \rho_1(-1)^{(n-1)(n-3)/8} 2^{y+u-2}, & \text{if } x = 0, y \geq 0, z = 0, u > 0, \\ \rho_2(-1)^{(n-1)/2} 2^{y+z-3}, & \text{if } x = 0, y \geq 0, z > 0, u = 0, \\ (2\rho_1 + \rho_2)(-1)^{(n-1)/2} 2^{y-3}, & \text{if } x = 0, y > 0, z = 0, u = 0, \end{cases}$$

where n is odd, $m = 1, 3, 5$ or 7, x, y, z, u are respectively the numbers of prime divisors of n of the form $8t + 1$, $8t - 1$, $8t + 3$, $8t - 3$, and

$$\rho_1 = (-1)^{(m-1)(m-3)/8}, \quad \rho_2 = (-1)^{(m-1)/2}.$$

For details see [Nagell, 1923] (cf. [Urbanowicz, 1983] and [Hardy and Williams, 1986]). Nagell [Nagell, 1923] has also shown that

$$\sum_{\substack{l \equiv -mn \,(\bmod\, r) \\ 1 \leq l \leq n, \gcd(l,n)=1}} 1 = A\left(\frac{(m+1)n}{r}, n\right) - A\left(\frac{mn}{r}, n\right), \tag{32}$$

if n is odd, $0 \leq m \leq r - 1$ and $r = 2, 4$ or 8.

Chapter II

CLASS NUMBER CONGRUENCES

In section 8.5 of Chapter I we observed that Gauss' theory of ambiguous classes shows that

$$h(d) \equiv \begin{cases} 0 \,(\bmod\, 2^{\nu-1}), & \text{if } d < 0, \text{ or } d > 0 \text{ and } N(\varepsilon_d) = -1, \\ 0 \,(\bmod\, 2^{\nu-2}), & \text{if } d > 0 \text{ and } N(\varepsilon_d) = 1, \end{cases} \quad (1)$$

where d denotes the discriminant of the quadratic field $\mathbb{Q}(\sqrt{d})$, ν the number of distinct prime divisors of d, and $N(\varepsilon_d)$ the norm of the fundamental unit ε_d (> 1) of $\mathbb{Q}(\sqrt{d})$ when $d > 0$. This congruence is a simple consequence of an algebraic fact. For negative d the congruence (1) can be obtained analytically from Dirichlet's class number formula, see Chapter VI. For positive d it is surprising that as yet no one has deduced this congruence analytically from Dirichlet's formula for $h(d)$. In the next sections we discuss $h(d)$ modulo powers of 2 when d has a small number of prime factors. In obtaining such congruences authors have traditionally used either analytic (complex or 2-adic) methods, i.e., those involving Dirichlet's class number formulae (or their 2-adic analogues), or algebraic methods, i.e., those using the structure of an appropriate class group. It is likely that Gauss' congruence can be obtained for any quadratic field using 2-adic analytic methods, see Chapter IV.

In this chapter we give a linear congruence relating class numbers obtained in [Hardy and Williams, 1986] and its generalization obtained in [Szmidt, Urbanowicz and Zagier, 1995].

1. IMAGINARY QUADRATIC FIELDS

1.1 *The Case $\nu = 1$*

If $d < 0$ and $\nu = 1$ we have $d = -4, -8$ or $-p$, where p is a prime with $p \equiv 3 \,(\bmod\, 4)$. Since $h(-4) = h(-8) = h(-3) = 1$ we need only consider

51

$h(-p)$, p prime, $p > 3$, $p \equiv 3 \ (\mathrm{mod}\, 4)$. From equation (17) in Chapter I, we see that

$$h(-p) = c \sum_{1 \le a \le (p-1)/2} \left(\frac{-p}{a}\right) \equiv \frac{p-1}{2} \equiv 1 \ (\mathrm{mod}\, 2),$$

where $c = 1$ if $p \equiv 7 \ (\mathrm{mod}\, 8)$ and $c = 1/3$ if $p \equiv 3 \ (\mathrm{mod}\, 8)$, so that $h(-p)$ is odd. The value of $h(-p)$ modulo 4 is given in the next theorem.

THEOREM 8 ([Mordell, 1961]) *Let p be a prime number with $p \equiv 3 \ (\mathrm{mod}\, 4)$ and $p > 3$. Then*

$$h(-p) \equiv \begin{cases} 1 \ (\mathrm{mod}\, 4), & \text{if } ((p-1)/2)! \equiv -1 \ (\mathrm{mod}\, p), \\ 3 \ (\mathrm{mod}\, 4), & \text{if } ((p-1)/2)! \equiv 1 \ (\mathrm{mod}\, p). \end{cases}$$

PROOF. Write $E = ((p-1)/2)!$. Denote by $\rho_1, \rho_2, \ldots$ the R quadratic residues of p less than $p/2$, and by $\nu_1, \nu_2, \ldots$ the N quadratic nonresidues less than $p/2$. Then the quadratic residues $\rho_1', \rho_2', \ldots$ greater than $p/2$ are given by $p - \nu_1, p - \nu_2, \ldots$ since $p \equiv 3 \ (\mathrm{mod}\, 4)$. Then

$$E = \rho_1 \rho_2 \cdots \nu_1 \nu_2 \cdots \equiv (-1)^N \rho_1 \rho_2 \cdots \rho_1' \rho_2' \cdots \equiv (-1)^N \ (\mathrm{mod}\, p) \text{ if } p > 3,$$

as

$$(-1)^N E \equiv g^{2+4+\cdots+(p-1)} = g^{(p^2-1)/4} = (g^{(p-1)/2})^{(p+1)/2} \equiv 1 \ (\mathrm{mod}\, p),$$

where g is a primitive root modulo p.

Now

$$R + N = \frac{p-1}{2},$$

and it is known from the class number formula that

$$R - N = \rho h(-p),$$

where

$$\rho = \begin{cases} 1, & \text{if } p \equiv 7 \ (\mathrm{mod}\, 8), \\ 3, & \text{if } p \equiv 3 \ (\mathrm{mod}\, 8), \ p > 3 \end{cases}$$

(see formula (17) of Chapter I). Hence

$$2N = \frac{p-1}{2} - \rho h(-p).$$

Then if $p \equiv 7 \ (\mathrm{mod}\, 8)$,

$$2N \equiv 3 - h(-p) \ (\mathrm{mod}\, 4),$$

and if $p \equiv 3 \,(\bmod\, 8)$,

$$2N \equiv 1 - 3h(-p) \,(\bmod\, 4).$$

The first congruence is

$$N \equiv \frac{1}{2}(-1 - h(-p)) \,(\bmod\, 2)$$

and the second is

$$N \equiv \frac{1}{2}(1 + h(-p)) \,(\bmod\, 2).$$

These are both included in

$$N \equiv \frac{1}{2}(1 + h(-p)) \,(\bmod\, 2). \qquad \blacksquare$$

For primes $p \equiv 3 \,(\bmod\, 4)$ with $p > 3$ the congruence $E \equiv (-1)^N \,(\bmod\, p)$ established in the proof above was first obtained by Dirichlet [Dirichlet, 1828]. Since Dirichlet's class number formula gives N in terms of $h(-p)$ (as shown above), Mordell's theorem is implicit in the work of Dirichlet. Kronecker [Kronecker, 1884] has combined these ideas of Dirichlet with some of his own formulae involving class numbers to determine the sign in the congruence $((p-1)/2)! \equiv +1$ or $-1 \,(\bmod\, p)$ in terms of the parity of the number of representations of p by a certain form.

The analogue of Mordell's theorem for $h(p)$ where p is a prime with $p \equiv 1 \,(\bmod\, 4)$ was proved by Chowla.

THEOREM 9 ([Chowla, 1961]) *Let p be a prime number with $p \equiv 1 \,(\bmod\, 4)$.*
Let

$$\varepsilon_p = \frac{t + u\sqrt{p}}{2}$$

be the fundamental unit (> 1) of $\mathbb{Q}(\sqrt{p})$. Then

$$h(p) \equiv \lambda \,(\bmod\, 4),$$

where

$$\left(\frac{p-1}{2}\right)! \equiv -\frac{\lambda t}{2} \,(\bmod\, p), \quad \lambda = \pm 1.$$

PROOF. From Dirichlet's formula we derive

$$\varepsilon_p^h \sqrt{p} = \prod_n (1 - \zeta^n),$$

where $\zeta = \exp(2\pi i/p)$ and n runs over the quadratic nonresidues modulo p in the interval $(0, p)$ (see section 8.2 of Chapter I). Working with integers in $\mathbb{Q}(\zeta)$, we note that

$$\sqrt{p} \equiv \left(\frac{p-1}{2}\right)! (1 - \zeta)^{(p-1)/2} \, (\bmod \, (1 - \zeta)^{(p+1)/2}),$$

$$\prod_n (1 - \zeta^n) \equiv (1 - \zeta)^{(p-1)/2} \, (\bmod \, (1 - \zeta)^{(p+1)/2}),$$

and the asserted congruence follows.   ∎

It appears to be an open problem to evaluate $h(-p)$ modulo 8 for $p \equiv 3$ $(\bmod \, 4)$. In [Williams, 1979, 1982], however, the author has related $h(-p)$ to $h(4p)$ modulo 4 and modulo 8, namely, he has shown that

$$h(-p) \equiv h(4p) + U + 1 \, (\bmod \, 4)$$

and

$$h(-p) \equiv -h(4p) + R + S + 2(-1)^{(p-3)/4} \, (\bmod \, 8),$$

if $R \equiv S \, (\bmod \, 4)$,

$$h(-p) \equiv h(4p) + R - S - 2 \, (\bmod \, 8),$$

if $R \equiv -S \, (\bmod \, 4)$, where

$$\varepsilon_p = T + U\sqrt{p} = \frac{1}{2}(R + S\sqrt{p})^2$$

is the fundamental unit (> 1) of $\mathbb{Q}(\sqrt{p})$.

These congruences were proved using Dirichlet's class number formulae and the values of the polynomials

$$F_-(X) = \prod_{\substack{1 \le n \le p \\ \left(\frac{n}{p}\right) = -1}} (X - \zeta^n), \qquad F_+(X) = \prod_{\substack{1 \le r \le p \\ \left(\frac{r}{p}\right) = 1}} (X - \zeta^r),$$

when X is a fourth root of unity.

1.2   *The Case $\nu = 2$*

Perhaps Hurwitz [Hurwitz, 1895] was the first to establish congruences in this case. When $d < 0$ and $\nu = 2$, we have

(i) $d = -4p$, where p is a prime with $p \equiv 1 \, (\bmod \, 4)$,

(ii) $d = -8p$, where p is an odd prime,

(iii) $d = -pq$, where p and q are primes with $p \equiv 1 \,(\bmod\, 4)$ and $q \equiv 3$ $(\bmod\, 4)$.

In these cases $h(d) \equiv 0 \,(\bmod\, 2)$ by (1).

Case (i). When $p \equiv 1 \,(\bmod\, 4)$ Brown [Brown, 1975] used Gauss' theory of ambiguous classes and genera to show that

$$h(-4p) \equiv \frac{p-1}{2} \,(\bmod\, 4),$$

see also [Glaisher, 1903a], [Lerch, 1905, p. 224]. This congruence also follows easily from equation (25) of Chapter I as

$$h(-4p) = 2 \sum_{1 \leq a \leq (p-1)/4} \left(\frac{p}{a}\right) \equiv 2 \sum_{1 \leq a \leq (p-1)/4} 1 = \frac{p-1}{2} \,(\bmod\, 4).$$

A congruence modulo 8 for $h(-4p)$ was given by Gauss in a letter to Dirichlet, see [Gauss, 1828, 1897]. It is well known that a prime $p \equiv 1 \,(\bmod\, 4)$ is uniquely expressible as the sum of two squares, i.e.,

$$p = a^2 + b^2,$$

where $a \equiv 1 \,(\bmod\, 4)$ and $b \equiv ((p-1)/2)!\, a \,(\bmod\, p)$. Gauss' congruence states that

$$h(-4p) \equiv -a + b + 1 \,(\bmod\, 8).$$

We now sketch a proof of Gauss' congruence. For Dedekind's original proof, see [Dedekind, 1863]. See also [Yamamoto, K., 1965] and [Barkan, 1975]. The starting point is formula (25) of Chapter I, that is

$$h(-4p) = \frac{p-1}{2} - 4N,$$

where N denotes the number of quadratic nonresidues modulo p in the interval $(0, p/4)$. Next, modifying the argument in [Yamamoto, K., 1965, Lemma 4] it can be shown that

$$2^{(p-1)/4} \equiv \begin{cases} (-1)^N \,(\bmod\, p), & \text{if } p \equiv 1 \,(\bmod\, 8), \\ (-1)^N ((p-1)/2)! \,(\bmod\, p), & \text{if } p \equiv 5 \,(\bmod\, 8), \end{cases}$$

which together with the above formula for $h(-4p)$ gives at once the congruence

$$2^{(p-1)/4}$$
$$\equiv \begin{cases} (-1)^{(p-1)/8 + h(-4p)/4} \,(\bmod\, p), & \text{if } p \equiv 1 \,(\bmod\, 8), \\ (-1)^{(p-5)/8 + (h(-4p)-2)/4} ((p-1)/2)! \,(\bmod\, p), & \text{if } p \equiv 5 \,(\bmod\, 8). \end{cases}$$

Then using the supplement to the law of biquadratic reciprocity in the form

$$2^{(p-1)/4} \equiv \begin{cases} 1 \,(\bmod p), & \text{if } b \equiv 0 \,(\bmod 8), \\ b/a \,(\bmod p), & \text{if } b \equiv 2 \,(\bmod 8), \\ -1 \,(\bmod p), & \text{if } b \equiv 4 \,(\bmod 8), \\ -b/a \,(\bmod p), & \text{if } b \equiv 6 \,(\bmod 8), \end{cases}$$

(see for instance [Kaplan, 1976]) we obtain Gauss' congruence in view of the simple congruence

$$p \equiv 2a + 1 - 2 \left(\frac{2}{p}\right) \,(\bmod 16).$$

When $p \equiv 1 \,(\bmod 8)$ a number of authors have investigated $h(-4p)$ modulo 8. Let a, b, c, d, e, f, g, h be integers such that $a \equiv 1 \,(\bmod 4)$, $b \equiv 0 \,(\bmod 4)$, $c \equiv 1 \,(\bmod 2)$, $d \equiv 0 \,(\bmod 2)$, $e > 0$, $f \equiv 0 \,(\bmod 4)$, $g > 0$, and

$$p = a^2 + b^2 = c^2 + 2d^2 = e^2 - 2f^2 = 2g^2 - h^2,$$

then

$$h(-4p) \equiv a + b - 1 \quad \text{(Gauss)}$$

$$\equiv 2d$$

$$\equiv 2e - 2$$

$$\equiv 2g - 2$$

$$\equiv 2 - 2 \left(\frac{g}{p}\right) \,(\bmod 8),$$

see [Barrucand and Cohn, 1969], [Hasse, 1969b], [Brown, 1972] and [Williams, 1976]. Barrucand and Cohn used the arithmetic of the biquadratic field $\mathbb{Q}(\sqrt{-1}, \sqrt{2})$ to prove their results, Hasse used norm residue symbols, whereas Brown used ambiguous classes and genera. A further result of Barrucand and Cohn is equivalent to the biquadratic reciprocity law

$$\left(\frac{p}{2}\right)_4 \left(\frac{2}{p}\right)_4 = (-1)^{h(-4p)/4},$$

where

$$\left(\frac{p}{2}\right)_4 = (-1)^{(p-1)/8}$$

and

$$\left(\frac{2}{p}\right)_4 = \left(\frac{w}{p}\right), \quad w^2 \equiv 2 \,(\bmod p).$$

From this law Kaplan [Kaplan, 1977a] deduced, using elementary arguments, the congruence

$$h(-4p) \equiv T \,(\bmod 8),$$

where $T + U\sqrt{p}$ is a unit of norm -1 in the real quadratic field $\mathbb{Q}(\sqrt{p})$. This congruence was first proved by E. Lehmer [Lehmer, 1971]. See also [Cohn and Cooke, 1976].

The determination of $h(-4p)$ modulo 16 was considered in [Williams, 1981a], where the author showed that

$$h(-4p) \equiv \begin{cases} T + (p-1) \,(\bmod 16), & \text{if } T \equiv 0 \,(\bmod 8), \\ T + (p-1) + 4(h(p)-1) \,(\bmod 16), & \text{if } T \equiv 4 \,(\bmod 8), \end{cases} \tag{2}$$

by a manipulation of Dirichlet's class number formula. Combining (2) with Theorem 9, and recalling that $t = 2T$, $u = 2U$ in this case, we deduce that

$$h(-4p) \equiv \begin{cases} T + (p-1) \,(\bmod 16), & \text{if } T \equiv 0 \,(\bmod 8), \\ T + (p-1) + 4(\lambda-1) \,(\bmod 16), & \text{if } T \equiv 4 \,(\bmod 8), \end{cases}$$

where

$$\lambda \equiv \left(\frac{p-1}{2}\right)! \, T \,(\bmod p), \quad \lambda = \pm 1.$$

When $T \equiv 0 \,(\bmod 8)$ in [Leonard and Williams, 1982] it was shown using genera that

$$h(-4p) \equiv 0 \,(\bmod 16) \Leftrightarrow \left(\frac{g}{p}\right)_4 = \left(\frac{2h}{g}\right).$$

When $p \equiv 5 \,(\bmod 8)$ we have $h(-4p) \equiv 2 \,(\bmod 4)$. In this case

$$T + U\sqrt{p} = \left(\frac{t + u\sqrt{p}}{2}\right)^\mu = \varepsilon_p^\mu, \quad \mu = 1 \text{ or } 3.$$

An easy calculation shows that

$$\frac{t}{2} \equiv (-1)^{(\mu-1)/2} T \,(\bmod p).$$

In [Williams, 1981c] it was shown analytically that

$$h(-4p) \equiv -\mu T h(p) \,(\bmod 8). \tag{3}$$

Appealing to Theorem 9, we see that

$$h(-4p) \equiv -\mu T \lambda \,(\bmod 8), \tag{4}$$

where

$$\lambda \equiv (-1)^{(\mu-1)/2} T \left(\frac{p-1}{2}\right)! \,(\bmod p), \quad \lambda = \pm 1. \tag{5}$$

In [Kaplan and Williams, 1982b] the congruence (3) was extended to

$$h(-4p) \equiv (4 - \mu)Th(p) \ (\bmod 16)$$

using Dirichlet's class number formulae and the values of $F_{\pm}(s)$ when s is an eighth root of unity.

Case (ii). We now turn to $h(-8p)$ where p is an odd prime. Using genera, Brown [Brown, 1972] has shown that

$$h(-8p) \equiv 1 - \left(\frac{2}{p}\right) \ (\bmod 4).$$

This can also be proved using short character sums (see [Lerch, 1905, p. 225]). From (17) of Chapter I we have

$$h(-8p) = \frac{1}{2} \sum_{1 \le l \le 4p} \left(\frac{-8p}{l}\right).$$

Making use of the basic properties of the Kronecker symbol, we obtain in an elementary way

$$h(-8p) = 2 \sum_{\substack{1 \le l \le p \\ l \equiv 1 \,(\bmod 4)}} \left(\frac{8p}{l}\right).$$

Hence

$$h(-8p) \equiv 2 \sum_{\substack{1 \le l \le p-1 \\ l \equiv 1 \,(\bmod 4)}} 1 = \frac{1}{2} \left(p - (-1)^{(p-1)/2}\right) \equiv 1 - \left(\frac{2}{p}\right) \ (\bmod 4).$$

We now discuss $h(-8p)$ modulo 8 in each of the four cases $p \equiv 1, 3, 5, 7$ $(\bmod 8)$.

When $p \equiv 1 \,(\bmod 8)$) we have $h(-8p) \equiv 0 \,(\bmod 4)$. In this case Pizer [Pizer, 1976] has employed a type number formula from the theory of quaternion algebras to prove that

$$h(-4p) + h(-8p) \equiv \frac{p-1}{2} \ (\bmod 8).$$

Appealing to Gauss' congruence for $h(-4p)$ modulo 8, we obtain

$$h(-8p) \equiv b \,(\bmod 8),$$

where $p = a^2 + b^2$, $a \equiv 1 \,(\bmod 4)$, $b \equiv 0 \,(\bmod 4)$, see [Barkan, 1975]. Hasse [Hasse, 1969a] has also determined $h(-8p)$ modulo 8 using the structure of the

class group of the imaginary quadratic field of discriminant $-8p$. He showed
that

$$h(-8p) \equiv 2 - 2 \left(\frac{-2}{e} \right) \, (\mathrm{mod}\, 8),$$

where $p = e^2 - 2f^2$, $e > 0$, $f > 0$. This congruence also follows from
the congruences of Pizer, and Barrucand and Cohn. Replacing (e, f) by
$(3e + 4f, 2e + 3f)$, if necessary, we can suppose that $e \equiv 1 \,(\mathrm{mod}\, 4)$. If
$h(-8p) \equiv 0 \,(\mathrm{mod}\, 8)$, so that $e \equiv 1 \,(\mathrm{mod}\, 8)$, then in [Leonard and Williams,
1982] the authors have shown that

$$h(-8p) \equiv 4 - 4 \left(\frac{e}{p} \right)_4 \, (\mathrm{mod}\, 16).$$

When $p \equiv 3 \,(\mathrm{mod}\, 8)$ we have $h(-8p) \equiv 2 \,(\mathrm{mod}\, 4)$. (A special case is
treated in [Brown, 1974c].) In this case Pizer [Pizer, 1976] has shown that

$$2h(-p) + h(-8p) \equiv \frac{p - 3}{2} \, (\mathrm{mod}\, 8).$$

Coupling this congruence with Theorem 8 we obtain

$$h(-8p) \equiv \frac{p - 3}{2} + 2\lambda \, (\mathrm{mod}\, 8),$$

where

$$\left(\frac{p-1}{2} \right)! \equiv \lambda \,(\mathrm{mod}\, p), \quad \lambda = \pm 1.$$

When $p \equiv 5 \,(\mathrm{mod}\, 8)$ we have $h(-8p) \equiv 2 \,(\mathrm{mod}\, 4)$. Pizer [Pizer, 1976]
has shown in this case that

$$h(-4p) + h(-8p) \equiv \frac{p - 13}{2} \, (\mathrm{mod}\, 8).$$

Making use of Gauss' congruence for $h(-4p)$ modulo 8 we obtain

$$h(-8p) \equiv b \, (\mathrm{mod}\, 8),$$

where $p = a^2 + b^2$, $a \equiv 1 \,(\mathrm{mod}\, 4)$, $b \equiv ((p - 1)/2)! \, a \,(\mathrm{mod}\, p)$, and
$b \equiv 2 \,(\mathrm{mod}\, 4)$, see [Barkan, 1975].

Alternatively, appealing to the congruence (4), we deduce that

$$h(-8p) \equiv \frac{p - 13}{2} - T\lambda \, (\mathrm{mod}\, 8),$$

where λ is given by (5). The explicit determination of $h(-8p)$ modulo 16
remains to be discovered.

When $p \equiv 7 \,(\mathrm{mod}\,8)$ we have $h(-8p) \equiv 0 \,(\mathrm{mod}\,4)$ and it follows from [Hasse, 1969a], [Berndt, 1976] and [Pizer, 1976] that

$$h(-8p) \equiv \begin{cases} 0\,(\mathrm{mod}\,8), & \text{if } p \equiv 15\,(\mathrm{mod}\,16), \\ 4\,(\mathrm{mod}\,8), & \text{if } p \equiv 7\,(\mathrm{mod}\,16). \end{cases}$$

Hasse used algebraic methods, Berndt used analytic methods, whereas Pizer used type numbers of Eichler orders. If $p \equiv 15 \,(\mathrm{mod}\,16)$ in [Leonard and Williams, 1982] the authors have shown algebraically that

$$h(-8p) \equiv 4 - 4\left(\frac{f}{e}\right) \,(\mathrm{mod}\,16),$$

where

$$p = e^2 - 2f^2, \quad e > 0, \quad f > 0, \quad e \equiv 1\,(\mathrm{mod}\,16).$$

The determination of $h(-8p)$ modulo 16 for $p \equiv 7 \,(\mathrm{mod}\,16)$ remains unknown.

Case (iii). Now we discuss the case $d = -pq$, where p and q are primes with $p \equiv 1 \,(\mathrm{mod}\,4)$ and $q \equiv 3 \,(\mathrm{mod}\,4)$. By (1) we have $h(-pq) \equiv 0 \,(\mathrm{mod}\,2)$. To examine $h(-pq)$ modulo 4 we appeal to Theorem 6 of Chapter I. We obtain

$$h(-pq) = 2 \sum_{1 \leq l \leq (q-1)/2} \left(\frac{-q}{l}\right) \sum_{1 \leq n \leq lp/q} \left(\frac{p}{n}\right)$$

so that

$$h(-pq) \equiv 2 \sum_{1 \leq l \leq (q-1)/2} \sum_{1 \leq n \leq lp/q} 1 \,(\mathrm{mod}\,4).$$

Therefore we have

$$h(-pq) \equiv \sum_{l=1}^{(q-1)/2} \left[\frac{lp}{q}\right] \equiv \begin{cases} 0\,(\mathrm{mod}\,4), & \text{if } \left(\dfrac{p}{q}\right) = 1, \\ 2\,(\mathrm{mod}\,4), & \text{if } \left(\dfrac{p}{q}\right) = -1. \end{cases}$$

The last congruence follows from a form of Gauss' lemma. For the congruence for $h(-pq)$ modulo 4 see also [Hasse, 1970a], [Berndt, 1976 ($q = 3$)] and [Pizer, 1976]. The determination of $h(-pq)$ modulo 8 was carried out in [Currie and Williams, 1982] by an extension of the above argument.

THEOREM 10 ([Currie and Williams, 1982]) *Let p and q be primes with $p \equiv -q \equiv 1 \,(\mathrm{mod}\,4)$. Define a and b by $p = a^2 + b^2$, $b \equiv ((p-1)/2)!\,a \,(\mathrm{mod}\,p)$, $a \equiv 1 \,(\mathrm{mod}\,4)$. Then*

(i) *if* $\left(\dfrac{p}{q}\right) = +1$ *(equivalently* $\left(\dfrac{q}{p}\right) = +1$) *we have*

$$h(-pq) \equiv \begin{cases} 0\,(\bmod 8), & \text{if } \left(\dfrac{a-bi}{a+bi}\right)^{(q+1)/4} \equiv 1\,(\bmod q), \\[3mm] 4\,(\bmod 8), & \text{if } \left(\dfrac{a-bi}{a+bi}\right)^{(q+1)/4} \equiv -1\,(\bmod q); \end{cases}$$

(ii) *if* $\left(\dfrac{p}{q}\right) = -1$ *(equivalently* $\left(\dfrac{q}{p}\right) = -1$) *we have*

$$h(-pq) \equiv \begin{cases} 2\,(\bmod 8), & \text{if } \left(\dfrac{a-bi}{a+bi}\right)^{(q+1)/4} \equiv -i\,(\bmod q), \\[3mm] 6\,(\bmod 8), & \text{if } \left(\dfrac{a-bi}{a+bi}\right)^{(q+1)/4} \equiv i\,(\bmod q); \end{cases}$$

if $q > 3$ *and* $h(-q) \equiv 1\,(\bmod 4)$, *and*

$$h(-pq) \equiv \begin{cases} 2\,(\bmod 8), & \text{if } \left(\dfrac{a-bi}{a+bi}\right)^{(q+1)/4} \equiv i\,(\bmod q), \\[3mm] 6\,(\bmod 8), & \text{if } \left(\dfrac{a-bi}{a+bi}\right)^{(q+1)/4} \equiv -i\,(\bmod q); \end{cases}$$

if $q > 3$ *and* $h(-q) \equiv 3\,(\bmod 4)$, *and*

$$h(-3p) \equiv \begin{cases} 2\,(\bmod 8), & \text{if } a \equiv -b\,(\bmod 3), \\ 6\,(\bmod 8), & \text{if } a \equiv b\,(\bmod 3). \end{cases}$$

if $q = 3$.

For $h(-pq)$ modulo 16 see [Leonard and Williams, 1983].

1.3 *The Case $\nu = 3$*

By (1) we have $h(d) \equiv 0\,(\bmod 4)$. A discussion of $h(d)$ modulo 8 or 16 requires examination of a large number of cases, see for example [Pumplün, 1965], [Brown, 1973], [Brown and Parry, 1973], [Kaplan, 1976], [Pizer, 1976], [Hardy and Williams, 1987]. We will therefore just focus on $h(-4pq)$ and $h(-8pq)$, where p and q are distinct primes with $p \equiv q\,(\bmod 4)$ in the first case.

THEOREM 11 *Let* $p \equiv q \equiv 1\,(\bmod 4)$ *be distinct primes. Then*

$$h(-4pq) \equiv \begin{cases} 4\,(\bmod 8), & \text{if } p \not\equiv q\,(\bmod 8) \text{ and } \left(\dfrac{p}{q}\right) = -1, \\[3mm] 0\,(\bmod 8), & \text{otherwise.} \end{cases}$$

Some special cases of Theorem 11 were proved by Berndt [Berndt, 1976] using short character sums, see section 9.6 of Chapter I. $h(-4pq)$ modulo 16 is treated in [Brown, 1974c].

THEOREM 12 *Let* $p \equiv q \equiv 3 \pmod 4$ *be distinct primes. Choose p and q so that* $\left(\dfrac{p}{q}\right) = 1$. *Then*

$$h(-4pq) \equiv \begin{cases} 4 \pmod 8, & \text{if } q \equiv 3 \pmod 8, \\ 0 \pmod 8, & \text{if } q \equiv 7 \pmod 8. \end{cases}$$

[Brown, 1974a] has given the value of $h(-4pq)$ modulo 16 in certain cases.

THEOREM 13 *Let p and q be distinct odd primes. Then*

$$h(-8pq) \equiv 4 \pmod 8$$

in the cases $p \equiv -q \equiv 3 \pmod 8$; $p \equiv q \equiv 5 \pmod 8$; $p \equiv 1 \pmod 8$, $q \equiv \pm 3$ *and* $\left(\dfrac{p}{q}\right) = -1$; $p \equiv \pm 3 \pmod 8$, $q \equiv 7 \pmod 8$ *and* $\left(\dfrac{p}{q}\right) = -1$; *and*

$$h(-8pq) \equiv 0 \pmod 8,$$

in all other cases.

PROOFS OF THEOREMS 11, 12 AND 13. These theorems can easily be deduced from the Hardy-Williams congruence discussed in section 1.5. The proofs are left to the reader as an exercise. Theorem 13 also follows from [Pizer, 1976, Corollary to Proposition 5]. ∎

When $h(-8pq) \equiv 0 \pmod 8$, Kaplan [Kaplan, 1976] has determined $h(-8pq)$ modulo 16 using ambiguous forms and genera. When $p \equiv q \equiv 5$ $\pmod 8$ and $\left(\dfrac{p}{q}\right) = 1$, in [Hardy and Williams, 1987] the authors have determined $h(-8pq)$ modulo 16.

1.4 $\nu > 3$

When $\nu > 3$ the situation is of course even more complicated. Some general results were obtained by Kaplan [Kaplan, 1976] using ambiguous forms and genera. We just cite one such result.

If $d = -4p_1 \ldots p_{\nu-1}$, where ν is odd and $p_1, \ldots, p_{\nu-1}$ are distinct primes with $p_i \equiv 1 \pmod 8$ $(1 \leq i \leq \nu - 1)$ and $\left(\dfrac{p_i}{p_j}\right) = -1$ $(1 \leq i < j \leq \nu - 1)$ then

$$h(-4p_1 \cdots p_{\nu-1}) \equiv 0 \pmod{2^\nu}.$$

For other results see [Plancherel, 1908], [Rédei, 1928], [Costa, 1993]. See also [Rédei and Reichardt, 1934] (cf. [Rédei, 1934a,b], [Lagarias, 1980], [Kisilevsky, 1982], [Hurrelbrink, 1994]).

1.5 Hardy and Williams Congruence

In [Hardy and Williams, 1986] the authors discovered a new congruence of linear type relating class numbers of imaginary quadratic fields. Since the class numbers of imaginary quadratic fields are in fact the first Bernoulli numbers (see section 8 of Chapter I), that is, the values of L-series at 0, the discovered congruence was the first example of a linear congruence between special values of L-series. Extensions of this type of linear congruence have been considered later by many authors. These authors extended the Hardy-Williams congruence to both classical L-function and 2-adic L-functions. We remark that H. Lang and Schertz [Lang and Schertz, 1976] have given a quadratic-type congruence between class numbers of both real and imaginary quadratic fields.

For any fundamental discriminants d and e with $e \mid d$, and for $\mathcal{D}$ from $\{-8e, -4e, e, 8e\}$, we set

$$H(d, \mathcal{D}) = \left(\frac{e}{|d/e|}\right) \prod_{\substack{p \mid |d/e| \\ p \, \text{prime}}} \left(1 - \left(\frac{\mathcal{D}}{p}\right)\right) h(\mathcal{D}),$$

where p runs through all prime divisors of $|d/e|$.

THEOREM 14 ([Hardy and Williams, 1986]) *Let d be an odd fundamental discriminant having ν prime factors. Then with above notation we have:*

$$\sum_{\substack{e \mid d, \, e > 1 \\ e \equiv 1 \, (\text{mod} \, 4)}} (\chi_e(2) H(d, -4e) + H(d, -8e))$$

$$+ \sum_{\substack{e \mid d, \, e < 0 \\ e \equiv 1 \, (\text{mod} \, 4)}} ((5 - \chi_e(2)) H(d, e) - H(d, 8e))$$

$$\equiv \frac{1}{2}(-1)^{\nu+1} \phi(|d|) + H_1(d) + H_2(d) \, (\text{mod} \, 2^{\nu+2}),$$

where

$$H_1(d) = \begin{cases} 2^{\nu-1}, & \text{if d is divisible only by primes} \equiv 3 \, (\text{mod} \, 4), \\ 0, & \text{otherwise}, \end{cases}$$

$$H_2(d) = \begin{cases} 0, & \text{if $3 \nmid d$, or $3 \mid d$ and $p \mid (d/3)$} \\ & \quad \text{for some prime $p \equiv 1 \, (\text{mod} \, 3)$}, \\ 2^{\nu+1}, & \text{otherwise}. \end{cases}$$

PROOF. This congruence is a special case (for $m = 1$ and $r = 8$) of a more general congruence proved in [Szmidt, Urbanowicz and Zagier, 1995] using Zagier's identity (Theorem 2 for $r\,|\,24$) and the same reasoning as used in [Hardy and Williams, 1986], where the authors derived their congruence from Theorem 3 using a Möbius inversion argument (for details see section 1.8). The form of the right hand side of the congruence of Theorem 14 follows from an explicit formula for the number of integers in the interval $(0, N/8)$ which are relatively prime to N given in (31), Chapter I.   ∎

The main theorem in [Hardy and Williams, 1986] is also a special case of more general congruences proved by Gras [Gras, 1989] and Uehara [Uehara, 1990]. These congruences are special cases of the congruence found in [Urbanowicz and Wójcik, 1995/1996], and [Wójcik, 1998]. Gras proved his congruence using 2-adic measure theory. Uehara's proof is more elementary but also obtained by 2-adic methods (without using 2-adic measures). In [Urbanowicz and Wójcik, 1995/1996] the authors applied Uehara's ideas as well as results of Coleman [Coleman, 1982]. We shall deal with these congruences in Chapter IV.

We remark that the (corrected) congruences of Kenku [Kenku, 1977] follow easily from the Hardy-Williams congruence. Kenku used the theory of elliptic curves. Costa [Costa, 1989, 1992], using the cuspidal behaviour of 2-adic modular forms, found some congruences which follow from the congruences of Gras [Gras, 1989] and Uehara [Uehara, 1990].

1.6   *Class Number Congruences for Imaginary Quadratic Fields Implied by the Short Character Sum Formulae*

Many of the congruences between class numbers of imaginary quadratic fields follow directly from the short character sum formulae cited in section 9.6 of Chapter I, or from the Hardy-Williams congruence for small ν. Berndt [Berndt, 1976] using only the short character sum formulae listed in section 9.6 found in an elementary way many congruences of the type

$$h(-sp) \equiv r \,(\,\mathrm{mod}\,2^{\alpha}\,),$$

where $-sp$ is a fundamental discriminant, $s\,|\,24$ or $s\,|\,40$, $\alpha \leq 3$, $0 \leq r \leq 2^{\alpha} - 1$ for primes p in certain arithmetic progressions.

1.7   *Class Number Congruences for Imaginary Quadratic Fields Implied by the Hardy-Williams Congruence*

In [Pizer, 1976] the following congruences are given. We leave it to the reader to show that they are simple consequences of the Hardy-Williams congruence.

THEOREM 15 *If p is a prime number not less than* 5, *we have the following congruences*:

(i) $\quad h(-4p) + h(-8p) \equiv \begin{cases} 0 \,(\bmod\, 8), & \text{if } p \equiv 1 \,(\bmod\, 16), \\ 4 \,(\bmod\, 8), & \text{if } p \equiv 9 \,(\bmod\, 16), \end{cases}$

(ii) $\quad 2h(-p) + h(-8p) \equiv \begin{cases} 0 \,(\bmod\, 8), & \text{if } p \equiv 3 \,(\bmod\, 16), \\ 4 \,(\bmod\, 8), & \text{if } p \equiv 11 \,(\bmod\, 16), \end{cases}$

(iii) $\quad h(-4p) + h(-8p) \equiv \begin{cases} 0 \,(\bmod\, 8), & \text{if } p \equiv 13 \,(\bmod\, 16), \\ 4 \,(\bmod\, 8), & \text{if } p \equiv 5 \,(\bmod\, 16), \end{cases}$

(iv) $\qquad\qquad\quad h(-8p) \equiv \begin{cases} 0 \,(\bmod\, 8), & \text{if } p \equiv 15 \,(\bmod\, 16), \\ 4 \,(\bmod\, 8), & \text{if } p \equiv 7 \,(\bmod\, 16). \end{cases}$

THEOREM 16 *If p and q are distinct primes* ≥ 3, *we have the following congruences*:

(i)
$$\left. \begin{array}{l} p \equiv 1 \,(\bmod\, 8) \\ q \equiv 1 \,(\bmod\, 8) \\ \left(\dfrac{p}{q}\right) = +1 \end{array} \right\} \Rightarrow h(-4pq) + h(-8pq) \equiv 0 \,(\bmod\, 16),$$

(ii)
$$\left. \begin{array}{l} p \equiv 1 \,(\bmod\, 8) \\ q \equiv 1 \,(\bmod\, 8) \\ \left(\dfrac{p}{q}\right) = -1 \end{array} \right\} \Rightarrow \begin{array}{l} h(-4pq) + h(-8pq) \\ \equiv \begin{cases} 0 \,(\bmod\, 16), & \text{if } pq \equiv 1 \,(\bmod\, 16), \\ 8 \,(\bmod\, 16), & \text{if } pq \equiv 9 \,(\bmod\, 16), \end{cases} \end{array}$$

(iii)
$$\left. \begin{array}{l} p \equiv 1 \,(\bmod\, 8) \\ q \equiv 3 \,(\bmod\, 8) \\ \left(\dfrac{p}{q}\right) = +1 \end{array} \right\} \Rightarrow \begin{array}{l} 2h(-4p) + 2h(-pq) + h(-8pq) \\ \equiv \begin{cases} 0 \,(\bmod\, 16), & \text{if } p \equiv 1 \,(\bmod\, 16), \\ 8 \,(\bmod\, 16), & \text{if } p \equiv 9 \,(\bmod\, 16), \end{cases} \end{array}$$

(iv)
$$\left. \begin{array}{l} p \equiv 1 \,(\bmod\, 8) \\ q \equiv 3 \,(\bmod\, 8) \\ \left(\dfrac{p}{q}\right) = -1 \end{array} \right\} \Rightarrow \begin{array}{l} 2h(-4p) + 2h(-pq) + h(-8pq) \\ \equiv \begin{cases} 0 \,(\bmod\, 16), & \text{if } q \equiv 3 \,(\bmod\, 16), \\ 8 \,(\bmod\, 16), & \text{if } q \equiv 11 \,(\bmod\, 16), \end{cases} \end{array}$$

(v)

$$p \equiv 1 \,(\mathrm{mod}\,8)$$
$$q \equiv 5 \,(\mathrm{mod}\,8)$$
$$\left.\left(\frac{p}{q}\right) = +1 \right\} \Rightarrow 2h(-8p) + h(-4pq) + h(-8pq)$$
$$\equiv 0 \,(\mathrm{mod}\,16),$$

(vi)

$$p \equiv 1 \,(\mathrm{mod}\,8)$$
$$q \equiv 5 \,(\mathrm{mod}\,8)$$
$$\left.\left(\frac{p}{q}\right) = -1 \right\} \Rightarrow 2h(-8p) + h(-4pq) + h(-8pq)$$
$$\equiv \begin{cases} 0 \,(\mathrm{mod}\,16), & \text{if } pq \equiv 13 \,(\mathrm{mod}\,16), \\ 8 \,(\mathrm{mod}\,16), & \text{if } pq \equiv 5 \,(\mathrm{mod}\,16), \end{cases}$$

(vii)

$$p \equiv 1 \,(\mathrm{mod}\,8)$$
$$q \equiv 7 \,(\mathrm{mod}\,8)$$
$$\left.\left(\frac{p}{q}\right) = +1 \right\} \Rightarrow h(-8pq) \equiv 0 \,(\mathrm{mod}\,16),$$

(viii)

$$p \equiv 1 \,(\mathrm{mod}\,8)$$
$$q \equiv 7 \,(\mathrm{mod}\,8)$$
$$\left.\left(\frac{p}{q}\right) = -1 \right\} \Rightarrow h(-8pq)$$
$$\equiv \begin{cases} 0 \,(\mathrm{mod}\,16), & \text{if } pq \equiv 15 \,(\mathrm{mod}\,16), \\ 8 \,(\mathrm{mod}\,16), & \text{if } pq \equiv 7 \,(\mathrm{mod}\,16), \end{cases}$$

(ix)

$$p \equiv 3 \,(\mathrm{mod}\,8)$$
$$q \equiv 3 \,(\mathrm{mod}\,8)$$
$$\left.\left(\frac{p}{q}\right) = +1 \right\} \Rightarrow 2h(-8p) + 2h(-8q) + h(-4pq) + h(-8pq)$$
$$\equiv \begin{cases} 4 \,(\mathrm{mod}\,16), & \text{if } q \equiv 11 \,(\mathrm{mod}\,16), \\ 12 \,(\mathrm{mod}\,16), & \text{if } q \equiv 3 \,(\mathrm{mod}\,16), \end{cases}$$

(x)

$$p \equiv 3 \,(\mathrm{mod}\,8)$$
$$q \equiv 5 \,(\mathrm{mod}\,8)$$
$$\left.\left(\frac{p}{q}\right) = +1 \right\} \Rightarrow 4h(-p) + 2h(-8q) + h(-8pq)$$
$$\equiv \begin{cases} 4 \,(\mathrm{mod}\,16), & \text{if } p \equiv 11 \,(\mathrm{mod}\,16) \\ & \quad \text{or } p = 3, \\ 12 \,(\mathrm{mod}\,16), & \text{if } p \equiv 3 \,(\mathrm{mod}\,16), \\ & \quad p \neq 3, \end{cases}$$

(xi)

$$\left.\begin{array}{l} p \equiv 3 \,(\mathrm{mod}\,8) \\ q \equiv 5 \,(\mathrm{mod}\,8) \\ \left(\dfrac{p}{q}\right) = -1 \end{array}\right\} \Rightarrow 4h(-p) + 2h(-8q) + h(-8pq)$$

$$\equiv \begin{cases} 4\,(\mathrm{mod}\,16), & \textit{if } q \equiv 5\,(\mathrm{mod}\,16),\ p \neq 3, \\ & \textit{or if } q \equiv 13\,(\mathrm{mod}\,16),\ p = 3, \\ 12\,(\mathrm{mod}\,16), & \textit{if } q \equiv 13\,(\mathrm{mod}\,16),\ p \neq 3, \\ & \textit{or if } q \equiv 5\,(\mathrm{mod}\,16),\ p = 3, \end{cases}$$

(xii)

$$\left.\begin{array}{l} p \equiv 3 \,(\mathrm{mod}\,8) \\ q \equiv 7 \,(\mathrm{mod}\,8) \\ \left(\dfrac{p}{q}\right) = +1 \end{array}\right\} \Rightarrow h(-4pq) + h(-8pq) \equiv 0\,(\mathrm{mod}\,16),$$

(xiii)

$$\left.\begin{array}{l} p \equiv 3 \,(\mathrm{mod}\,8) \\ q \equiv 7 \,(\mathrm{mod}\,8) \\ \left(\dfrac{p}{q}\right) = -1 \end{array}\right\} \Rightarrow h(-4pq) + h(-8pq)$$

$$\equiv \begin{cases} 0\,(\mathrm{mod}\,16), & \textit{if } pq \equiv 13\,(\mathrm{mod}\,16), \\ 8\,(\mathrm{mod}\,16), & \textit{if } pq \equiv 5\,(\mathrm{mod}\,16), \end{cases}$$

(xiv)

$$\left.\begin{array}{l} p \equiv 5 \,(\mathrm{mod}\,8) \\ q \equiv 5 \,(\mathrm{mod}\,8) \\ \left(\dfrac{p}{q}\right) = +1 \end{array}\right\} \Rightarrow 2h(-4p) + 2h(-8q) + h(-4pq) + h(-8pq)$$

$$\equiv \begin{cases} 4\,(\mathrm{mod}\,16), & \textit{if } p \equiv 5\,(\mathrm{mod}\,16), \\ 12\,(\mathrm{mod}\,16), & \textit{if } p \equiv 13\,(\mathrm{mod}\,16), \end{cases}$$

(xv)

$$\left.\begin{array}{l} p \equiv 5 \,(\mathrm{mod}\,8) \\ q \equiv 5 \,(\mathrm{mod}\,8) \\ \left(\dfrac{p}{q}\right) = -1 \end{array}\right\} \Rightarrow 2h(-4q) + 2h(-8p) + h(-4pq) + h(-8pq)$$

$$\equiv \begin{cases} 4\,(\mathrm{mod}\,16), & \textit{if } p \equiv 5\,(\mathrm{mod}\,16), \\ 12\,(\mathrm{mod}\,16), & \textit{if } p \equiv 13\,(\mathrm{mod}\,16), \end{cases}$$

(xvi)

$$\left.\begin{array}{l} p \equiv 5 \,(\mathrm{mod}\,8) \\ q \equiv 7 \,(\mathrm{mod}\,8) \\ \left(\dfrac{p}{q}\right) = +1 \end{array}\right\} \Rightarrow 2h(-pq) + h(-8pq)$$

$$\equiv \begin{cases} 0\,(\mathrm{mod}\,16), & \textit{if } q \equiv 15\,(\mathrm{mod}\,16), \\ 8\,(\mathrm{mod}\,16), & \textit{if } q \equiv 7\,(\mathrm{mod}\,16), \end{cases}$$

(xvii)
$$\left.\begin{array}{l} p \equiv 5\,(\bmod 8) \\ q \equiv 7\,(\bmod 8) \\ \left(\dfrac{p}{q}\right) = -1 \end{array}\right\} \Rightarrow \begin{array}{l} 2h(-pq) + h(-8pq) \\ \equiv \begin{cases} 0\,(\bmod 16), & if\, p \equiv 13\,(\bmod 16), \\ 8\,(\bmod 16), & if\, p \equiv 5\,(\bmod 16), \end{cases} \end{array}$$

(xviii)
$$\left.\begin{array}{l} p \equiv 7\,(\bmod 8) \\ q \equiv 7\,(\bmod 8) \\ \left(\dfrac{p}{q}\right) = +1 \end{array}\right\} \Rightarrow \begin{array}{l} h(-4pq) + h(-8pq) \\ \equiv \begin{cases} 0\,(\bmod 16), & if\, q \equiv 15\,(\bmod 16), \\ 8\,(\bmod 16), & if\, q \equiv 7\,(\bmod 16). \end{cases} \end{array}$$

THEOREM 17 *If p, q, r are distinct odd primes such that* $pqr \equiv 7\,(\bmod 8)$, *we have*:

If not all p, q, r are $\equiv 3\,(\bmod 4)$, *then*

$$h(-pqr) \equiv 0\,(\bmod 8), if\, \left(\frac{p}{q}\right) + \left(\frac{p}{r}\right) + \left(\frac{q}{r}\right) > 0,$$

$$\equiv 4\,(\bmod 8), if\, \left(\frac{p}{q}\right) + \left(\frac{p}{r}\right) + \left(\frac{q}{r}\right) < 0.$$

If all p, q, r are $\equiv 3\,(\bmod 4)$, *then*

$$h(-pqr) \equiv 4\,(\bmod 8), if\, \left(\left(\frac{p}{q}\right), \left(\frac{p}{r}\right), \left(\frac{q}{r}\right) \right) = (1, -1, 1)\, or\, (-1, 1, -1),$$

$$\equiv 0\,(\bmod 8), \textit{otherwise}.$$

Many other congruences of the above type also follow from the Hardy-Williams congruence. For example, by considering cases depending upon the values of p, q, r modulo 8 and the values of the Legendre symbols $\left(\dfrac{p}{q}\right), \left(\dfrac{q}{r}\right), \left(\dfrac{r}{p}\right)$, we could obtain from the Hardy-Williams congruence congruences involving $h(-4pqr)$ and $h(-8pqr)$ modulo 32 for distinct primes p, q, r. However there are too many cases to make it practical to give a complete analysis. For example in the case $p \equiv q \equiv r \equiv 1\,(\bmod 4)$ it is necessary to consider 20 cases and in the case $p \equiv q \equiv 1\,(\bmod 4)$, $r \equiv 3\,(\bmod 4)$ 40 cases are required. As an example we give the following theorem.

THEOREM 18 *Let p, q, r be distinct odd primes.*

(i) *If* $p \equiv q \equiv r \equiv 1\,(\bmod 8)$ *and*

$$\left(\frac{p}{q}\right) = \left(\frac{q}{r}\right) = \left(\frac{r}{p}\right) = -1,$$

we have

$$h(-4pqr) + h(-8pqr) \equiv 2p + 2q + 2r - 6 \,(\mathrm{mod}\,32).$$

(ii) *If* $p \equiv 1 \,(\mathrm{mod}\,8)$, $q \equiv 5 \,(\mathrm{mod}\,8)$, $r \equiv 7 \,(\mathrm{mod}\,8)$ *and*

$$\left(\frac{p}{q}\right) = -1, \quad \left(\frac{q}{r}\right) = 1, \quad \left(\frac{r}{p}\right) = -1,$$

we have

$$2h(-pqr) + h(-8pqr) \equiv 2q + 2r + 8 \,(\mathrm{mod}\,32).$$

(iii) *If* $p \equiv 1 \,(\mathrm{mod}\,8)$, $q \equiv 3 \,(\mathrm{mod}\,8)$, $q > 3$, $r \equiv 7 \,(\mathrm{mod}\,8)$ *and*

$$\left(\frac{p}{q}\right) = -1, \quad \left(\frac{q}{r}\right) = 1, \quad \left(\frac{r}{p}\right) = -1,$$

we have

$$h(-8pqr) - h(-4pqr) \equiv 2p + 2q - 2r + 6 \,(\mathrm{mod}\,32).$$

(iv) *If* $p \equiv q \equiv r \equiv 7 \,(\mathrm{mod}\,8)$ *and*

$$\left(\frac{p}{q}\right) = \left(\frac{q}{r}\right) = \left(\frac{r}{p}\right) = -1,$$

we have

$$4h(-pqr) - h(-8pqr) \equiv 2p + 2q + 2r - 10 \,(\mathrm{mod}\,32).$$

PROOF. This is an immediate consequence of the Hardy-Williams congruence. ∎

1.8   *Linear Congruence Relations*

In this section we give a new congruence relating the values of Dirichlet L-series attached to quadratic characters at nonpositive integers (or equivalently, between the numbers $B_{m,\chi_d}/m$) modulo powers of 2 (or 3). This congruence generalizes the Hardy-Williams congruence since class numbers of imaginary quadratic fields are the first Bernoulli numbers up to sign. For $r \in \mathbb{Z}$ denote by $\mathcal{T}_r$ the set of all fundamental discriminants dividing r. For example, for the divisors of 24 we have $\mathcal{T}_1 = \mathcal{T}_2 = \{1\}$, $\mathcal{T}_3 = \mathcal{T}_6 = \{-3, 1\}$, $\mathcal{T}_4 = \{-4, 1\}$, $\mathcal{T}_8 = \{-8, -4, 1, 8\}$, $\mathcal{T}_{12} = \{-4, -3, 1, 12\}$, and $\mathcal{T}_{24} = \mathcal{T}_8 \cup \mathcal{T}_{12} \cup \{-24, 24\}$.

If χ is a Dirichlet character modulo M and d is any nonzero integer, then for $m \geq 0$ we set

$$B_{m,\chi}^{[d]} = \prod_{\substack{p|d \\ p \text{ prime}}} (1 - \chi(p)p^{m-1}) \cdot B_{m,\chi} \,.$$

We show next that the quantity $B_{m,\chi}^{[d]}$ is just the generalized Bernoulli number $B_{m,\chi'}$ for the character χ' modulo $M|d|$ induced by χ. As the character χ modulo $M|d|$ is induced from the character χ modulo M, we have

$$B_{m,\chi'} = B_{m,\chi} \sum_{e|(M|d|)} \mu(e)\chi(e)e^{m-1}$$

$$= B_{m,\chi} \prod_{\substack{p|(M|d|) \\ p \text{ prime}}} (1 - \chi(p)p^{m-1})$$

$$= B_{m,\chi} \prod_{\substack{p \,|\, |d| \\ p \text{ prime}}} (1 - \chi(p)p^{m-1})$$

$$= B_{m,\chi}^{[d]} \,.$$

This assertion also follows from formula (10) of Chapter I and the first formula of section 4.5 of this chapter, or else from (23) of Chapter I and a Möbius inversion argument

$$\mathcal{L}_\chi(s) = \sum_{\substack{n \geq 1 \\ \gcd(n,M)=1}} \chi(n)\exp(ns)$$

$$= \sum_{n \geq 1} \chi(n)\exp(ns) \sum_{e|\gcd(n,M)} \mu(e)$$

$$= \sum_{e|M} \mu(e)\chi(e)\mathcal{L}_\chi(es) \,.$$

Furthermore, for any nonzero integer d we define a modification of the generalized Bernoulli polynomial by

$$B_{m,\chi}^{[d]}(X) = \sum_{n=0}^{m} \binom{m}{n} B_{n,\chi}^{[d]} X^{m-n},$$

which has the property

$$B_{m,\chi}^{[d]}(-X) = (-1)^m \chi(-1) B_{m,\chi}^{[d]}(X) \tag{6}$$

unless $M = m = d = 1$ (see (6) of Chapter I). We shall use the above formula later.

THEOREM 19 ([Szmidt, Urbanowicz and Zagier, 1995]) *Let d be a fundamental discriminant, and r and c integers prime to d with $r > 1$ and $r \mid 24$. Then for any integer $m \geq 1$ the number*

$$r^{m-1}\phi(r) \sum_{e \in T_d} \chi_e(c) B_{m,\chi_e}^{[d]} - \sum_{\tau \in T_r} \chi_\tau(-d) \sum_{e \in T_d} \chi_e(rc) B_{m,\chi_{e\tau}}^{[d]}(d) \qquad (7)$$

is an integer divisible by $2^{\nu+\varepsilon} r^{m-1}\phi(r)m$, where ν denotes the number of distinct prime factors of d and $\varepsilon = 1$ if $8 \mid d$ and 0 otherwise.

PROOF. We shall apply Theorem 2 in the case when r is a divisor of 24. We also restrict χ to be a quadratic character. Then all the characters ψ modulo r are quadratic and so all the generalized Bernoulli numbers occurring in Theorem 2 are attached to quadratic characters. Specifically, we take two coprime fundamental discriminants K and d and let χ range over the characters modulo $M = |Kd|$ induced by χ_{Ke} with $e \in T_d$. Multiplying both sides of the equation of Theorem 2 by $\phi(r)\chi_e(c)$ for a fixed integer c prime to M and summing over all such characters, we find that

$$\sum_{e \in T_d} \chi_e(c)\left(-r^{m-1}\phi(r) B_{m,\chi_{Ke}}^{[d]} + \chi_{Ke}(r) \sum_{\tau \in T_r} \chi_\tau(-N) B_{m,\chi_{Ke\tau}}^{[d]}(N)\right)$$

$$= m r^{m-1}\phi(r) \sum_{\substack{0<n<(N/r) \\ \gcd(n,d)=1}} \chi_K(n) n^{m-1} \sum_{e \in T_d} \chi_e(nc)$$

and the left hand side is divisible by $m r^{m-1}\phi(r)2^{\nu+\varepsilon}$ because

$$\sum_{e \in T_d} \chi_e(nc) = \prod_{\substack{p|d, p>2 \\ p \text{ prime}}} \left(1 + \left(\frac{nc}{p}\right)\right) \cdot \left(1 + \left(\frac{-4}{nc}\right)\right)_{\text{if } 4 \mid d}$$

$$\cdot \left(1 + \left(\frac{8}{nc}\right)\right)_{\text{if } 8 \mid d} \equiv 0 \,(\mathrm{mod}\, 2^{\nu+\varepsilon}).$$

To obtain the theorem, we take $N = M = |d|$ and if $d < 0$, we use (6), i.e., apply the evenness or oddness of $B_{m,\chi}^{[d]}(X)$ to replace the argument $|d|$ of the modified generalized Bernoulli polynomials by d. ∎

Since $B_{m,\chi}$ for quadratic characters χ is almost always integral (see Chapter I, section 3.3) the essential statement of the theorem is that a certain integer is divisible by a power of 2 and, if $3 \mid r$, by a power of 3. For example, for $r = 24$ it says that the quotient of (7) by m is divisible by $2^{3m+\nu}3^{m-1}$. These congruences extend those discovered in [Hardy and Williams, 1986] ($m = 1$ and $r = 8$), those in [Urbanowicz, 1990b] ($m = 2$ and $r = 8$) and in [Szmidt and Urbanowicz, 1994] (here $m \geq 1$, $r = 8$ and the congruence is modulo

$2^{3m-1+\nu}$). They are of the same general linear type as those in [Gras, 1989], [Uehara, 1990], [Urbanowicz and Wójcik, 1995/1996], and [Wójcik, 1998]. In these four papers the authors used 2-adic L-functions. We shall deal with this approach in Chapter IV.

The congruence of Theorem 19 for $r = 8$ was discovered and proved in [Szmidt and Urbanowicz, 1994] using a different method. Zagier found the simpler method of proof presented here. The key of his proof was the identity of Theorem 2. The starting point for the proof of [Szmidt and Urbanowicz, 1994] was Berndt's formula (stated in Theorem 1), more precisely the equation of Theorem 5. The remainder of the proof is the same as that of Theorem 19. The theorem of [Szmidt, Urbanowicz and Zagier, 1995] with $r = 8$ is a special case of the congruence of Wójcik [Wójcik, 1998].

2.   REAL QUADRATIC FIELDS

2.1   *The Case $\nu = 1$*

If $d > 0$ and $\nu = 1$ we have $d = 8$ or $d = p$, where p is a prime with $p \equiv 1 \,(\bmod\, 4)$. Since $h(8) = 1$ we need only consider $h(p)$, p (prime) $\equiv 1$ $(\bmod\, 4)$. In this case Gauss' congruence (1) does not give any information. We show that $h(p)$ is odd. We consider the real number

$$\eta = \frac{\prod\limits_{n} \sin(\pi n/p)}{\prod\limits_{r} \sin(\pi r/p)} \,,$$

where r and n run through the quadratic residues and nonresidues modulo p respectively in the interval $(0, p/2)$. An easy calculation shows that η is a unit of the real quadratic field $\mathbb{Q}(\sqrt{p})$ of norm -1. By Dirichlet's class number formula

$$\eta = \varepsilon_p^{h(p)} \,,$$

where ε_p is the fundamental unit (> 1) of $\mathbb{Q}(\sqrt{p})$. Hence $\eta > 1$, $N(\varepsilon_p) = -1$, and $h(p)$ is odd.

We remark that more generally Dirichlet's formula (see section 8 of Chapter I) asserts that for a real quadratic field of discriminant d

$$\frac{\prod\limits_{\substack{0<n<d/2 \\ \left(\frac{d}{n}\right)=-1}} \sin(\pi n/d)}{\prod\limits_{\substack{0<r<d/2 \\ \left(\frac{d}{r}\right)=1}} \sin(\pi r/d)} = \varepsilon_d^{h(d)} \,,$$

where ε_d is the fundamental unit (> 1) of this field. For details see [Borevich and Shafarevich, 1966, p. 344].

The determination of $h(p)$ modulo 4 due to Chowla was given in Theorem 9. When $p \equiv 5 \pmod 8$ a different evaluation of $h(p)$ modulo 4 has been given in [Williams, 1981c] making use of Chowla's result. The author found the following congruences. If $\varepsilon_p = \frac{1}{2}(t + u\sqrt{p})$ is the fundamental unit of $\mathbb{Q}(\sqrt{p})$ (so that $t \equiv u \pmod 2$ and $t^2 - pu^2 = -4$) and a, b are the unique integers such that $p = a^2 + b^2$, $a \equiv 1 \pmod 4$, $b \equiv 2 \pmod 4$, $b \equiv ((p-1)/2)! \, a$ $\pmod p$, then

$$
h(p) \equiv
\begin{cases}
\dfrac{1}{2}(-2t + u + b + 1) \pmod 4, & \text{if } t \equiv u \equiv 1 \pmod 2, \\[2mm]
\dfrac{1}{4}(t + u + 2b + 2) \pmod 4, & \text{if } t \equiv u \equiv 0 \pmod 2.
\end{cases}
$$

It is still an unsolved problem to determine $h(p)$ modulo 8 when $p \equiv 1$ $\pmod 4$.

2.2 The Case $\nu = 2$

In this case we have

(i) $d = 4p$, where p is a prime with $p \equiv 3 \pmod 4$, or

(ii) $d = 8p$, where p is an odd prime, or

(iii) $d = pq$, where p and q are distinct primes with $p \equiv q \pmod 4$.

(i) When $d = 4p$, $p \equiv 3 \pmod 4$, since $h(12) = 1$ we may suppose that $p > 3$, the norm of the fundamental unit $\varepsilon_{4p} \, (> 1)$ of $\mathbb{Q}(\sqrt{p})$ is 1, and Gauss' congruence (1) does not give any information. As $N(\varepsilon_{4p}) = 1$, it does not seem possible to prove that $h(4p)$ is odd directly from Dirichlet's class number formula for $h(4p)$. Instead we consider $F_+(i)$, where the polynomial $F_+(X)$ was defined in section 1.1. By means of Dirichlet's formulae for $h(4p)$ and $h(-p)$ (see section 8 of Chapter I) one can show that

$$
F_+(i) =
\begin{cases}
w^3(-1)^{(h(-p)+1)/2}\varepsilon_{4p}^{-h(4p)/2}, & \text{if } p \equiv 3 \pmod 8, \\[2mm]
w^5\varepsilon_{4p}^{-h(4p)/2}, & \text{if } p \equiv 7 \pmod 8,
\end{cases}
\tag{8}
$$

where $w = (1+i)/\sqrt{2}$, see [Williams, 1982]. Thus

$$
\sqrt{2}F_+(i) = i^b(1+i)\varepsilon_{4p}^{-h(4p)/2}
$$

for some rational integer b. If $h(4p)$ is even, clearly

$$
\sqrt{2}F_+(i) \in \mathbb{Q}(i, \sqrt{p}).
$$

On the other hand, $F_+(X)$ is a polynomial in X with coefficients in $\mathbb{Q}(\sqrt{-p})$. Thus

$$F_+(i) \in \mathbb{Q}(i, \sqrt{-p}) = \mathbb{Q}(i, \sqrt{p}).$$

Hence $\sqrt{2} \in \mathbb{Q}(i, \sqrt{p})$, which is a contradiction. Thus $h(4p)$ must be odd. See also [Brown, 1974b], where this result is obtained using ambiguous forms and genera.

In [Williams, 1979] it is shown using (8) that

$$h(-p) \equiv h(4p) + U + 1 \,(\bmod\, 4),$$

where $\varepsilon_{4p} = T + U\sqrt{p} = \frac{1}{2}(R + S\sqrt{p})^2$ $(R, S, T, U \in \mathbb{Z})$. Thus, appealing to Theorem 8, we see that

$$h(4p) \equiv \lambda - U - 1 \,(\bmod\, 4),$$

where

$$\left(\frac{p-1}{2}\right)! \equiv -\lambda \,(\bmod\, p), \quad \lambda = \pm 1.$$

As for the value of $h(4p)$ modulo 8, in [Williams, 1982] the author showed that

$$h(4p)$$
$$\equiv \begin{cases} R + S + 2(-1)^{(p-3)/4} - h(-p) \,(\bmod\, 8), & \text{if } R \equiv S \,(\bmod\, 4), \\ -R + S + 2 + h(-p) \,(\bmod\, 8), & \text{if } R \equiv -S \,(\bmod\, 4). \end{cases}$$

Nothing is known about $h(4p)$ modulo 16.

(ii) Now suppose $d = 8p$, where p is an odd prime. Brown [Brown, 1974b] using ambiguous forms and genera showed that

$$h(8p) \equiv \begin{cases} 0 \,(\bmod\, 2), & \text{if } p \equiv 1 \,(\bmod\, 8), \ N(\varepsilon_{8p}) = +1, \\ 0 \,(\bmod\, 4), & \text{if } p \equiv 1 \,(\bmod\, 8), \ N(\varepsilon_{8p}) = -1, \\ 2 \,(\bmod\, 4), & \text{if } p \equiv 5 \,(\bmod\, 8), \\ 1 \,(\bmod\, 2), & \text{if } p \equiv 3 \,(\bmod\, 4). \end{cases}$$

We leave it to the reader to derive these congruences from values of the polynomials $F_\pm(s)$.

In [Friesen and Williams, 1985] the value of $h(8p)$ is determined modulo 8 when $p \equiv 5 \,(\bmod\, 8)$. It is shown that

$$h(8p) \equiv 2T + b + 2 \,(\bmod\, 8),$$

where $\varepsilon_{8p} = T + U\sqrt{2p}$ and

$$p = a^2 + b^2, \quad a \equiv 1 \,(\bmod\, 4), \quad b \equiv \left(\frac{p-1}{2}\right)! \, a \,(\bmod\, p).$$

This result is a simple corollary from Gauss' congruence for $h(-4p)$ modulo 8 (see section 1.2), Theorem 15, and a congruence relating $h(-8p)$ and $h(8p)$ modulo 8 given in [Williams, 1981b]. A congruence relating $h(-8p)$ and $h(8p)$ modulo 16 is given in [Kaplan and Williams, 1982a].

When $p \equiv 1 \,(\bmod\, 8)$ there are a great many results concerning $h(8p)$ modulo 8 and 16. We refer the reader to [Kaplan, 1973a], [Brown, 1974b, 1983], [Kaplan and Williams, 1982b, 1984], [Hardy, Kaplan and Williams, 1986] for further details.

The question of determining $h(8p)$ modulo 4 when $p \equiv 3 \,(\bmod\, 4)$ appears to be open.

(iii) Next we discuss briefly the case $d = pq$, where p and q are distinct primes with $p \equiv q \,(\bmod\, 4)$. Brown [Brown, 1974b] has shown that:

$$h(pq) \equiv 1 \,(\bmod\, 2), \text{ if } p \equiv q \equiv 3 \,(\bmod\, 4),$$

$$h(pq) \equiv 2 \,(\bmod\, 4), \text{ if } p \equiv q \equiv 1 \,(\bmod\, 4), \left(\frac{p}{q}\right) = -1,$$

and if $p \equiv q \equiv 1 \,(\bmod\, 4)$, $\left(\frac{p}{q}\right) = 1$

$$h(pq) \equiv 4 \,(\bmod\, 8) \text{ if } \left(\frac{p}{q}\right)_4 = \left(\frac{q}{p}\right)_4 = -1,$$

$$h(pq) \equiv 2 \,(\bmod\, 4), \text{ if } \left(\frac{p}{q}\right)_4 \left(\frac{q}{p}\right)_4 = -1,$$

$$h(pq) \equiv 0 \,(\bmod\, 8), \text{ if } \left(\frac{p}{q}\right)_4 = \left(\frac{q}{p}\right)_4 = 1, \ N(\varepsilon_{pq}) = -1,$$

$$h(pq) \equiv 0 \,(\bmod\, 4), \text{ if } \left(\frac{p}{q}\right)_4 = \left(\frac{q}{p}\right)_4 = 1, \ N(\varepsilon_{pq}) = 1,$$

where we write $\left(\frac{p}{q}\right)_4 = 1$ or -1 according as p is or is not a biquadratic residue of q. See also [Rédei and Reichardt, 1934], [Scholz, 1935] and [Kaplan, 1972, 1973a,b].

2.3   *The Case $\nu \geq 3$*

Brown [Brown, 1981] has also obtained some divisibility properties for $h(d)$ when $d = 4pq$ ($p \not\equiv q \,(\bmod\, 4)$), $d = 8pq$, and $d = pqr$. He used ambiguous forms and genera to obtain similar congruences to those in section 2.2 (iii).

For other results concerning the 2-part of the group of classes of ideals in a quadratic field see [Rédei, 1932a,b, 1934c, 1936, 1939a,b], [Reichardt, 1934, 1970], [Damey and Payan, 1970], [Hasse, 1970b], [Bauer, 1971, 1972],

[Koch and Zink, 1972], [Endô, 1973a,b], [Gras, 1973], [Waterhouse, 1973], [Kaplan, 1974, 1977b], [Halter-Koch, 1984], [Kaplan and Williams, 1986], [Yamamoto, Y., 1988], [Uehara, 1989], [Kohno and Nakahara, 1993] and [Sueyoshi, 1995, 1997]. See also [Szymiczek, 1996].

2.4   Congruences between Class Numbers of Real and Imaginary Quadratic Fields

Some such congruences have already been given earlier in this chapter. For further such congruences, see [Lang and Schertz, 1976], [Oriat, 1977, 1978], [Kaplan, 1981], [Morton, 1983], [Lang, H., 1985], [Hikita, 1986a,b], [Desnoux, 1987, 1988], [Zhang, 1989], [Pioui, 1990], [Stevenhagen, 1988, 1993]. All these congruences follow from congruences of Gras [Gras, 1989] and Uehara [Uehara, 1990]. We give just two examples.

Let d be an odd discriminant of a real quadratic field having ν prime factors, all of which are congruent to 1 modulo 4, and $T + U\sqrt{2d}$ is the fundamental unit (> 1) of $\mathbb{Q}(\sqrt{2d})$. Hikita [Hikita, 1986a] proved the following congruence

$$h(-8d) \equiv \left(\frac{2}{d}\right) TUh(8d) \,(\bmod\, 2^{\nu+2})\,.$$

H. Lang [Lang, H., 1985] showed that if d is an odd discriminant of a real quadratic field having ν distinct prime factors all of which are congruent to 1 modulo 8, then

$$h(-4d) \equiv \lambda(d)h(d) \,(\bmod\, 2^{\nu+2})\,,$$

where

$$\lambda(d) = \begin{cases} Tr_{\mathbb{Q}(\sqrt{d})/\mathbb{Q}}\left(\dfrac{\varepsilon}{2\sqrt{d}} + \varepsilon - 1\right), & \text{if } N(\varepsilon) = 1, \\[2ex] \dfrac{1}{2} Tr_{\mathbb{Q}(\sqrt{d})/\mathbb{Q}}(\varepsilon), & \text{if } N(\varepsilon) = -1, \end{cases}$$

and ε is the fundamental unit (> 1) of $\mathbb{Q}(\sqrt{d})$.

2.5   Hasse's Classical Klassenzahlbericht

Conner and Hurrelbrink [Conner and Hurrelbrink, 1988, Chapter III] made use of Hasse's approach given in [Hasse, 1952, 1985]. They combined some cohomologically derived results with some explicit calculations of Hasse that follow from the analytic class number formula for abelian fields. Their principal tool is an exact hexagon which simplifies for quadratic fields on incorporating standard results from Gauss' genus theory. Many of the classical results given in Chapter II of this book are derived as examples illustrating the general results of [Conner and Hurrelbrink, 1988].

Chapter III

CONGRUENCES
BETWEEN THE ORDERS OF K_2-GROUPS

This chapter will focus on the results which appear in [Urbanowicz, 1990a,b] and [Urbanowicz, 1990/1991a,b]. In section 7.2 of Chapter I we observed that Browkin and Schinzel's theorem shows that

$$k_2(d) \equiv \begin{cases} 0 \, (\bmod \, 2^{\nu+s}), & \text{if } d > 0 \text{ is odd,} \\ 0 \, (\bmod \, 2^{\nu+s-1}), & \text{if } d > 0 \text{ is even, or } d < 0 \text{ is odd,} \quad (1) \\ 0 \, (\bmod \, 2^{\nu+s-2}), & \text{if } d < 0 \text{ is even,} \end{cases}$$

where d denotes the discriminant of the quadratic field $F = \mathbb{Q}(\sqrt{d})$, ν the number of distinct prime divisors of d, and 2^s the number of elements of the set $\{\pm 1, \pm 2\}$ that are norms of an element of F. In view of identity (19) of Chapter I we have the same congruence for B_{2,χ_d}. Recently in [Fox, Urbanowicz and Williams, 1999] the authors have found a congruence of this type for generalized Bernoulli numbers of higher order, namely they proved that

$$(B_{k,\chi_d}/k) \equiv 0 \, (\bmod \, 2^{\nu-1}),$$

if $\chi_d(-1) = (-1)^k$. For $k = 2$ and positive d the above congruence coincides with

$$k_2(d) \equiv 0 \, (\bmod \, 2^{\nu}),$$

which implies congruence (1). In the case when d is even and $s = 0$ the above congruence also follows from [Browkin and Schinzel, 1982, Corollary 3]. In this case the 4-rank of the group K_2O_F, i.e. the number of cyclic summands of its Sylow 2-subgroup of orders divisible by 4, is positive. This was proved by Browkin and Schinzel under more general assumptions.

1.   REAL QUADRATIC FIELDS

We start with the case of real quadratic fields. Congruences obtained in this case are implied by those for class numbers of imaginary quadratic fields stated in Chapter II, §1.

1.1   *Congruences between $k_2(d)$ and Class Numbers of Appropriate Imaginary Quadratic Fields (Elementary Approach)*

In [Urbanowicz, 1983] the author, using only short character sum formulae, found congruences between the orders of K_2-groups of the integers of real quadratic fields and class numbers of appropriate imaginary quadratic fields. He considered the right hand side of the equation in the Birch-Tate conjecture (see section 7.5 of Chapter I).

Let d be the discriminant of a quadratic field. Throughout the section let $D\ (> 0)$ denote the squarefree part of d. We have

$$D = \begin{cases} |d|, & \text{if } d \text{ odd}, \\ |d|/4, & \text{if } d \text{ even}. \end{cases}$$

We follow the notation of section 10.3, Chapter I. If d is odd, appealing to (20) of Chapter I, we have

$$k_2(d) \equiv 4 \sum_{1 \le a \le (d-1)/2} \left(\frac{d}{a}\right) a$$

$$\equiv 4 \sum_{1 \le a \le (d-1)/4} \left(\frac{a}{d}\right) \equiv 4A(d/4, d) \, (\bmod 8). \tag{2}$$

If $d\ (d > 0, d \ne 8)$ is even, making use of the basic properties of the Jacobi and Kronecker symbols, we obtain in an elementary way

$$k_2(d) = 2 \sum_{1 \le a \le (D/2)} \left(\frac{D}{2a-1}\right) (D - 2a + 1).$$

Therefore for even and positive d we obtain

$$k_2(d) = \begin{cases} 4 \displaystyle\sum_{1 \le a \le (D-1)/2} \left(\dfrac{D}{D - 2a}\right) a, & \text{if } 4 \,\|\, d, \\[3mm] 2 \displaystyle\sum_{1 \le a \le (D/2)} \left(\dfrac{D}{D - 2a + 1}\right) (2a - 1), & \text{if } 8 \,|\, d, \ d \ne 8. \end{cases} \tag{3}$$

On the other hand, appealing to (17) of Chapter I we obtain

$$h(d) = \sum_{\substack{1 \le a \le D \\ a\,\text{odd}}} (-1)^{(a-1)/2} \left(\frac{D}{a}\right) = 2 \sum_{\substack{1 \le a \le D \\ a \equiv 1\,(\text{mod}\,4)}} \left(\frac{D}{a}\right), \qquad (4)$$

if d ($d < 0$, $d \ne -8$) is divisible by 8. (Note that the first equality in (4) also holds when $d = -8$.)

Coupling identity (4) with (3) we deduce in an elementary way that

$$k_2(d) = -\frac{1}{2}dh(-D)\left(\left(\frac{2}{D}\right) - 1\right) + 16 \sum_{1 \le a \le (D-3)/4} \left(\frac{a}{D}\right) a, \qquad (5)$$

if $d > 0$ and $4 \,\|\, d$, and

$$k_2(d) = -16 \sum_{\substack{1 \le a \le (D/2) \\ a \equiv 3\,(\text{mod}\,4)}} \left(\frac{D}{a}\right)\frac{a+1}{4}$$

$$+ 2(D+2) \sum_{\substack{1 \le a \le (D/2) \\ a \equiv 3\,(\text{mod}\,4)}} \left(\frac{D}{a}\right) + Dh(-d), \qquad (6)$$

if $d > 0$, $8 \,|\, d$ and $d \ne 8$. For details see [Urbanowicz, 1983].

In the five theorems below d denotes the discriminant of a quadratic field and D has the same meaning as above. Following [Urbanowicz, 1983] we have:

THEOREM 20 *Let d ($\ne 8$) be positive and x (resp. y) be the number of prime factors of d of the form $4t + 1$ (resp. $4t + 3$). Then*

$$k_2(d) \equiv \begin{cases} \phi(D)\,(\text{mod}\,8), & \text{if } x > 0, \\ \phi(D) + 2^y\,(\text{mod}\,8), & \text{if } x = 0. \end{cases}$$

PROOF. If d is odd then the congruence follows easily from (2) and formula (30) of Chapter I.

If $4 \,\|\, d$ then from (3) it follows that

$$k_2(d) \equiv 4 \sum_{1 \le a \le (D+1)/4} \left(\frac{2a-1}{D}\right)$$

$$\equiv 4\left(\sum_{1 \le a \le (D-1)/2} \left(\frac{a}{D}\right) - \sum_{1 \le a \le (D-3)/4} \left(\frac{a}{D}\right)\right)$$

$$\equiv 4\,(A(D/2, D) - A(D/4, D))\,(\text{mod}\,8).$$

Thus the congruence of the theorem follows from (30) of Chapter I.

If $8 \mid d$ then we make use of formulae (4), (5) and (32) of Chapter I. We have

$$k_2(d) \equiv Dh(-d) \equiv 2D \sum_{\substack{1 \leq a \leq (D/2) \\ a \equiv 1 \,(\mathrm{mod}\,4)}} \left(\frac{D}{a}\right) \equiv 4 \sum_{\substack{1 \leq a \leq (D/2) \\ a \equiv 1 \,(\mathrm{mod}\,4)}} \left(\frac{D}{a}\right)$$

$$\equiv 4\left(A((m+1)D/8, D/2) - A(mD/8, D/2)\right) \,(\mathrm{mod}\,8),$$

where $m \equiv -(D/2) \,(\mathrm{mod}\,4)$, $m = 1$ or 3. ∎

THEOREM 21 *If d is negative and odd, then*

$$k_2(-4d) \equiv 2dh(d)(\chi_d(2) - 1) \,(\mathrm{mod}\,16).$$

PROOF. The congruence is an immediate consequence of (5). ∎

THEOREM 22 *Let d be negative and $d \equiv 1 \,(\mathrm{mod}\,8)$. Let x, y, z, u denote respectively the numbers of prime divisors of d of the form $8t+1$, $8t-1$, $8t+3$ and $8t-3$. Then we have*

$$k_2(-4d) - 2\phi(|d|) \equiv \begin{cases} 0 \,(\mathrm{mod}\,32), & \text{if } x > 0 \\ & \text{or } x = 0,\, y \geq 0,\, z > 0,\, u > 0, \\ 2^{y+u+2} \,(\mathrm{mod}\,32), & \text{if } x = 0,\, y \geq 0,\, z = 0,\, u > 0, \\ -2^{y+z+1} \,(\mathrm{mod}\,32), & \text{if } x = 0,\, y \geq 0,\, z > 0,\, u = 0, \\ 2^{y+1} \,(\mathrm{mod}\,32), & \text{if } x = 0,\, y > 0,\, z = 0,\, u = 0. \end{cases}$$

PROOF. In view of (5) we have

$$k_2(-4d) = 16 \sum_{1 \leq a \leq (|d|-3)/4} \chi_d(a)a \equiv 16 \sum_{1 \leq a \leq (|d|+1)/8} \chi_d(2a-1)$$

$$\equiv 16\left(A(|d|/4, |d|) - A(|d|/8, |d|)\right) \,(\mathrm{mod}\,32).$$

Thus the theorem follows from formulae (30) and (31) of Chapter I. ∎

THEOREM 23 *If d $(d \neq 5)$ is positive and odd, then*

$$k_2(d) \equiv 2h(-4d) \,(\mathrm{mod}\,16).$$

PROOF. Appealing to formulae (20) and (25) of Chapter I we have

$$k_2(d) = \frac{-8}{4 - \chi_d(2)}\left(\sum_{1 \leq a \leq (d-1)/4} \chi_d(2a)a + \sum_{1 \leq a \leq (d-1)/4} \chi_d(2a-1)a\right)$$

$$- \frac{2\chi_d(2)}{4 - \chi_d(2)}h(-4d).$$

Therefore we obtain

$$k_2(d) \equiv \frac{-2\chi_d(2)}{4 - \chi_d(2)} h(-4d) \,(\bmod\, 16)$$

because in view of formula (32) of Chapter I we have

$$\sum_{1 \leq a \leq (d-1)/4} \chi_d(2a)a + \sum_{1 \leq a \leq (d-1)/4} \chi_d(2a-1)a$$

$$\equiv \sum_{\substack{1 \leq a \leq d \\ a \equiv 4 \,(\bmod\, 8)}} \chi_d(a) - \sum_{\substack{1 \leq a \leq d \\ a \equiv 2 \,(\bmod\, 8)}} \chi_d(a)$$

$$\equiv (A(5d/8, d) - A(4d/8, d)) - (A(7d/8, d) - A(6d/8, d))$$

$$\equiv 0 \,(\bmod\, 2). \qquad\qquad\qquad\qquad\qquad\blacksquare$$

THEOREM 24 *Let d be positive and $8 \mid d$ $(d \neq 8)$. Let x (resp. y) denote the number of prime divisors of d of the form $4t + 1$ (resp. $4t + 3$). Then we have*

$$k_2(d) \equiv \begin{cases} \left(\dfrac{d}{8} + 1\right)\phi\left(\dfrac{d}{8}\right) + \dfrac{d}{4}h(-d) \,(\bmod\, 16), & \text{if } x > 0, \\[3mm] \left(\dfrac{d}{8} + 1\right)\phi\left(\dfrac{d}{8}\right) + \dfrac{d}{4}h(-d) - \left(\dfrac{d}{8} + 1\right)2^y \,(\bmod\, 16), & \text{if } x = 0. \end{cases}$$

PROOF. The theorem follows from (6) and formulae (31), (32) of Chapter I as

$$k_2(d) - Dh(-d) \equiv 2(D+2) \sum_{\substack{1 \leq a \leq (D/2) \\ a \equiv 3 \,(\bmod\, 4)}} \left(\frac{D}{a}\right)$$

$$\equiv 2(D+2)\left(A\left(\frac{(m+1)D}{8}, \frac{D}{2}\right) - A\left(\frac{mD}{8}, \frac{D}{2}\right)\right) \,(\bmod\, 16),$$

where $m \equiv (D/2) \,(\bmod\, 4)$, $m = 1$ or 3. $\blacksquare$

In the next sections we discuss $k_2(d)$ modulo powers of 2 when $d\,(> 0)$ has a small number of prime factors. In obtaining such congruences authors have traditionally used either analytic (complex or 2-adic) methods, i.e., those involving the right hand side of the equation in the Birch-Tate conjecture, see section 7.5 of Chapter I, or algebraic methods, i.e., those using the structure of the group $K_2 O_F$, where F is a real quadratic field with discriminant d. All the results in sections 1.2 and 1.3 can also be obtained algebraically (for details see [Browkin and Schinzel, 1982], [Conner and Hurrelbrink, 1988, 1989a,b], [Candiotti and Kramer, 1989], [Qin, 1994a, 1995b], [Hurrelbrink and Kolster, 1998]).

1.2   *Corollaries to Theorems 20–24*

In the six theorems below d denotes a positive discriminant and p, q denote distinct odd primes. Following [Urbanowicz, 1983] we have:

THEOREM 25   *We have*

$$8 \mid k_2(d)$$

unless $d = 8$, p, $4p$ *or* $8p$, *where* $p \equiv \pm 3 \,(\,\mathrm{mod}\,8\,)$, *in which case* $4 \parallel k_2(d)$.

PROOF. The theorem follows easily from Theorem 20 by considering cases. If $d = 8$ then $k_2(d) = 4$. If $d = p$, $4p$ or $8p$ then

(i) for $p \equiv 1\,(\,\mathrm{mod}\,4\,)$ we have in Theorem 20 $x = 1$, $y = 0$, hence

$$k_2(d) \equiv p - 1 \equiv \begin{cases} 0\,(\,\mathrm{mod}\,8\,), & \text{if } p \equiv 1\,(\,\mathrm{mod}\,8\,), \\ 4\,(\,\mathrm{mod}\,8\,), & \text{if } p \equiv -3\,(\,\mathrm{mod}\,8\,), \end{cases}$$

(ii) for $p \equiv 3\,(\,\mathrm{mod}\,4\,)$ we have in Theorem 20 $x = 0$, $y = 1$, hence

$$k_2(d) \equiv p + 1 \equiv \begin{cases} 0\,(\,\mathrm{mod}\,8\,), & \text{if } p \equiv -1\,(\,\mathrm{mod}\,8\,), \\ 4\,(\,\mathrm{mod}\,8\,), & \text{if } p \equiv 3\,(\,\mathrm{mod}\,8\,). \end{cases}$$

If d has at least two odd prime factors then from Theorem 20 it follows easily that $8 \mid k_2(d)$. ∎

THEOREM 26   *Let* $d \equiv 1\,(\,\mathrm{mod}\,8\,)$, *or* $4 \parallel d$ *and* $(d/4) \equiv -1\,(\,\mathrm{mod}\,8\,)$. *Then we have*

$$16 \mid k_2(d)$$

unless $d = p = e^2 - 2f^2$, $e > 0$, $e \equiv 3\,(\,\mathrm{mod}\,4\,)$, $f \equiv 0\,(\,\mathrm{mod}\,4\,)$, *or* $d = pq$, $p \equiv q \equiv 3\,(\,\mathrm{mod}\,8\,)$, *in which case* $8 \parallel k_2(d)$.

PROOF. If $(d/4) \equiv -1\,(\,\mathrm{mod}\,8\,)$ the theorem is an immediate consequence of Theorem 21.

If $d \equiv 1\,(\,\mathrm{mod}\,8\,)$ and the number ν of prime factors of d is less than 3 it is sufficient to apply Theorem 23 together with results of Barrucand and Cohn [Barrucand and Cohn, 1969], Brown [Brown, 1974a] and Pizer [Pizer, 1976] (see section 1.2 of Chapter II). When $\nu \geq 3$ then we use congruence (1) of Chapter II. ∎

THEOREM 27   *If* $(d/4) \equiv -1\,(\,\mathrm{mod}\,8\,)$ *then*

$$32 \mid k_2(d)$$

unless $d = 4p$, $p \equiv 7\,(\,\mathrm{mod}\,16\,)$, *or* $d = 4pq$, $p \equiv -q \equiv 3\,(\,\mathrm{mod}\,8\,)$, *in which case* $16 \parallel k_2(d)$.

PROOF. We adopt the notation of Theorem 22.

Let $(d/4) = p \equiv -1 \,(\text{mod}\,8)$. Then $x = z = u = 0$, $y = 1$, and from Theorem 22 it follows that

$$k_2(d) \equiv 2\phi(d/4) + 4 \equiv 2(p+1) \,(\text{mod}\,32).$$

Thus $16 \,\|\, k_2(d)$ if $p \equiv 7 \,(\text{mod}\,16)$, and $32 \,|\, k_2(d)$ if $p \equiv -1 \,(\text{mod}\,16)$.

Let $(d/4) = pq \equiv -1 \,(\text{mod}\,8)$. Then evidently $x = y = 1$, $z = u = 0$, or $x = y = 0$, $z = u = 1$. In both the cases from Theorem 22 we obtain

$$k_2(d) \equiv 2\phi(d/4) \equiv 2(p-1)(q-1) \,(\text{mod}\,32).$$

Thus if $x = y = 1$ then $32 \,|\, k_2(d)$, and if $z = u = 1$ then $16 \,\|\, k_2(d)$.

Finally, if $d/4$ has at least three prime factors, i.e., $x + y + z + u \geq 3$ then $2^{3x+y+z+2u} \,|\, \phi(d/4)$, and so $32 \,|\, 2\phi(d/4)$ with one possible exception $x = u = 0$, $y + z = 3$, when $16 \,\|\, 2\phi(d/4)$. In this case by Theorem 22 we have

$$k_2(d) \equiv 2\phi(d/4) - 2^{y+z+1} \equiv 0 \,(\text{mod}\,32).$$

In all the other cases, the congruence $k_2(d) \equiv 0 \,(\text{mod}\,32)$ easily follows. ■

THEOREM 28 *Suppose that $(d/4) \equiv 3 \,(\text{mod}\,8)$ and d has at least two odd prime factors. Then we have*

$$16 \,|\, k_2(d)$$

unless $d = 4pq$, $\left(\dfrac{p}{q}\right) = -1$, in which case $8 \,\|\, k_2(d)$.

PROOF. By Theorem 21 it follows that

$$k_2(d) \equiv dh(-D) \,(\text{mod}\,16),$$

where $D = (d/4)$. Since, by (1) of Chapter II we have $2^{\nu-1} \,|\, h(-D)$, where ν denotes the number of odd prime factors of discriminant d, it is sufficient to consider the case when $\nu = 2$. Let $d = 4pq$. From results of Hasse [Hasse, 1970a] (see section 1.2 of Chapter II) it follows that $h(-D) \equiv 0 \,(\text{mod}\,4)$ if and only if $\left(\dfrac{p}{q}\right) = 1$. This proves the theorem. ■

THEOREM 29 *Suppose that $d \equiv 5 \,(\text{mod}\,8)$ and d has at least two prime factors. Then we have*

$$16 \,|\, k_2(d)$$

unless $d = pq$, $\left(\dfrac{p}{q}\right) = -1$, $p \not\equiv 7 \,(\text{mod}\,8)$, in which case $8 \,\|\, k_2(d)$.

PROOF. By (1) of Chapter II we have $2^{\nu} \mid h(-4d)$, where ν is the number of prime factors of d and so by virtue of Theorem 23 it suffices to consider the case $d = pq$. If $p \equiv q \equiv 1 \,(\mathrm{mod}\,4)$ then, by virtue of results of Brown [Brown, 1974a] and Pizer [Pizer, 1976] (see section 1.3 of Chapter II), it follows that $\left(\dfrac{p}{q}\right) = -1$, $p \equiv -1\,(\mathrm{mod}\,8)$ imply the congruence $h(-4d) \equiv 0\,(\mathrm{mod}\,8)$, and $\left(\dfrac{p}{q}\right) = -1$, $p \equiv 3\,(\mathrm{mod}\,8)$ imply the congruence $h(-4d) \equiv 4\,(\mathrm{mod}\,8)$. Thus the theorem follows from Theorem 23. ∎

In order to simplify the formulation of results in case $8 \mid d$ we divide this case into two subcases when d has only one odd prime factor or d has at least two distinct odd prime factors.

THEOREM 30 *Let $d = 8p$, where $p \equiv \pm 1\,(\mathrm{mod}\,8)$. Then we have*

$$16 \mid k_2(d)$$

unless $p \equiv 7\,(\mathrm{mod}\,16)$, or $p = e^2 - 2f^2$, $e > 0$, $e \equiv 5,7\,(\mathrm{mod}\,8)$, f even, in which case $8 \| k_2(d)$.

THEOREM 31 *Suppose that $8 \mid d$ and d has at least two odd prime factors. Then we have*

$$16 \mid k_2(d)$$

unless $d = 8pq$, $p \equiv -q \equiv 3\,(\mathrm{mod}\,8)$, or $p \equiv q \equiv 5\,(\mathrm{mod}\,8)$, or $p \equiv 1\,(\mathrm{mod}\,8)$, $q \equiv 3,5\,(\mathrm{mod}\,8)$, $\left(\dfrac{p}{q}\right) = -1$, or $p \equiv 3,5\,(\mathrm{mod}\,8)$, $q \equiv 7\,(\mathrm{mod}\,8)$, $\left(\dfrac{p}{q}\right) = -1$, in which case $8 \| k_2(d)$.

PROOFS OF THEOREMS 30 AND 31. Both theorems are immediate consequences of Theorem 24 and results of Hasse [Hasse, 1969a], Pizer [Pizer, 1976] and Berndt [Berndt, 1976] (see section 1.2 of Chapter II). ∎

1.3 *The Cases $\nu \leq 3$*

Let d be a positive discriminant. Denote by ν the number of prime factors of d, and by p, q distinct odd prime numbers. In the next sections we discuss $k_2(d)$ modulo powers of 2 when d has a small number of prime factors. We now consider the cases when $\nu = 1, 2$ or 3.

If $\nu = 1$ we have $d = 8$ or $d = p$, with $p \equiv 1\,(\mathrm{mod}\,4)$. Since $k_2(8) = 4$ we need only consider $k_2(p)$, $p \equiv 1\,(\mathrm{mod}\,4)$. In this case appealing to Theorems

25 and 26 we have

$$k_2(p) \equiv \begin{cases} 4\,(\bmod\,8), & \text{if } p \equiv 5\,(\bmod\,8), \\ 8\,(\bmod\,16), & \text{if } p = e^2 - 2f^2, e > 0, e \equiv 3\,(\bmod\,4), \\ & \qquad f \equiv 0\,(\bmod\,4), \\ 0\,(\bmod\,16), & \text{otherwise}. \end{cases}$$

The determination of $k_2(p)$ modulo 16 and 32 if $p \equiv 5\,(\bmod\,8)$ and modulo 32 if $p \equiv 1\,(\bmod\,8)$ will be given in Theorem 34, section 1.5. We shall relate the numbers $k_2(d)$ modulo powers of 2 to class numbers of appropriate imaginary quadratic fields.

If $\nu = 2$ we have

(i) $d = 4p$ with $p \equiv 3\,(\bmod\,4)$,

(ii) $d = 8p$,

(iii) $d = pq$ with $p \equiv q\,(\bmod\,4)$.

Case (i). In this case appealing to Theorems 25, 26 and 27 we obtain

$$k_2(4p) \equiv \begin{cases} 4\,(\bmod\,8), & \text{if } p \equiv 3\,(\bmod\,8), \\ 16\,(\bmod\,32), & \text{if } p \equiv 7\,(\bmod\,16), \\ 0\,(\bmod\,32), & \text{if } p \equiv 15\,(\bmod\,16). \end{cases}$$

The determination of $k_2(4p)$ modulo 32 if $p \equiv 3\,(\bmod\,8)$ will be given in Theorem 35, section 1.5.

Case (ii). In this case appealing to Theorem 25 and 30 we have

$$k_2(8p) \equiv \begin{cases} 4\,(\bmod\,8), & \text{if } p \equiv 3 \text{ or } 5\,(\bmod\,8), \\ 8\,(\bmod\,16), & \text{if } p \equiv 7\,(\bmod\,16), \text{ or } p = e^2 - 2f^2, e > 0, \\ & \qquad e \equiv 5 \text{ or } 7\,(\bmod\,8), f \equiv 0\,(\bmod\,2), \\ 0\,(\bmod\,16), & \text{otherwise}. \end{cases}$$

The determination of $k_2(8p)$ modulo 8, 16 and 32 if $p \equiv 3$ or $5\,(\bmod\,8)$ and modulo 32 in the other cases will be given in Theorems 34 and 35, section 1.5.

Case (iii). Appealing to Theorems 26 and 29 in this case we obtain

$$k_2(pq) \equiv \begin{cases} 8\,(\bmod\,16), & \text{if } p \equiv q \equiv 3\,(\bmod\,8), \text{ or } pq \equiv 5\,(\bmod\,8), \\ & \qquad \left(\dfrac{p}{q}\right) = -1, p \not\equiv 7\,(\bmod\,8), \\ 0\,(\bmod\,16), & \text{otherwise}. \end{cases}$$

For relations between the numbers $k_2(pq)$ modulo 32 and the class numbers of appropriate imaginary fields see Theorem 34, section 1.5.

When $\nu = 3$ the situation is of course more complicated. In this case we refer the reader to Theorems 27, 28 and 31.

1.4   Congruences Implied by Theorem 4

Theorem 4 implies many new, surprising congruences between class numbers of imaginary quadratic fields as well as the orders of K_2-groups of the integers of appropriate real quadratic fields. In the five theorems below d denotes an odd discriminant of a quadratic field, and p, q, r denote distinct odd primes.

THEOREM 32 ([Urbanowicz, 1990a, Corollary 1 to Theorem 1]) *For $d > 5$ we have*:

(i) $k_2(d) \equiv 2h(-4d) + 2\phi(d) + \varepsilon \,(\mathrm{mod}\,32\,)$,

   where $\varepsilon = 0$ unless $d = p \equiv -3 \,(\mathrm{mod}\,8\,)$ or $d = pq$, where $p \equiv q \not\equiv 1$ $(\mathrm{mod}\,8\,)$ or $p \equiv q+4 \equiv 3\,(\mathrm{mod}\,8\,)$. In these cases $\varepsilon = 16$ if $p \equiv q \equiv -3$ $(\mathrm{mod}\,8\,)$, $\varepsilon = -8$, if $p \equiv q \equiv -1\,(\mathrm{mod}\,8\,)$ and $\varepsilon = 8$, otherwise,

(ii) $k_2(d) \equiv 6h(-4d) - 4(2 - \chi_d(2))h(-8d)\,(\mathrm{mod}\,32\,)$,

(iii) $k_2(8d) \equiv -2(2 - \chi_d(2))(2h(-4d) - \chi_d(2)h(-8d))\,(\mathrm{mod}\,32\,)$,

(iv) $k_2(8d) + (\chi_d(2) - 34)k_2(d)$
$$\equiv -2(2\chi_d(2) - 1)(\chi_d(2)h(-4d) + h(-8d))\,(\mathrm{mod}\,64\,),$$

$k_2(8d) + 3(3\chi_d(2) - 2)k_2(d)$
$$\equiv -2(2\chi_d(2) - 1)((\chi_d(2) - 4)h(-4d) + h(-8d))\,(\mathrm{mod}\,64\,),$$

$k_2(8d) - 3(3\chi_d(2) - 2)k_2(d)$
$$\equiv -2(2\chi_d(2) - 1)((3\chi_d(2) + 4)h(-4d) - 3h(-8d))\,(\mathrm{mod}\,64\,),$$

$k_2(8d) + 15(\chi_d(2) - 2)k_2(d)$
$$\equiv -6(2\chi_d(2) - 1)(\chi_d(2)h(-4d) - h(-8d))\,(\mathrm{mod}\,64\,),$$

(v) *if $d = p = 8t + 1$ or $8t - 3$ then*:

$$k_2(d) \equiv 2h(-4d) + 16t\,(\mathrm{mod}\,32\,),$$

$$k_2(d) \equiv 32\alpha + 2\beta(-(2 + \chi_d(2))h(-4d) + 2h(-8d))\,(\mathrm{mod}\,64\,),$$

   where $\alpha = 1$ if $p \equiv -3\,(\mathrm{mod}\,16\,)$ and $\alpha = 0$ otherwise, and $\beta = -1$, resp. 5 if $p \equiv 1\,(\mathrm{mod}\,8\,)$, $p \equiv 5\,(\mathrm{mod}\,16\,)$, resp. $p \equiv -3\,(\mathrm{mod}\,16\,)$,

$$k_2(8d) \equiv 32\alpha + 2\beta(2\chi_d(2)h(-4d) - h(-8d))\,(\mathrm{mod}\,64\,),$$

   where $\alpha = 0$ if $p \equiv 1\,(\mathrm{mod}\,8\,)$ and $\alpha = 1$ otherwise, and $\beta = -1, -3$, resp. 5 if $p \equiv 1\,(\mathrm{mod}\,8\,)$, $p \equiv -3\,(\mathrm{mod}\,16\,)$, resp. $p \equiv 5\,(\mathrm{mod}\,16\,)$.

PROOF. (i) If d has at least two prime factors then by (1) of Chapter II we have $h(-4d) \equiv 0\,(\mathrm{mod}\,4\,)$. Otherwise $d = p\,(\text{prime}) \equiv 1\,(\mathrm{mod}\,4\,)$ and

$h(d) = h(-4p) \equiv (p-1)/2 \,(\bmod\, 4)$, see section 1.2 of Chapter II. Thus $4\,|\,h(-4d)$ unless $d = p \equiv 5\,(\bmod\, 8)$, in which case $2\,\|\,h(-4d)$. Hence we obtain

$$-\frac{2}{45}(2\chi_d(2) - 7)dh(-4d) \equiv 2h(-4d) \,(\bmod\, 32).$$

Appealing to Theorem 4 we obtain

$$k_2(d) = \frac{16}{45}(2\chi_d(2) - 7)(s_1(8,1,\chi_d)$$

$$+ \, s_1(8,2,\chi_d)) - \frac{2}{45}(2\chi_d(2) - 7)dh(-4d) \tag{7}$$

(cf. [Urbanowicz, 1990a, Theorem 1(i)]). Next we see that

$$s_1(8,1,\chi_d) + s_1(8,2,\chi_d) \equiv \sum_{1 \le a \le d/4} \chi_d(a) - \sum_{1 \le a \le d/8} \chi_d(a)$$

$$\equiv A(d/4, d) - A(d/8, d) \,(\bmod\, 2).$$

Thus to complete the proof of (i), it is sufficient to use formulae (30) and (31) of Chapter I.

(ii) This part is an immediate consequence of an appropriate equation of Theorem 4. The theorem gives the formula

$$k_2(d) = -\frac{32}{25}(\chi_d(2) + 4)(s_1(8,1,\chi_d) + s_1(8,3,\chi_d))$$

$$- \, \frac{2}{75}(\chi_d(2) + 4)d((\chi_d(2) + 2)h(-4d) - 2h(-8d)).$$

The required congruence now follows as

$$\frac{1}{75}(\chi_d(2) + 4)d \equiv \chi_d(2) - 2 \,(\bmod\, 8). \tag{8}$$

(iii) This part also follows from Theorem 4, more precisely from the formula

$$k_2(8d) = -32(s_1(8,2,\chi_d) + s_1(8,3,\chi_d)) - 2d(2\chi_d(2)h(-4d) - h(-8d)).$$

(iv) This part is also an immediate consequence of Theorem 4 in view of

$$\pm\chi_d(2)h(-4d) \pm h(-8d) \equiv 0 \,(\bmod\, 4). \tag{9}$$

Indeed, from (17), Chapter I we have

$$\pm\chi_d(2)h(-4d) \pm 4(-8d) = \pm 2\chi_d(2) \sum_{\substack{1 \le a \le d \\ a \equiv 0\,(\bmod\, 4)}} \chi_d(a) \pm 2 \sum_{\substack{1 \le a \le d \\ a \equiv 1\,(\bmod\, 4)}} \chi_{8d}(a)$$

$$= 2\chi_d(2) \sum_{\substack{1 \le a \le d \\ a \equiv 0\,(\bmod\, 4)}} \left(\pm 1 \pm (-1)^{a/4}\right)\chi_d(a).$$

(v) The first congruence is a particular case of part (i) of this theorem since

$$s_1(8,1,\chi_d) + s_1(8,3,\chi_d)$$

is even except when $p \equiv -3 \,(\,\mathrm{mod}\,16\,)$ and

$$s_1(8,2,\chi_d) + s_1(8,3,\chi_d)$$

is odd except when $p \equiv 1\,(\,\mathrm{mod}\,16\,)$. The remaining congruences of (v) follow from (8) and the congruence $h(-4d) \equiv 0\,(\,\mathrm{mod}\,4\,)$ valid for $d \equiv 1\,(\,\mathrm{mod}\,8\,)$. ∎

THEOREM 33 ([Urbanowicz, 1990a, Corollary 1 to Theorem 2]) *For $d < -3$ we have*:

(i) $k_2(-4d) \equiv 6(1 - \chi_d(2))h(d) + 2\phi(|d|) + \varepsilon\,(\,\mathrm{mod}\,32\,)$,

   where $\varepsilon = 0$ unless $|d| = p \equiv 3\,(\,\mathrm{mod}\,4\,)$, or $|d| = pq$, where $p \equiv q + 2 \equiv -1\,(\,\mathrm{mod}\,8\,)$, or $|d| = pqr$, where $p \equiv q \equiv r \equiv -1, 3\,(\,\mathrm{mod}\,8\,)$ or $p \equiv q \equiv -1$, resp. $3\,(\,\mathrm{mod}\,8\,)$ and $r \equiv 3$, resp. $-1\,(\,\mathrm{mod}\,8\,)$. In these excluded cases $\varepsilon = 4$ if $|d| = p \equiv -1\,(\,\mathrm{mod}\,8\,)$, $\varepsilon = -4$ if $|d| = p \equiv 3\,(\,\mathrm{mod}\,8\,)$, and $\varepsilon = 16$ otherwise,

(ii) $k_2(-4d) \equiv 6\chi_d(2)(7(\chi_d(2) - 1)h(d) + 2h(8d))\,(\,\mathrm{mod}\,32\,)$,

   (in particular if $|d| \equiv -1\,(\,\mathrm{mod}\,8\,)$, we have

   $k_2(-4d) \equiv -4h(8d)\,(\,\mathrm{mod}\,32\,)\,)$,

(iii) $k_2(-8d) \equiv 2(1 - 2\chi_d(2))(6(1 - \chi_d(2))h(d) - h(8d))\,(\,\mathrm{mod}\,32\,)$,

   (in particular if $|d| \equiv -1\,(\,\mathrm{mod}\,8\,)$, we have

   $k_2(-8d) \equiv 2h(8d)\,(\,\mathrm{mod}\,32\,)\,)$,

(iv) $k_2(-8d) + \chi_d(2)k_2(-4d)$
   $\equiv -2(2\chi_d(2) - 1)((\chi_d(2) - 1)h(d) + h(8d))\,(\,\mathrm{mod}\,64\,)$,

   $k_2(-8d) + (\chi_d(2) - 4)k_2(-4d)$
   $\equiv -2(2\chi_d(2) - 1)(5(\chi_d(2) - 1)h(d) + h(8d))\,(\,\mathrm{mod}\,64\,)$,

   $k_2(-8d) - (\chi_d(2) + 4)k_2(-4d)$
   $\equiv -2(2\chi_d(2) - 1)(7(\chi_d(2) - 1)h(d) - 3h(8d))\,(\,\mathrm{mod}\,64\,)$,

   $k_2(-8d) - \chi_d(2)k_2(-4d)$
   $\equiv 2(2\chi_d(2) - 1)(13(\chi_d(2) - 1)h(d) + 3h(8d))\,(\,\mathrm{mod}\,64\,)$,

(v) *if* $|d| = p = 8t - 1$ *or* $8t + 3$ *then*

$$k_2(-4d) \equiv -6h(d)(\chi_d(2) - 1) + 16t \ (\mathrm{mod}\ 32),$$

$$k_2(-4d) \equiv 32\alpha + 2\beta\chi_d(2)(7(\chi_d(2) - 1)h(d) + 2h(8d)) \ (\mathrm{mod}\ 64),$$

$$k_2(-8d) \equiv 32\alpha + 2\beta(13(1 - \chi_d(2))h(d) + h(8d)) \ (\mathrm{mod}\ 64),$$

where $\alpha = 1$ *if* $p \equiv 7 \,(\mathrm{mod}\ 16)$ *and* $\alpha = 0$ *otherwise, and* $\beta = -1, 3,$ *or* 11 *if* $p \equiv -1 \,(\mathrm{mod}\ 8)$, $p \equiv 3 \,(\mathrm{mod}\ 16)$, *or* $p \equiv 11 \,(\mathrm{mod}\ 16)$ *respectively.*

PROOF. (i) By Theorem 4 we have

$$k_2(-4d) = 16(s_1(8, 1, \chi_d) + s_1(8, 2, \chi_d)) - 2|d|(\chi_d(2) - 1)h(d)$$

(cf. formula (5)). Further

$$s_1(8, 1, \chi_d) + s_1(8, 2, \chi_d) \equiv A(|d|/4, |d|) - A(|d|/8, |d|) \ (\mathrm{mod}\ 2).$$

Appealing to formulae (30), (31) of Chapter I the required congruence for $k_2(-4d)$ follows.

(ii) This part is a consequence of the formula

$$k_2(-4d) = 32\chi_d(2)(s_1(8, 1, \chi_d) + s_1(8, 3, \chi_d))$$
$$+ 2|d|\chi_d(2)(7(\chi_d(2) - 1)h(d) + 2h(8d))$$

(see Theorem 4).

(iii) This part is an immediate corollary to the following formula of Theorem 4

$$k_2(-8d) = 32(s_1(8, 1, \chi_d) - s_1(8, 4, \chi_d)) - 2|d|(6(\chi_d(2) - 1)h(d) + h(8d)),$$

as $|d| \equiv 1 - 2\chi_d(2) \ (\mathrm{mod}\ 8)$ and $2 \,|\, h(8d)$.

(iv) This part is also a consequence of Theorem 4 in view of the congruence $h(8d) \equiv 0 \,(\mathrm{mod}\ 4)$ for $|d| \equiv -1 \,(\mathrm{mod}\ 8)$, and

$$2h(d) \pm h(8d) \equiv 0 \,(\mathrm{mod}\ 4) \text{ for } |d| \equiv 3 \,(\mathrm{mod}\ 8).$$

(v) The first congruence of (v) is a particular case of (i). Since

$$s_1(8, 1, \chi_d) \pm s_1(8, 4, \chi_d)$$

is even except when $p \equiv 7 \,(\mathrm{mod}\ 16)$ the remaining cases of (v) follow at once. ∎

1.5   *Corollaries to Theorems 32 and 33*

In this section we shall extend results of section 1.2.

THEOREM 34 ([Urbanowicz, 1990a, Corollary 2 to Theorem 1]) *In the notation of Theorem 32, we have*:

(i) $4 \| k_2(d) \Leftrightarrow 2 \| h(-4d) \Leftrightarrow 2 \| h(-8d) \Leftrightarrow 4 \| k_2(8d)$
$\qquad \Leftrightarrow d = p \equiv -3 \,(\bmod\, 8\,),$

(ii) $8 \| k_2(d) \Leftrightarrow 4 \| h(-4d),$

$8 \| k_2(8d) \Leftrightarrow 4 \| h(-8d),$

(*for* (i) *and* (ii) *see also section* 1.2),

(iii) $16 \| k_2(d) \Leftrightarrow (8 \| h(-4d) \ and \ 16 \,|\, \phi(d) + \varepsilon/2)$
$\qquad or \ (16 \,|\, h(-4d) \ and \ 8 \| \phi(d) + \varepsilon/2)$
$\qquad \Leftrightarrow (8 \| h(-4d) \ and \ 8 \,|\, h(-8d)) \ or \ (16 \,|\, h(-4d) \ and \ 4 \| h(-8d)),$

$16 \| k_2(8d) \Leftrightarrow (8 \| h(-8d) \ and \ 8 \,|\, h(-4d)) \ or \ (16 \,|\, h(-8d)$
$\qquad and \ 4 \| h(-4d)),$

$32 \,|\, k_2(d), \ 32 \,|\, k_2(8d) \ otherwise, \ .$

(iv) *if* $d = p \equiv 1 \,(\bmod\, 8\,)$ *then*:

$16 \| k_2(d) \Leftrightarrow (8 \| h(-4d) \ and \ p \equiv 1 \,(\bmod\, 16\,))$
$\qquad\qquad or \ (16 \,|\, h(-4d) \ and \ p \equiv 9 \,(\bmod\, 16\,)),$

$32 \| k_2(d)$
$\qquad \Leftrightarrow (8 \,|\, h(-4d) \ and \ \dfrac{h(-4d)}{8} + \dfrac{h(-8d)}{4} \equiv 2 \,(\bmod\, 4\,))$
$\qquad \Leftrightarrow (8 \| h(-4d) \ and \ 4 \| h(-8d) \ and \ \dfrac{h(-4d)}{8} \equiv \dfrac{h(-8d)}{4} \,(\bmod\, 4\,)),$
$\qquad or \ (16 \| h(-4d) \ and \ 16 \,|\, h(-8d)),$
$\qquad or \ (32 \,|\, h(-4d) \ and \ 8 \| h(-8d)),$

$64 \,|\, k_2(d) \Leftrightarrow (8 \,|\, h(-4d) \ and \ \dfrac{h(-4d)}{8} + \dfrac{h(-8d)}{4} \equiv 0 \,(\bmod\, 4\,)).$

PROOF. To prove (i) and (ii) of the theorem it is sufficient to use the congruence (i) of Theorem 32 modulo 16, i.e., the congruence

$$k_2(d) \equiv 2h(-4d) \,(\bmod\, 16\,)$$

(cf. Theorem 23) and congruence (9). To prove (iii) of the theorem, suppose $8 \,|\, h(-4d)$. Then, it suffices to apply (i) of Theorem 32. The second part of (iii) for $k_2(d)$ is an immediate consequence of the congruence (ii) of this theorem. The part of (iii) concerning $k_2(8d)$ follows from Theorem 32 (iii). Part (iv) follows easily from part (v) of this theorem.   ■

THEOREM 35 ([Urbanowicz, 1990a, Corollary 2 to Theorem 2]) *In the notation of Theorem 33, we have:*

(i) *if* $|d| \equiv -1\,(\bmod\,8)$ *then* $16\,|\,k_2(-4d)$.

   Moreover in this case

   $$16\,\|\,k_2(-4d) \Leftrightarrow (8\,\|\,\phi(|d|) + \varepsilon/2) \Leftrightarrow 4\,\|\,h(8d),$$

   $$32\,|\,k_2(-4d) \Leftrightarrow (16\,|\,\phi(|d|) + \varepsilon/2) \Leftrightarrow 8\,|\,h(8d),$$

   $$16\,\|\,k_2(-8d) \Leftrightarrow 8\,\|\,h(8d),$$

   $$32\,|\,k_2(-8d) \Leftrightarrow 16\,|\,h(8d).$$

(ii) *if* $|d| = p \equiv -1\,(\bmod\,8)$ *then:*

   $$16\,\|\,k_2(-4d) \Leftrightarrow p \equiv 7\,(\bmod\,16),$$

   $$32\,\|\,k_2(-4d) \Leftrightarrow p \equiv -1\,(\bmod\,16)\ \text{and}\ 8\,\|\,h(8d),$$

   $$64\,\|\,k_2(-4d) \Leftrightarrow p \equiv -1\,(\bmod\,16)\ \text{and}\ 16\,|\,h(8d),$$

   $$32\,\|\,k_2(-8d) \Leftrightarrow (p \equiv 7\,(\bmod\,16)\ \text{and}\ 32\,|\,h(8d))$$
   $$\text{or}\ (p \equiv -1\,(\bmod\,16)\ \text{and}\ 16\,\|\,h(8d)),$$

   $$64\,|\,k_2(-8d) \Leftrightarrow (p \equiv 7\,(\bmod\,16)\ \text{and}\ 16\,\|\,h(8d))$$
   $$\text{or}\ (p \equiv -1\,(\bmod\,16)\ \text{and}\ 32\,|\,h(8d)).$$

(iii) *if* $|d| \equiv 3\,(\bmod\,8)$ *then:*

   $$4\,\|\,k_2(-4d) \Leftrightarrow 2 \nmid h(d) \Leftrightarrow 2\,\|\,h(8d) \Leftrightarrow 4\,\|\,k_2(-8d)$$
   $$\Leftrightarrow |d| = p \equiv 3\,(\bmod\,8),$$

   $$8\,\|\,k_2(-4d) \Leftrightarrow 2\,\|\,h(d),$$

   $$8\,\|\,k_2(-8d) \Leftrightarrow 4\,\|\,h(8d),$$

   $$16\,\|\,k_2(-4d) \Leftrightarrow (4\,\|\,h(d)\ \text{and}\ 16\,|\,\phi(|d|) + \varepsilon/2)$$
   $$\text{or}\ (8\,|\,h(d)\ \text{and}\ 8\,\|\,\phi(|d|) + \varepsilon/2)$$
   $$\Leftrightarrow (4\,\|\,h(d)\ \text{and}\ 8\,|\,h(8d))\ \text{or}\ (8\,|\,h(d)\ \text{and}\ 4\,\|\,h(8d)),$$

   $$32\,|\,k_2(-4d) \Leftrightarrow (4\,\|\,h(d)\ \text{and}\ 8\,\|\,\phi(|d|) + \varepsilon/2)$$
   $$\text{or}\ (8\,|\,h(d)\ \text{and}\ 16\,|\,\phi(|d|) + \varepsilon/2),$$

$$16 \parallel k_2(-8d) \Leftrightarrow (16 \mid h(8d) \; and \; 2 \parallel h(d)) \; or \; (8 \parallel h(8d) \; and \; 4 \mid h(d)),$$

$$32 \mid k_2(-8d) \Leftrightarrow (16 \mid h(8d) \; and \; 4 \mid h(d)) \; or \; (8 \parallel h(8d) \; and \; 2 \parallel h(d)).$$

PROOF. (i) If $|d| \equiv -1 \, (\bmod \, 8)$ then by Theorem 33 we have

$$k_2(-4d) \equiv 2\phi(|d|) + \varepsilon \, (\bmod \, 32).$$

From this congruence part (i) of theorem follows easily.

(ii) The first of equivalence of (ii) follows from the first equivalence of (i). The remaining equivalences are consequences of the last two congruences of Theorem 33. If $|d| \equiv 3 \, (\bmod \, 8)$ then by Theorem 33(i) we have

$$k_2(-4d) \equiv 12h(d) \, (\bmod \, 16),$$

and, from Theorem 33 (ii), (iii) with $|d| \equiv 3 \, (\bmod \, 8)$, part (iii) of Theorem 35 follows at once.   ∎

Theorems 32 and 33 also imply congruences modulo higher powers of 2. In the theorem below we present ones modulo 128.

THEOREM 36 ([Urbanowicz, 1990b, Corollary]) *Let $p > 5$ be a prime number and let $f_s(x)$ for $s \in \{1, 3, 5, 7\}$ be the polynomial defined as follows:*

$$f_s(x) = ax^2 + bx + c,$$

where

$$a = -\frac{1}{2},$$

$$b = -\left(\frac{-4}{s}\right)\left(1 + 2\left(\frac{8}{s}\right)\right),$$

$$c = \frac{1}{2}\left(11 - 4\left(\frac{8}{s}\right)\right).$$

For $p \equiv s \, (\bmod \, 8)$, $s \in \{1, 3, 5, 7\}$ we have:

(i) *if $p \equiv 1 \, (\bmod \, 4)$ then*

$$-k_2(8p) + \left(34 - \left(\frac{8}{p}\right)\right)k_2(p)$$

$$\equiv 2p\left(\left(\frac{8}{p}\right)h(-4p) + h(-8p)\right) + f_s(p) \, (\bmod \, 128),$$

(ii) *if* $p \equiv 3 \pmod 4$ *then*

$$-k_2(8p) - \left(\frac{8}{p}\right) k_2(4p)$$

$$\equiv 2p\left(\left(1 - \left(\frac{8}{p}\right)\right) h(-p) - h(-8p)\right) + f_s(p) \pmod{128}.$$

PROOF. The theorem is an immediate consequence of Theorems 32 and 33 (it can also be deduced from Theorem 37 for $\nu = 1$). ∎

1.6   *Linear Congruence Relations for* $k_2(d)$

In [Urbanowicz, 1990b] the Hardy-Williams congruence was extended to a linear congruence involving not only class numbers of imaginary quadratic fields but also the orders of K_2-groups of the integers of real quadratic fields.

Let d and e be fundamental discriminants with $e \mid d$. Then for any

$$\mathcal{D} \in \{-8e, -4e, e, 8e\}$$

set

$$K(d, \mathcal{D}) = -\left(\frac{e}{|d/e|}\right) \prod_{\substack{p \mid |d/e| \\ p \text{ prime}}} \left(p - \left(\frac{\mathcal{D}}{p}\right)\right) k_2(\mathcal{D}),$$

where p runs through all prime divisors of $|d/e|$.

THEOREM 37 ([Urbanowicz, 1990b]) *Let* d *be an odd discriminant of a quadratic field having* ν *prime factors. Then, in the notation of Theorem 14, we have:*

$$\left(\sum_{\substack{e \mid d, e>1 \\ e \equiv 1 \,(\text{mod}\, 4)}} (K(d, 8e) - (34 - \chi_e(2))K(d, e))\right.$$

$$\left. + \sum_{\substack{e \mid d, e<0 \\ e \equiv 1 \,(\text{mod}\, 4)}} (K(d, -8e) + \chi_e(2)K(d, -4e))\right)$$

$$- 2|d|\left(\sum_{\substack{e \mid d, e>1 \\ e \equiv 1 \,(\text{mod}\, 4)}} (\chi_e(2)H(d, -4e) + H(d, -8e))\right.$$

$$\left. + \sum_{\substack{e \mid d, e<0 \\ e \equiv 1 \,(\text{mod}\, 4)}} ((1 - \chi_e(2))H(d, e) - H(d, 8e))\right)$$

$$\equiv -K_1(d) + 2|d|K_3(d) \pmod{2^{\nu+6}},$$

where

$$K_1(d) = \frac{1}{2}(11\phi(|d|) + K(d,8) + K_2(d)),$$

$$K_2(d) = \begin{cases} 56K(d,5), & if\ 5\,|\,d, \\ 0, & otherwise, \end{cases}$$

and

$$K_3(d) = \frac{1}{4}(-1)^\nu \phi(|d|) + K_4(d) + K_5(d),$$

$$K_4(d) = \begin{cases} -\frac{13}{3}H(d,-3), & if\ 3\,|\,d, \\ 0, & otherwise, \end{cases}$$

$$K_5(d) = \begin{cases} 0, & if\ x > 0\ or\ x = 0,\ y \geq 0,\ z > 0,\ u > 0, \\ 2^\nu, & if\ x = 0,\ y \geq 0,\ z = 0,\ u > 0, \\ 2^{\nu-1}, & if\ x = 0,\ y \geq 0,\ z > 0,\ u = 0, \\ 3 \cdot 2^{\nu-1}, & if\ x = 0,\ y > 0,\ z = 0,\ u = 0, \end{cases}$$

and x, y, z, u denote respectively the numbers of prime factors of $|d|$ of the form $8t + 1$, $8t - 1$, $8t + 3$, $8t - 3$.

PROOF. The theorem follows from Theorem 19 with $m = 2$ and $r = 8$. ∎

In [Urbanowicz, 1990b] the author derived the congruence of Theorem 37 from Theorem 4. The form of the right hand side of the congruence of the theorem follows from an explicit formula for the sum

$$\sum_{\substack{1 \leq a \leq (N/8) \\ \gcd(a,N)=1}} a^m$$

with $m = 1$. For arbitrary m the formula was determined in [Szmidt and Urbanowicz, 1994]. It can be also obtained from Theorem 2.

1.7   *Imaginary Quadratic Fields*

In contrast to the case of real quadratic fields, all results on $k_2(d)$ modulo powers of 2 were obtained only with algebraic methods, i.e., those using the structure of the group $(K_2O_F)_2$. It appears to be an open problem to evaluate the right hand side of the conjectured equation (21) of Chapter I modulo powers of 2. In Chapter IV we shall look more closely at this case.

In this section we review some of the standard facts on $k_2(d)$ modulo powers of 2 when d has a small number of prime factors. We consider the cases when $\nu = 1, 2$ or 3.

If $d < 0$ and $\nu = 1$ we have $d = -4$, -8 or $-p$, where p is a prime with $p \equiv 3 \,(\bmod 4)$. Since $k_2(-4) = k_2(-8) = 1$ we need only consider $k_2(-p)$, p (prime) $\equiv 3 \,(\bmod 4)$. In this case Browkin and Schinzel [Browkin and Schinzel, 1982, Corollaries 8 and 9] used their formula for the 2-rank of the group $K_2 O_F$ (see section 7.2 of Chapter I) to show that

$$k_2(-p) \equiv \begin{cases} 1 \,(\bmod 2), & \text{if } p \equiv 3 \,(\bmod 8), \\ 2 \,(\bmod 4), & \text{if } p \equiv -1 \,(\bmod 8). \end{cases}$$

The problems of evaluating $k_2(-p)$ modulo 4 if $p \equiv 3 \,(\bmod 8)$ and $k_2(-p)$ modulo 8 if $p \equiv -1 \,(\bmod 8)$ appear to be open problems.

If $d < 0$ and $\nu = 2$, we have

(i) $d = -4p$, where p is a prime with $p \equiv 1 \,(\bmod 4)$,

(ii) $d = -8p$, where p is an odd prime,

(iii) $d = -pq$, where p and q are primes with $p \equiv 1 \,(\bmod 4)$ and $q \equiv 3$ $(\bmod 4)$.

(i) In this case, Browkin and Schinzel [Browkin and Schinzel, 1982, Corollaries 8 and 9], and Conner and Hurrelbrink [Conner and Hurrelbrink, 1989a, Example 6.3, 1989b, Example 7.2] have shown that

$$k_2(-4p) \equiv \begin{cases} 1 \,(\bmod 2), & \text{if } p \equiv -3 \,(\bmod 8), \\ 0 \,(\bmod 4), & \text{if } p \equiv 1 \,(\bmod 8). \end{cases}$$

The determination of $k_2(-4p)$ modulo 4 for $p \equiv -3 \,(\bmod 8)$ remains unknown. Hurrelbrink has conjectured that

$$k_2(-4p) \equiv \begin{cases} 0 \,(\bmod 8), & \text{if } p \equiv 1 \,(\bmod 16), \\ 4 \,(\bmod 8), & \text{if } p \equiv 9 \,(\bmod 16). \end{cases}$$

The conjecture was confirmed in [Qin, 1998]. The question of evaluating $k_2(-4p)$ modulo 16 appears to be an open problem.

(ii) In this case Browkin and Schinzel [Browkin and Schinzel, 1982, Corollaries 8, 9], and Conner and Hurrelbrink [Conner and Hurrelbrink, 1989a, Example 6.6, 1989b, Example 8.2] have shown that $k_2(-8p) \equiv 0 \,(\bmod 2)$ if $p \not\equiv \pm 3 \,(\bmod 8)$, and

$$k_2(-8p) \equiv \begin{cases} 1 \,(\bmod 2), & \text{if } p \equiv \pm 3 \,(\bmod 8), \\ 2 \,(\bmod 4), & \text{if } p \equiv 7 \,(\bmod 8) \text{ or } p \equiv 9 \,(\bmod 16), \\ 0 \,(\bmod 4), & \text{if } p \equiv 1 \,(\bmod 16). \end{cases}$$

The questions of determining $k_2(-8p)$ modulo 8 for $p \equiv 7 \pmod 8$ and $p \equiv 9 \pmod{16}$ appear to be open. The best general reference here is [Conner and Hurrelbrink, 1989a,b].

(iii) In this case by congruence (1) we have $k_2(-pq) \equiv 0 \pmod 2$. Conner and Hurrelbrink [Conner and Hurrelbrink, 1989a, Example 6.4, 1989b, Examples 8.5, 8.6 and 8.8] have shown that

$$k_2(-pq) \equiv 2 \pmod 4$$

for $p \equiv 5 \pmod 8$ (see also [Browkin and Schinzel, 1982, Corollary 9]), and if $p \equiv 1 \pmod 8$, $q \equiv 3 \pmod 8$

$$k_2(-pq) \equiv \begin{cases} 2 \pmod 4, & \text{if } \left(\dfrac{p}{q}\right) = -1, \\ 0 \pmod 4, & \text{if } \left(\dfrac{p}{q}\right) = +1, \end{cases}$$

and if $p \equiv 1 \pmod 8$, $q \equiv 7 \pmod 8$, $\left(\dfrac{p}{q}\right) = -1$

$$k_2(-pq) \equiv 4 \pmod 8.$$

The following deeper result for $p \equiv 1 \pmod 8$, $q \equiv 7 \pmod 8$, $\left(\dfrac{p}{q}\right) = +1$ is provided by [Conner and Hurrelbrink, 1995, Theorem 2 p. 60, Corollary p. 65]. Putting $\mu = h(-8q)/4$ we have

$$k_2(-pq) \equiv \begin{cases} 4 \pmod 8, & \text{if } p = x^2 + 32y^2 \text{ and } p^\mu = 2a^2 + qb^2 \text{ with} \\ & b \not\equiv 0 \pmod p \text{ either both have integral} \\ & \text{solutions, or neither one has an integral solution,} \\ 0 \pmod 8, & \text{if exactly one of } p = x^2 + 32y^2 \text{ or } p^\mu = 2a^2 + qb^2 \\ & \text{with } b \not\equiv 0 \pmod p \text{ has an integral solution}. \end{cases}$$

The determination $k_2(-pq)$ modulo higher powers of 2 in the above cases remains to be discovered.

The cases $\nu = 3$ and 4 were considered in [Qin, 1995b] and [Conner and Hurrelbrink, 1989a]. These cases require examination of a large number of subcases.

2.   CONGRUENCES FOR HIGHER BERNOULLI NUMBERS

2.1   *Kummer Congruences modulo Powers of* 2

Both class numbers of imaginary quadratic fields and the orders of K_2-groups of the integers of real quadratic fields are up to sign generalized Bernoulli numbers of small indices. Namely we have $h(d) = -B_{1,\chi_d}$ for

$d < -4$ and $k_2(d) = B_{2,\chi_d}$ for $d > 8$. In this section we discuss the case of higher generalized Bernoulli numbers. For further discussion see [Gras, 1987], [Lang, H., 1988], [Pioui, 1990, 1992] and [Goren, 1999].

Let d be an odd discriminant of a quadratic field. Write

$$\mathcal{R}_d = \{-8d, -4d, d, 8d\}.$$

Let $\mathcal{R}_d^+$ be the subset of $\mathcal{R}_d$ consisting of positive discriminants, and $\mathcal{R}_d^-$ be the subset of $\mathcal{R}_d$ consisting of negative discriminants. We have

$$\mathcal{R}_d^+ = \{d, 8d\} \text{ and } \mathcal{R}_d^- = \{-8d, -4d\}$$

if $d > 0$, and

$$\mathcal{R}_d^+ = \{-8d, -4d\} \text{ and } \mathcal{R}_d^- = \{d, 8d\}$$

if $d < 0$.

Following [Urbanowicz, 1990/1991a,b] we shall relate the numbers $B_{k,\chi_{ed}}/k$ with $e \in \mathcal{T}_8$ to linear integral combinations of $h(\mathcal{D})$ ($\mathcal{D} \in \mathcal{R}_d^-$) and $k_2(\mathcal{D})$ ($\mathcal{D} \in \mathcal{R}_d^+$) modulo 64. More precisely, when $e = 1$ we shall express the numbers $B_{k,\chi_d}/k$ as linear combinations of $h(-4d)$ and $k_2(d)$ if $d > 0$, or $h(d)$ and $k_2(-4d)$ if $d < 0$. When $e \neq 1$ we shall express the numbers $B_{k,\chi_{ed}}/k$ as linear combinations of two class numbers $h(\mathcal{D})$, where $\mathcal{D} \in \mathcal{R}_d^-$ and one of the numbers $k_2(\mathcal{D})$, where $\mathcal{D} \in \mathcal{R}_d^+$. We shall treat the case of $e = 1$ in detail and indicate how the techniques used may be applied to the other cases.

Some special cases of our congruences can be derived from the generalized Kummer congruences (see section 3.4 of Chapter I). Namely, when $k \equiv 1$ $(\bmod 32)$, $k \geq 3$ and $d < 0$, $d \neq -4$ the rational numbers $B_{k,\chi_d}/k$ are 2-integral and we have

$$(B_{k,\chi_d}/k) \equiv \epsilon_k(d)h(d) \,(\bmod 64),$$

where

$$\epsilon_k(d) = -\frac{1 - \chi_d(2)}{1 - \chi_d(2)2^{k-1}}.$$

If $k \equiv 2 \,(\bmod 32)$, $k \geq 2$ and $d > 0$ the numbers $B_{k,\chi_d}/k$ are also 2-integral and we have

$$(B_{k,\chi_d}/k) \equiv \frac{1}{2}\rho_k(d)k_2(d) \,(\bmod 64),$$

where

$$\rho_k(d) = \frac{1 - \chi_d(2)2}{1 - \chi_d(2)2^{k-1}}.$$

Here $\chi_d(-1) = (-1)^k$.

When $d < 0$ and $\chi_d(2) = 1$ we have $\epsilon_k(d) = 0$. Thus if $k \equiv 1 \,(\bmod\, 2^a)$, $k \geq 3$, $a \geq 1$ then we obtain

$$(B_{k,\chi_d}/k) \equiv 0 \,(\bmod\, 2^{a+1}).$$

In other words, for given $d < 0$ odd with $\chi_d(2) = 1$ there exist infinitely many odd k such that $B_{k,\chi_d}/k$ is divisible by any given power of 2.

When $d < 0$, $\chi_d(2) = -1$ and $k \geq a+2$, we have $\epsilon_k(d) \equiv -2 \,(\bmod\, 2^{a+1})$. Thus if $k \equiv 1 \,(\bmod\, 2^a)$, $k \geq 3$, $a \geq 1$ then we find that

$$(B_{k,\chi_d}/k) \equiv -2h(d) \,(\bmod\, 2^{a+1}).$$

Hence $(B_{k,\chi_d}/k) \equiv 0 \,(\bmod\, 2^{a+1})$ if $a \leq \mathrm{ord}_2 h(d)$, and $(B_{k,\chi_d}/k) \not\equiv 0$ $(\bmod\, 2^{a+1})$ if $a > \mathrm{ord}_2 h(d)$.

When $d > 0$, $\chi_d(2) = 1$ (resp. -1) and $k \geq a + 2$, we have $\rho_k(d) \equiv -1$ (resp. 3) $(\bmod\, 2^{a+1})$. Thus if $k \equiv 2 \,(\bmod\, 2^a)$, $k \geq 2$, $a \geq 1$, we have

$$(B_{k,\chi_d}/k) \equiv \begin{cases} -k_2(d)/2 \,(\bmod\, 2^{a+1}), & \text{if } \chi_d(2) = 1, \\ 3k_2(d)/2 \,(\bmod\, 2^{a+1}), & \text{if } \chi_d(2) = -1. \end{cases}$$

Hence $(B_{k,\chi_d}/k) \equiv 0 \,(\bmod\, 2^{a+1})$ if $a \leq \mathrm{ord}_2 k_2(d) - 2$, and $(B_{k,\chi_d}/k) \not\equiv 0$ $(\bmod\, 2^{a+1})$ if $a > \mathrm{ord}_2 k_2(d) - 2$.

2.2   Some Useful Elementary Congruences

The main results of section 2 will be derived from formula (4) of Chapter I and some elementary congruences for powers of integers a^k modulo $2^{\mathrm{ord}_2 k + 6}$.

It is easily seen that for odd natural a and even k we have

$$a^k \equiv \frac{k}{2} a^2 - \frac{k}{2} + 1 \,(\bmod\, 64). \tag{10}$$

Moreover the above congruence implies

$$a^k \equiv \frac{k-1}{2} a^3 - \frac{k-1}{2} a + a \,(\bmod\, 64), \tag{11}$$

if k is odd.

For even $k \geq 4$ and odd a with $a \equiv r \,(\bmod\, 8)$ we have a more general congruence

$$a^k \equiv r^k - \frac{k}{2} r^2 + \frac{k}{2} a^2 \,(\bmod\, 2^{\mathrm{ord}_2 k + 6}). \tag{12}$$

Indeed, we have

$$a^k = (a - r + r)^k = \sum_{i=0}^{k} \binom{k}{i}(a - r)^i r^{k-i} \equiv r^k + k(a - r)r^{k-1}$$

$$+ \binom{k}{2}(a - r)^2 r^{k-2} + (a - r)^k \ (\mathrm{mod}\ 2^{\mathrm{ord}_2 \, k + 6}),$$

because for $i \geq 3$

$$\mathrm{ord}_2 \left(\frac{(a - r)^i}{i!} \right) \geq \mathrm{ord}_2 \left(\frac{2^{3i}}{i!} \right) > 2i \geq 6.$$

Consequently, in the case of even k, $k \geq 4$ the congruence $a \equiv r \ (\mathrm{mod}\ 8)$ with odd r leads to

$$a^k \equiv r^k + k(a - r)r + \frac{k}{2}(a - r)^2 \ (\mathrm{mod}\ 2^{\mathrm{ord}_2 \, k + 6}),$$

because for $k \geq 3$ we have

$$\mathrm{ord}_2(a - r)^k \geq 3k \geq \mathrm{ord}_2 \, k + 6.$$

Hence (12) follows easily.

Congruences (10), (11) and (12) give the following useful lemmas. We list them following [Urbanowicz, 1990/1991a,b].

LEMMA 1 *Let* X *be a subset of the set of odd natural numbers. For any quadratic Dirichlet character* χ_d *we have:*

$$\sum_{a \in X} \chi_d(a)a^k \equiv \frac{k - \mu}{2} \sum_{a \in X} \chi_d(a)a^{\mu+2} - \frac{k - \mu - 2}{2} \sum_{a \in X} \chi_d(a)a^{\mu} \ (\mathrm{mod}\ 64),$$

where $k \equiv \mu \ (\mathrm{mod}\ 2)$, $\mu \in \{0, 1\}$.

LEMMA 2 *For any character* χ_d *we have:*

$$\sum_{\substack{1 \leq a \leq |d| \\ a \, \mathrm{odd}}} \chi_d(a)a^k$$

$$\equiv \begin{cases} \dfrac{k}{2} \displaystyle\sum_{\substack{1 \leq a \leq |d| \\ a \, \mathrm{odd}}} \chi_d(a)a^2 - \dfrac{k - 2}{2} \displaystyle\sum_{\substack{1 \leq a \leq |d| \\ a \, \mathrm{odd}}} \chi_d(a) \ (\mathrm{mod}\ 64), & \text{if } 2 \mid k, \\[2em] \dfrac{k - 1}{2} \displaystyle\sum_{\substack{1 \leq a \leq |d| \\ a \, \mathrm{odd}}} \chi_d(a)a^3 - \dfrac{k - 3}{2} \displaystyle\sum_{\substack{1 \leq a \leq |d| \\ a \, \mathrm{odd}}} \chi_d(a)a \ (\mathrm{mod}\ 64), & \text{if } 2 \nmid k. \end{cases}$$

LEMMA 3 *Let X be a subset of the set of odd natural numbers. For any quadratic Dirichlet character χ_d and even k we have:*

$$\sum_{a \in X} \chi_d(a) a^k \equiv \frac{k}{2} \sum_{a \in X} \chi_d(a) a^2 + \left(1 - \frac{k}{2}\right) \sum_{\substack{a \in X \\ a \equiv \pm 1 \,(\mathrm{mod}\, 8)}} \chi_d(a)$$

$$+ 9 \left(3^{k-2} - \frac{k}{2}\right) \sum_{\substack{a \in X \\ a \equiv \pm 3 \,(\mathrm{mod}\, 8)}} \chi_d(a) \,(\mathrm{mod}\, 2^{\mathrm{ord}_2 k + 6}).$$

LEMMA 4 *For any character χ_d and even k we have:*

$$\sum_{\substack{1 \le a \le |d| \\ a \,\mathrm{odd}}} \chi_d(a) a^k \equiv \frac{k}{2} \sum_{\substack{1 \le a \le |d| \\ a \,\mathrm{odd}}} \chi_d(a) a^2 + \left(1 - \frac{k}{2}\right) \sum_{\substack{1 \le a \le |d| \\ a \equiv \pm 1 \,(\mathrm{mod}\, 8)}} \chi_d(a)$$

$$+ 9 \left(3^{k-2} - \frac{k}{2}\right) \sum_{\substack{1 \le a \le |d| \\ a \equiv \pm 3 \,(\mathrm{mod}\, 8)}} \chi_d(a) \,(\mathrm{mod}\, 2^{\mathrm{ord}_2 k + 6}).$$

2.3   Some Useful Formulae for $B_{k,\chi}$

We shall couple the lemmas of the previous section with formula (4) of Chapter I. Throughout the next sections we follow the notation of section 10 of Chapter I. Putting $\chi = \chi_d$, $N = 2|d|$ and $k \ge 2$ with $\chi(-1) = (-1)^k$, we obtain

$$B_{k,\chi_d} = \sum_{i=0}^{k-2} \binom{k}{i} (2B_{k-i})(2|d|)^{k-1-i} S_i(|d|, \chi_d)$$

$$- k S_{k-1}(|d|, \chi_d) + \frac{1}{|d|} S_k(|d|, \chi_d).$$

Hence we have

$$B_{k,\chi_d} + k S_{k-1}(|d|, \chi_d) - \frac{1}{|d|} S_k(|d|, \chi_d)$$

$$= \begin{cases} \displaystyle\sum_{i=0}^{(k-3)/2} \binom{k}{2i+1} (2B_{k-2i-1})(2|d|)^{k-2i-2} S_{2i+1}(|d|, \chi_d), \\ \qquad\qquad\qquad\qquad\qquad\qquad\qquad \text{if } d < 0,\, 2 \nmid k, \quad (13) \\ \displaystyle\sum_{i=0}^{(k/2)-1} \binom{k}{2i} (2B_{k-2i})(2d)^{k-2i-1} S_{2i}(d, \chi_d), \text{ if } d > 0,\, 2 \mid k. \end{cases}$$

2.4   *The Determination of $S_k(|d|, 1, 2, \chi_d)$ modulo 64 for $k \leq 3$*

The basic idea is to evaluate the sums $S_k(|d|, 1, 2, \chi_d)$ with $k \leq 3$ modulo $2^{\mathrm{ord}_2 k + 6}$ and then to use Lemmas 1, 2, 3, 4 of section 2.2 in conjunction with formula (13).

The numbers $h(d)$ and $-B_{1,\chi_d}$ for $d = -3$, -4 (resp. $k_2(d)$ and B_{2,χ_d} for $d = 5$, 8) differ by a factor $\xi(d)$ (resp. $\eta(d)$). We have $\xi(-3) = \frac{1}{3}$, $\xi(-4) = \frac{1}{2}$ (resp. $\eta(5) = \frac{1}{5}$, $\eta(8) = \frac{1}{2}$). We shall need the following lemmas concerning the sums $S_k(|d|, 1, 2, \chi_d)$:

LEMMA 5 ([Urbanowicz, 1990/1991a]) *Let d be an odd discriminant of an imaginary quadratic field. Then we have:*

(i) $S_0(|d|, 1, 2, \chi_d) = (1 - 2\chi_d(2))h(d)\xi(d)$,

(ii) $S_1(|d|, 1, 2, \chi_d) = \chi_d(2)dh(d)\xi(d)$,

(iii) $S_2(|d|, 1, 2, \chi_d) \equiv C_1(d)h(d)\xi(d) + k_2(-4d) \pmod{64}$,

   where
$$C_1(d) := -2\chi_d(2)d + 2\chi_d(2) - 1,$$

(iv) $S_3(|d|, 1, 2, \chi_d) \equiv C_2(d)h(d)\xi(d) \pmod{64}$,

   where
$$C_2(d) := 3\chi_d(2)d + 2\chi_d(2) - 4.$$

LEMMA 6 ([Urbanowicz, 1990/1991b]) *Let d be an odd discriminant of a real quadratic field. Then we have:*

(i) $S_0(|d|, 1, 2, \chi_d) = 0$,

(ii) $S_1(|d|, 1, 2, \chi_d) = \frac{1}{2}(4\chi_d(2) - 1)k_2(d)\eta(d)$,

(iii) $S_2(|d|, 1, 2, \chi_d) = 2d\chi_d(2)k_2(d)\eta(d)$,

(iv) $S_3(|d|, 1, 2, \chi_d) \equiv \frac{3}{2}C_3(d)k_2(d)\eta(d) + 2dh(-4d) \pmod{64}$,

   where
$$C_3(d) := 2d - 4 + \chi_d(2).$$

PROOF OF LEMMA 5. (i) This case follows from (17) of Chapter I.

(ii) This case follows from (16) of Chapter I and the formulae

$$S_1(|d|/2, \chi_d) = -\frac{1}{2}d(1 - \chi_d(2))h(d)\xi(d) \qquad (14)$$

(see for example [Urbanowicz, 1990a, p. 255]) and

$$S_k(|d|, 1, 2, \chi_d) = S_k(|d|, \chi_d) - 2^k \chi_d(2) S_k(|d|/2, \chi_d). \tag{15}$$

(iii) For any discriminant d we have

$$S_2(|d|, 1, 2, \chi_d)$$
$$\equiv 2 \sum_{\substack{1 \le r \le 7 \\ r \,\text{odd}}} r S_1(|d|, r, 8, \chi_d) - \sum_{\substack{1 \le r \le 7 \\ r \,\text{odd}}} r^2 S_0(|d|, r, 8, \chi_d) \,(\text{mod}\, 64).$$

When d is negative and odd, in view of Lemma 1(ii) of section 10, Chapter I, after an easy computation, the above congruence yields

$$S_2(|d|, 1, 2, \chi_d) \equiv 2S_1(|d|, 1, 2, \chi_d) - S_0(|d|, 1, 2, \chi_d)$$
$$+ 4(10 - |d|) S_0(|d|/4, \chi_d) + 16 S_1(|d|/4, \chi_d)$$
$$+ 8(1 + \chi_d(2)) S_0(|d|/2, \chi_d) \,(\text{mod}\, 64).$$

Thus the case (iii) of the lemma follows from Lemma 5(i), (ii), formula (17) of Chapter I and Theorems 3 and 4.

(iv) For any discriminant d we have

$$S_3(|d|, 1, 2, \chi_d) \equiv 3 \sum_{\substack{1 \le r \le 7 \\ r \,\text{odd}}} r S_2(|d|, r, 8, \chi_d) - 3 \sum_{\substack{1 \le r \le 7 \\ r \,\text{odd}}} r^2 S_1(|d|, r, 8, \chi_d)$$
$$+ \sum_{\substack{1 \le r \le 7 \\ r \,\text{odd}}} r^3 S_0(|d|, r, 8, \chi_d) \,(\text{mod}\, 2^9).$$

Therefore in view of the congruence

$$S_2(|d|, r, 8, \chi_d) \equiv 2r S_1(|d|, r, 8, \chi_d) - r^2 S_0(|d|, r, 8, \chi_d) \,(\text{mod}\, 64),$$

we deduce that

$$S_3(|d|, 1, 2, \chi_d)$$
$$\equiv 3 \sum_{\substack{1 \le r \le 7 \\ r \,\text{odd}}} r^2 S_1(|d|, r, 8, \chi_d) - 2 \sum_{\substack{1 \le r \le 7 \\ r \,\text{odd}}} r^3 S_0(|d|, r, 8, \chi_d) \,(\text{mod}\, 64).$$

Consequently in virtue of Lemma 1 (ii) of section 10, Chapter I, the above congruence gives

$$S_3(|d|, 1, 2, \chi_d) \equiv 3S_1(|d|, 1, 2, \chi_d) - 2S_0(|d|, 1, 2, \chi_d)$$
$$+ 12 S_0(|d|/4, \chi_d) - 8(1 + \chi_d(2)) S_0(|d|/2, \chi_d) \,(\text{mod}\, 64).$$

Thus the case (iv) of the lemma follows from Lemma 5(i), (ii) and formula (25) of Chapter I. ∎

PROOF OF LEMMA 6. The case (i) of the lemma is obvious and the case (ii) follows immediately from formulae (20) of Chapter I and (15).

Coupling formula (15) with the obvious identities $S_2(d, \chi_d) = dk_2(d)\eta(d)$ (see section 8 of Chapter I) and

$$S_2(d, \chi_d) = 2S_2(d/2, \chi_d) - 2dS_1(d/2, \chi_d)$$

we obtain

$$S_2(d, 1, 2, \chi_d) = (1 + 2\chi_d(2) - \chi_d(2)^2)dk_2(d)\eta(d),$$

which gives (iii) of the lemma.

The congruence (iv) follows from identity (15) with $k = 3$. In view of of the obvious equation

$$B_{3, \chi_d} = 0$$

and formula (4) of Chapter I, we obtain

$$S_3(d, \chi_d) = \frac{3}{2}d^2 k_2(d)\eta(d).$$

Therefore it suffices to use the congruence

$$S_3(d/2, \chi_d) \equiv S_1(d/2, 1, 2, \chi_d) \equiv S_1(d/2, \chi_d) - 2\chi_d(2)S_1(d/4, \chi_d) \,(\mathrm{mod}\,8)$$

together with formula (20) of Chapter I and the appropriate equation of Theorem 4.   ∎

REMARKS In [Urbanowicz, 1990/1991,a,b] the author proved analogous formulae for even discriminants d. When $d < 0$ we have:

$$S_0(|d|, 1, 2, \chi_d) = 0,$$

$$S_1(|d|, 1, 2, \chi_d) = dh(d)\xi(d),$$

$$S_2(|d|, 1, 2, \chi_d) = -d^2 h(d)\xi(d),$$

$$S_3(|d|, 1, 2, \chi_d) \equiv \begin{cases} 4(15\xi(d) - 14)h(d) \,(\mathrm{mod}\,64), & \text{if } 4\,\|\,d, \\ 8(-1)^{(|d|-8)/16}h(d) \,(\mathrm{mod}\,64), & \text{if } 8\,|\,d. \end{cases}$$

The last congruence is too weak to allow us to determine the numbers $B_{k, \chi_d}/k$ modulo 64. We only obtain a congruence modulo $2^{6-\mathrm{ord}_2 d}$. In [Urbanowicz, 1990/1991a, Lemma 2] the quotients $S_3(|d|, 1, 2, \chi_d)/|d|$ modulo 64 were related to some linear combinations of $k_2(d^*)$, $h(-4d^*)$ and $h(-8d^*)$ if $d = -4d^*$ or $d = -8d^*$, and of $k_2(-4d^*)$, $h(8d^*)$ and $h(d^*)$ if $d = 8d^*$. Here d^* denotes an odd fundamental discriminant.

When $d > 0$ we have:

$$S_0(d, 1, 2, \chi_d) = S_1(d, 1, 2, \chi_d) = 0,$$

$$S_2(d, 1, 2, \chi_d) = dk_2(d)\eta(d),$$

$$S_3(d, 1, 2, \chi_d) = \frac{3}{2}d^2 k_2(d)\eta(d).$$

2.5   *The Determination of $S_k(|d|, \chi_d)$ modulo 64 for any k*

In this section we shall couple Lemmas 1, 2, 3 and 4 with Lemmas 5, 6 and 7. Following [Urbanowicz, 1990/1991a] we have:

LEMMA 7 *Let d be an odd discriminant of an imaginary quadratic field. Then we have:*

(i) *if $k \geq 4$ is even*

$$S_k(|d|, \chi_d) \equiv A_1(d, k)h(d)\xi(d) + \frac{k}{2}k_2(-4d) \ (\mathrm{mod}\,64),$$

where

$$A_1(d, k) := -3 \cdot 2^{k-1}(1 - \chi_d(2)) + k(2\chi_d(2) - 1 - d\chi_d(2)) - 2\chi_d(2) + 1,$$

(ii) *if $k \geq 3$ is odd*

$$S_k(|d|, \chi_d) \equiv A_2(d, k)h(d)\xi(d) - 2^{k-3}k_2(-4d) \ (\mathrm{mod}\,64),$$

where

$$A_2(d, k) := 2^{k-2}d(1 - \chi_d(2)) + (k - 1)(\chi_d(2) - 2 + d\chi_d(2)) + d\chi_d(2).$$

LEMMA 8 *Let d be the discriminant of a real quadratic field. Then we have:*

(i) *if $k \geq 4$ is even*

$$S_k(d, \chi_d) \equiv (kd\chi_d(2) - 2^{k-2})k_2(d)\eta(d) \ (\mathrm{mod}\,64),$$

(ii) *if $k \geq 3$ is odd*

$$S_k(d, \chi_d) \equiv \frac{1}{4}A_3(d, k)k_2(d)\eta(d) + \left(k - 1 - 2^{k-2}\right)dh(-4d) \ (\mathrm{mod}\,64),$$

where

$$A_3(d, k) := 3 \cdot 2^{k-1}(3 - 2\chi_d(2)) - 2(k - 1)(d + 2) + 2(4\chi_d(2) - 1).$$

LEMMA 9 *Let d be the even discriminant of a real quadratic field and let $k \geq 4$ be an even natural number. Write $\lambda_k = 1$ if $k \leq 8$, and $\lambda_k = 0$ otherwise. Then we have:*

$$S_k(d, \chi_d)$$
$$\equiv \frac{1}{2}(1 - 3^k - \lambda_k 2^k + 4k)h(-4d) + kd\chi_d(2)k_2(d)\eta(d) \,(\bmod\ 2^{\mathrm{ord}_2\, k+6}\,).$$

PROOF OF LEMMA 7. We start with the following congruence implied by (15). If $k \geq 3$ we have

$$S_k(|d|, \chi_d) \equiv 2^k \chi_d(2)S_\sigma(|d|/2, 1, 2, \chi_d) + S_k(|d|, 1, 2, \chi_d) \,(\bmod\ 64), \quad (16)$$

where $k \equiv \sigma \,(\bmod\ 2), \sigma \in \{0, 1\}$. Hence from Lemma 2 it follows

$$S_k(|d|, \chi_d) \equiv 2^k \chi_d(2)S_0(|d|/2, 1, 2, \chi_d) + S_k(|d|, 1, 2, \chi_d)$$
$$\equiv 2^k \chi_d(2)(S_0(|d|/2, \chi_d) - \chi_d(2)S_0(|d|/4, \chi_d))$$
$$+ \frac{k}{2}S_2(|d|, 1, 2, \chi_d) - \frac{k-2}{2}S_0(|d|, 1, 2, \chi_d) \,(\bmod\ 64),$$

if $k \geq 4$ is even. Therefore (i) of the lemma follows immediately from formula (17) of Chapter I, Theorem 3 and Lemma 5(i), (iii).

Similarly, when $k \geq 3$ is odd we obtain

$$S_k(|d|, \chi_d) \equiv 2^k \chi_d(2)S_1(|d|/2, 1, 2, \chi_d) + S_k(|d|, 1, 2, \chi_d)$$
$$\equiv 2^k \chi_d(2)(S_1(|d|/2, \chi_d) - 2\chi_d(2)S_1(|d|/4, \chi_d))$$
$$+ \frac{k-1}{2}S_3(|d|, 1, 2, \chi_d) - \frac{k-3}{2}S_1(|d|, 1, 2, \chi_d) \,(\bmod\ 64).$$

Therefore (ii) of the lemma is an obvious consequence of formula (14), Theorem 4 and Lemma 5(i), (iv). ■

PROOF OF LEMMA 8. As in the proof of Lemma 7 we shall use formula (16). Since $S_0(d, \chi_d) = S_0(d/2, \chi_d) = 0$, formula (16) together with Lemma 2 give

$$S_k(d, \chi_d) \equiv 2^k \chi_d(2)S_0(d/2, 1, 2, \chi_d) + S_k(d, 1, 2, \chi_d)$$
$$\equiv -2^k \chi_d^2(2)S_0(d/4, \chi_d) + \frac{k}{2}S_2(d, 1, 2, \chi_d) \,(\bmod\ 64),$$

if $k \geq 4$ is even. Thus in view of formula (25) of Chapter I and Lemma 6(iii) we have

$$S_k(d, \chi_d) \equiv -2^{k-1}h(-4d) + kd\chi_d(2)k_2(d)\eta(d) \,(\bmod\ 64).$$

Consequently, Lemma 8(i) follows from Theorem 23.

We now turn to Lemma 8(ii). For odd $k \geq 3$ formula (16) and Lemma 2 imply

$$S_k(d, \chi_d) \equiv 2^k \chi_d(2) S_1(d/2, 1, 2, \chi_d) + S_k(d, 1, 2, \chi_d)$$

$$\equiv 2^k \chi_d(2)(S_1(d/2, \chi_d) - 2\chi_d(2)S_1(d/4, \chi_d))$$

$$+ \frac{k-1}{2} S_3(d, 1, 2, \chi_d) - \frac{k-3}{2} S_1(d, 1, 2, \chi_d) \, (\bmod 64).$$

Hence, from Lemma 5(ii),(iv), Theorem 4 and formula (20) of Chapter I, Lemma 8(ii) follows. ∎

PROOF OF LEMMA 9. In virtue of formula (15), formula (25) of Chapter I and the obvious congruences

$$2^k \equiv 0 \, (\bmod 2^{\mathrm{ord}_2 k + 6} , \, 2^{\mathrm{ord}_2 k + 5} , \, \mathrm{resp.} \, 2^{\mathrm{ord}_2 k + 2}),$$

if $k \geq 10$, $k = 8$ or 6, resp. $k = 4$, we obtain

$$S_k(d, \chi_d) \equiv S_k(d, 1, 2, \chi_d) - 2^{k-1} \lambda_k h(-4d) \, (\bmod 2^{\mathrm{ord}_2 k + 6})$$

because

$$S_k(d/2, \chi_d) \equiv S_0(d/2, 1, 2, \chi_d) = -\chi_d(2) S_0(d/4, \chi_d) \, (\bmod 2)$$

if $k \geq 6$. Note that the above congruence holds modulo 16 if $k = 4$.

On the other hand, applying Lemma 1(i) of section 10, Chapter I gives

$$S_0(d, \pm 1, 8, \chi_d) = -S_0(d, \pm 3, 8, \chi_d) = S_0(d/4, \chi_d).$$

Hence, from Lemma 4, formula (25) of Chapter I and Lemma 6(iii), Lemma 9 follows. ∎

REMARKS In [Urbanowicz, 1990/1991a,b] the author has shown analogous congruences for even discriminants d. When $d < 0$ and $k \geq 3$ we have:

$$S_k(|d|, \chi_d) \equiv \begin{cases} -\dfrac{1}{2} k d^2 h(d) \xi(d) \, (\bmod 64), & \text{if } 2 \mid k, \\ A_2(d, k) h(d) \xi(d) \, (\bmod 64), & \text{if } 2 \nmid k, \end{cases}$$

where

$$A_2(d, k) = \begin{cases} \dfrac{k-1}{2} (\xi(d)(60 - d) + 8) + d\xi(d), & \text{if } 4 \| d, \\ \dfrac{k-1}{2} (8(-1)^{(d-8)/16} - d) + d, & \text{if } 8 \mid d. \end{cases}$$

These congruences are too weak to allow us to determine $B_{k,\chi_d}/k$ modulo 64. We only find a congruence modulo $2^{6-\mathrm{ord}_2 d}$. In [Urbanowicz, 1990/1991a, Lemma 5] the quotients $S_k(|d|,\chi_d)/|d|$ modulo 64 were related to some linear combinations of class numbers and the orders of K_2-groups of the integers of appropriate quadratic fields of the same form as those obtained for $S_3(|d|,1,2,\chi_d)$ but now with coefficients of the form $Ak^3 + Bk^2 + Ck + D$, where A, B, C, D depend only upon d and $A,B,C,D \in \frac{1}{4}\mathbb{Z}$. The author made use of Lemmas 1 and 2. See the remark after Lemma 6.

When $d > 0$ we have

$$
S_k(d,\chi_d) \equiv
\begin{cases}
\dfrac{k}{2}dk_2(d)\eta(d)\,(\bmod\,64)\,, & \text{if } 2\,|\,k\,, \\[2mm]
\dfrac{3}{4}(k-1)d^2 k_2(d)\eta(d)\,(\bmod\,64)\,, & \text{if } 2\nmid k\,.
\end{cases}
$$

These congruences allow us to determine the numbers B_{k,χ_d} (but not their quotients $B_{k,\chi_d}/k$) modulo $2^{6-\mathrm{ord}_2 d}$. In [Urbanowicz, 1990/1991b, Lemma 5] the quotients $S_k(d,\chi_d)/d$ modulo $2^{6+\mathrm{ord}_2 k}$ were related to some linear combinations of the numbers

$$k_2(-d^*),\ h(d^*),\ h(8d^*),\ \text{if}\ d = -4d^*\,,$$

$$k_2(8d^*),\ h(-8d^*),\ h(-4d^*),\ \text{if}\ d = 8d^*\,,$$

or

$$k_2(-8d^*),\ h(8d^*),\ h(d^*),\ \text{if}\ d = -8d^*\,.$$

Recall that d^* is an odd discriminant. These combinations have coefficients of the from $Ak^2 + Bk + C$, where A, B, C depend only upon d and $A,B,C \in \frac{1}{4}\mathbb{Z}$. In this case the author made use of Lemmas 3 and 4.

In the case when d and k are even, the situation is much more complicated and we refer the reader to [Urbanowicz, 1990/1991a,b]. We just prove a weaker version of results obtained in this paper.

LEMMA 10 ([Urbanowicz, 1990/1991b, Lemma 6]) *Let d be an even discriminant of a real quadratic field and let $k \geq 4$ be an even natural number. Then we have*:

$$
S_k(d,\chi_d) \equiv \frac{kd}{2}k_2(d)\eta(d)\,(\bmod\,2^{\mathrm{ord}_2 k+6})\,.
$$

PROOF. In view of Lemma 3 for any d we have

$$
S_k(d,\chi_d) - \frac{k}{2}S_2(d,\chi_d)
$$

$$\equiv \left(1 - \frac{k}{2}\right) S_0(d, \pm 1, 8, \chi_d) + 9\left(3^{k-2} - \frac{k}{2}\right) S_0(d, \pm 3, 8, \chi_d)$$

$$\equiv \left(1 - \frac{k}{2}\right) S_0(d, \chi_d) + \left(3^k - 4k - 1\right) S_0(d, \pm 3, 8, \chi_d)$$

$$(\bmod\, 2^{\mathrm{ord}_2\, k+6}). \qquad (17)$$

On the other hand we have

$$S_0(d, \pm 3, 8, \chi_d) = S_0(d/4, \pm 3, 8, \chi_d) + \sum_{\substack{0 \le a \le d/4 \\ a \equiv (d/2)\pm 3\,(\bmod\, 8)}} \chi_d((d/2) - a)$$

$$+ \sum_{\substack{0 \le a \le d/4 \\ a \equiv (-d/2)\pm 3\,(\bmod\, 8)}} \chi_d((d/2) + a) + \sum_{\substack{0 \le a \le d/4 \\ a \equiv \pm 3\,(\bmod\, 8)}} \chi_d(d - a)$$

$$= T_1 + T_2 + T_3 + T_4,$$

where T_i denotes the ith summand of the middle term of the above equation. Thus in view of

$$T_1 + T_4 = \begin{cases} S_0(d/4, \chi_d), & \text{if } 4 \,\|\, d, \\ 2S_0(d/4, \pm 3, 8, \chi_d), & \text{if } 8 \,|\, d, \end{cases}$$

and

$$T_2 + T_3 = \begin{cases} -S_0(d/4, \chi_d), & \text{if } 4 \,\|\, d, \\ -2S_0(d/4, \pm 1, 8, \chi_d), & \text{if } 8 \,|\, d, \end{cases}$$

we obtain

$$S_0(d, \pm 3, 8, \chi_d) = 0.$$

Therefore (17) implies the congruence

$$S_k(d, \chi_d) \equiv \frac{k}{2} S_2(d, \chi_d) \,(\bmod\, 2^{\mathrm{ord}_2\, k+6}),$$

and by virtue of the identity $S_2(d, \chi_d) = dk_2(d)\eta(d)$ Lemma 10 follows. ∎

Now we are ready to prove the main results of section 2 of Chapter III.

2.6 The Determination of the Quotient $B_{k,\chi_d}/k$ modulo 64 for Negative d

Our purpose is to relate the numbers $B_{k,\chi_d}/k$ modulo 64 to linear combinations of class numbers and the orders of K_2-groups of the integers of appropriate quadratic fields. Following [Urbanowicz, 1990/1991a] we have:

THEOREM 38 *Let d be an odd discriminant of an imaginary quadratic field and let $k \geq 3$ be an odd natural number. The numbers $B_{k,\chi_d}/k$ are 2-integral and the following congruence holds:*

$$(B_{k,\chi_d}/k) \equiv -k\mu(1 - \chi_d(2))h(d)\xi(d) - \frac{k-1}{2}\nu k_2(-4d) \, (\bmod\, 64),$$

where $\mu := \mu_k(d)$, $\nu := \nu_k(d)$, and $\mu_3 = -1 - (2/d)$, $\mu_5 = -15$, $\nu_3 = 5/d$, and $\mu_k, \nu_k = 1$ otherwise.

PROOF. We start with formula (13). In view of the von Staudt-Clausen theorem for $p = 2$ and Lemma 7 we see that the numbers B_{k,χ_d} are 2-integral.

We apply formula (13) in the case when d and k are odd with $k \geq 7$. Then we have

$$|d|B_{k,\chi_d} + k|d|S_{k-1}(|d|, \chi_d) - S_k(|d|, \chi_d) \tag{18}$$

$$\equiv \sum_{i=(k-3)/2-2}^{(k-3)/2} \binom{k}{2i+i} (2B_{k-2i-1})2^{k-2i-2}|d|^{k-2i-1} S_{2i+1}(|d|, \chi_d)$$

$$(\bmod\, 64).$$

Hence by virtue of Lemma 7 when $k \geq 9$ we have

$$|d|B_{k,\chi_d}$$

$$\equiv h(d)\xi(d) \sum_{i=(k-3)/2-2}^{(k-3)/2} \binom{k}{2i+1} (2B_{k-2i-1})2^{k-2i-2}|d|^{k-2i-1} A_2(d, 2i+1)$$

$$- k_2(-4d)2^{k-4} \sum_{i=(k-3)/2-2}^{(k-3)/2} \binom{k}{2i+1} (2B_{k-2i-1})|d|^{k-2i-1}$$

$$- k|d| \left(A_1(d, k-1)h(d)\xi(d) + \frac{k-1}{2}k_2(-4d) \right)$$

$$+ A_2(d, k)h(d)\xi(d) \, (\bmod\, 64).$$

Thus in view of $k_2(-4d) \equiv 0 \, (\bmod\, 4)$ and $k \geq 9$, it follows that

$$|d|B_{k,\chi_d} \equiv A_4(d, k)h(d)\xi(d) + d\binom{k}{2}k_2(-4d) \, (\bmod\, 64),$$

where

$$A_4(d,k) := 32\binom{k}{6} + 8d\binom{k}{4}\chi_d(2)$$

$$+ \frac{2d^3}{3}\binom{k}{2}\chi_d(2) + dkA_1(d,k-1) + A_2(d,k).$$

The task is now to find a simpler form of $A_4(d,k)$ modulo 64. An easy computation shows that

$$A_4(d,k) \equiv \begin{cases} 0 & (\bmod\,64), & \text{if } \chi_d(2) = 1, \\ 2(k^2-1) + 2d & (\bmod\,64), & \text{if } \chi_d(2) = -1. \end{cases}$$

Consequently, if $k \geq 9$ we have

$$|d|B_{k,\chi_d} \equiv (k^2 - 1 + d)(1 - \chi_d(2))h(d)\xi(d) + d\binom{k}{2}k_2(-4d)\,(\bmod\,64).$$

A trivial verification gives Theorem 38 for $k \geq 9$ immediately.

We are left with the task of determining $B_{k,\chi_d}/k$ modulo 64 for $k = 3, 5$ and 7.

(i) If $k = 3$ then by formula (13) we have

$$|d|B_{3,\chi_d} = 2d^2S_1(|d|,\chi_d) + 3dS_2(|d|,\chi_d) + S_3(|d|,\chi_d).$$

Therefore by Lemma 7 (recall that $S_1(|d|,\chi_d) = dh(d)\xi(d)$ and $S_2(|d|,\chi_d) = -d^2h(d)\xi(d)$) we obtain

$$|d|B_{3,\chi_d} \equiv (-d^3 + A_2(d,3))h(d)\xi(d) - k_2(-4d)\,(\bmod\,64).$$

Hence Theorem 38 for $k = 3$ follows.

(ii) If $k = 5$ then formula (13) and Lemma 7 imply (recall that $k_2(-4d) \equiv 0$ $(\bmod\,4)$)

$$|d|B_{5,\chi_d} \equiv A_5(d)h(d)\xi(d) + 10dk_2(-4d)\,(\bmod\,64),$$

where

$$A_5(d) := -24d + 28(d-1)A_2(d,3) + 5dA_1(d,4) + A_2(d,5).$$

An easy computation gives Theorem 38 for $k = 5$ immediately.

(iii) If $k = 7$ then by (13) and Lemma 7 we have

$$|d|B_{7,\chi_d} \equiv A_6(d)h(d)\xi(d) + 5dk_2(-4d)\,(\bmod\,64),$$

where

$$A_6(d) := 32 + 24A_2(d,3) + 14d^2 A_2(d,4) + 7dA_1(d,6) + A_2(d,7).$$

This now yields Theorem 38 for $k = 7$. ∎

REMARKS In [Urbanowicz, 1990/1991a] analogous congruences for d negative and even were found. The author has shown that the numbers $B_{k,\chi_d}/k$ are 2-integral (unless $d = -4$ when $\mathrm{ord}_2(B_{k,\chi_d}/k) = -1$) and

$$(B_{k,\chi_d}/k) \equiv \begin{cases} \alpha_1 k_2(d^*) + \beta_1 h(-4d^*) + \gamma_1 h(-8d^*) \,(\bmod\, 64), \\ \qquad\qquad\qquad\qquad\qquad\qquad \text{if } d = -4d^*, \\ \alpha_2 k_2(-4d^*) + \beta_2 h(8d^*) + \gamma_2 h(d^*) \,(\bmod\, 64), \\ \qquad\qquad\qquad\qquad\qquad\qquad \text{if } d = 8d^*, \\ \alpha_3 k_2(d^*) + \beta_3 h(-4d^*) + \gamma_3 h(-8d^*) \,(\bmod\, 64), \\ \qquad\qquad\qquad\qquad\qquad\qquad \text{if } d = -8d^*, \end{cases}$$

where α_i, β_i, γ_i are of the form $Ak + B$, where A, B are explicitly determined integers depending upon d and k, expressed in terms of the Jacobi symbols

$$\left(\frac{a}{b}\right), \qquad a = -1 \text{ or } 2, \text{ and } b = k \text{ or } d^*.$$

In the case when d and $k \geq 3$ are even, formula (13) implies the congruence

$$|d|B_{k,\chi_d} \equiv 6\binom{k}{2} d^2 S_{k-2}(|d|, \chi_d) - k|d|S_{k-1}(|d|, \chi_d)$$
$$+ S_k(|d|, \chi_d) \,(\bmod\, 64|d|).$$

In [Urbanowicz, 1990/1991a] the author has shown that

$$6\binom{k}{2} d^2 S_{k-2}(|d|, \chi_d) \equiv 0 \,(\bmod\, 64|d|).$$

Therefore we obtain the congruence

$$|d|B_{k,\chi_d} \equiv -k|d|S_{k-1}(|d|, \chi_d) + S_k(|d|, \chi_d) \,(\bmod\, 64|d|).$$

This congruence gives a weaker version of congruences obtained in [Urbanowicz, 1990/1991a]:

THEOREM 39 *Let $d \neq -4$ be an even discriminant of an imaginary quadratic field and let $k \geq 3$ be an odd natural number. Then the numbers B_{k,χ_d} are 2-integral and we have:*

$$(B_{k,\chi_d}/k) \equiv -h(d) \,(\bmod\, 2^{6-\mathrm{ord}_2 d}).$$

2.7   *The Determination of the Quotient $B_{k,\chi_d}/k$ modulo 64 for Positive d*

In the case of positive d, following [Urbanowicz, 1990/1991b] we have:

THEOREM 40 *Let d be an odd discriminant of a real quadratic field and let $k \geq 4$ be an even natural number. Then the numbers $B_{k,\chi_d}/k$ are 2-integral and the following congruence holds:*

$$
(B_{k,\chi_d}/k) \equiv \left(2k\chi_d(2) + k + 2\right) \mu h(-4d)
$$

$$
+ \frac{3}{2} \left(-k - 2\chi_d(2) + 1\right) \nu k_2(d)\eta(d) \pmod{64},
$$

where $\mu := \mu_k(d)$, $\nu := \nu_k(d)$, and

$$
\mu_4 = -d + 10 + 4\chi_d(2), \quad \mu_6 = 8 + 5\chi_d(2),
$$
$$
\nu_4 = 2d + 8\chi_d(2) + 7, \quad \nu_6 = -4\chi_d(2) - 11,
$$

and $\mu_k, \nu_k = 1$ otherwise.

PROOF. We start with formula (13). By the von Staudt-Clausen theorem for $p = 2$ and Lemma 8 we can deduce that for any d the numbers B_{k,χ_d} are 2-integral. By Lemma 9 we see that also the numbers $B_{k,\chi_d}/k$ are 2-integral in view of

$$
\text{ord}_2 \left(1 - 3^k - \lambda_k 2^k\right) \geq \text{ord}_2 k.
$$

In view of formula (13) in the case when $d > 0$ is odd and $k \geq 8$ is even, we have

$$
dB_{k,\chi_d} + kdS_{k-1}(d, \chi_d) - S_k(d, \chi_d)
$$

$$
\equiv \sum_{i=k/2-2}^{k/2-1} \binom{k}{2i} (2B_{k-2i})2^{k-2i-1} d^{k-2i} S_{2i}(d, \chi_d) \pmod{2^{\text{ord}_2 k+6}}
$$

because (recall that $S_2(d, \chi_d) \equiv 0 \pmod 4$ and $S_{2i}(d, \chi_d) \equiv 0 \pmod 8$ if $i \geq 2$)

$$
\text{ord}_2(2^s/s) \geq 4
$$

if $s = k - 2i \geq 6$, and

$$
\text{ord}_2 \left(\binom{k}{2i}2^{k-2i-1} S_{2i}(d, \chi_d)\right) \geq \text{ord}_2 k + 6
$$

if $i \leq \frac{k}{2}$, and

$$
\text{ord}_2 \left(\binom{k}{2}2^{k-3}\right) \geq \text{ord}_2 k + 6
$$

if $k \geq 8$. If $k \geq 8$ then we obtain

$$dB_{k,\chi_d} \equiv -\frac{8d^4}{15}\binom{k}{4}S_{k-4}(d,\chi_d) + \frac{2d^2}{3}\binom{k}{2}S_{k-2}(d,\chi_d)$$

$$- dkS_{k-1}(d,\chi_d) + S_k(d,\chi_d)\,(\bmod\,2^{\mathrm{ord}_2\,k+6}).$$

Therefore by Lemma 8 (applied to the sums $S_{k-4}(d,\chi_d)$, $S_{k-2}(d,\chi_d)$, $S_{k-1}(d,\chi_d)$) and Lemma 9 (applied to the sum $S_k(d,\chi_d)$) we find that (recall that $k_2(-4d) \equiv 0\,(\bmod\,4)$, $h(-4d) \equiv 0\,(\bmod\,2)$ and $h(-4d) \equiv 0\,(\bmod\,4)$ if $\chi_d(2) = 1$)

$$(B_{k,\chi_d}/k) \equiv H_k(d)k_2(d)\eta(d) - \frac{1}{2}(2\chi_d(2) - 1)k_2(d)\eta(d)$$

$$+ \left(-k + 20 + 9\left(\frac{1 - 3^k - \lambda_k 2^k}{2k}\right)\right)dh(-4d)\,(\bmod\,64),$$

where

$$H_k(d) := \frac{1}{2}(2(1 + \chi_d(2)) + d + 8\lambda_k)(k - 2)$$

because for even k

$$\mathrm{ord}_2(1 - 3^k) = \mathrm{ord}_2 k + 2,$$

and for $k = 8$

$$\mathrm{ord}_2(2^k/k) = 5.$$

On the other hand when $k \geq 8$ we have

$$\frac{1 - 3^k}{2k} = -\frac{1}{k}\sum_{i=1}^{k}\binom{k}{i}2^{i-1}$$

$$\equiv -\frac{1}{k}\sum_{\substack{1 \leq i \leq 8 \\ i \neq 7}}\binom{k}{i}2^{i-1} \equiv 2\left(2k - 1 + 4(-1)^{k/2}\right)\,(\bmod\,32).$$

and

$$H_k(d) \equiv \frac{1}{2}\left(5(k - 2) - ((-1)^{k/2} - 1)(2\chi_d(2) - 3 + d) + 8\lambda_k\right)(\bmod\,16).$$

Combining the above with Theorem 23 gives Theorem 40 for $k \geq 8$ immediately.

We prove the rest of the theorem by considering cases.

(i) If $k = 4$ then by formula (13) we have

$$B_{4,\chi_d} \equiv \frac{1}{2}(2d^2 + 2d + 10\chi_d(2) - 13)k_2(d)\eta(d) - \frac{10}{d}h(-4d)\,(\bmod\,64).$$

Hence Theorem 40 for $k = 4$ follows immediately.

(ii) If $k = 6$ then from (13), by Lemma 8 (applied to the sums $S_4(d, \chi_d)$ and $S_5(d, \chi_d)$) and Lemma 9 (applied to the sum $S_6(d, \chi_d)$) we find that

$$dB_{6,\chi_d} \equiv \frac{3d}{2}(20\chi_d(2) - A_1(d, 4))k_2(d)\eta(d) + 24d^2h(-4d) \ (\text{mod } 128d).$$

Hence Theorem 40 for $k = 6$ follows at once.   ∎

REMARKS In [Urbanowicz, 1990/1991b] analogous congruences for d positive and even were found. The author has shown that the numbers $B_{k,\chi_d}/k$ are 2-integral and

$$(B_{k,\chi_d}/k) \equiv \begin{cases} \alpha_1 k_2(-4d^*) + \beta_1 h(d^*) + \gamma_1 h(8d^*), & \text{if } d = -4d^*, \\ \alpha_2 k_2(8d^*) + \beta_2 h(-8d^*) + \gamma_2 h(-4d^*), & \text{if } d = 8d^*, \\ \alpha_3 k_2(-8d^*) + \beta_3 h(d^*) + \gamma_3 h(8d^*), & \text{if } d = -8d^*, \end{cases}$$

where α_i, β_i, γ_i are of a similar form to the coefficients in the congruence for $B_{k,\chi_d}/k$ in the case when d is negative and even (see the remarks after the proof of Theorem 38).

Lemma 10 together with the congruence

$$dB_{k,\chi_d} \equiv -kdS_{k-1}(d, \chi_d) + S_k(d, \chi_d) \ (\text{mod } 2^{\text{ord}_2 \, k+6}d)$$

(implied for d even by formula (13)) gives a weaker version of the congruences obtained in [Urbanowicz, 1990/1991b] in this case:

THEOREM 41 *Let d be an even, positive discriminant of a quadratic field and let $k \geq 4$ be an even natural number. Then the numbers $B_{k,\chi_d}/k$ are 2-integral and*

$$(B_{k,\chi_d}/k) \equiv \frac{1}{2}k_2(d)\eta(d) \ (\text{mod } 2^{6-\text{ord}_2 \, d}).$$

2.8   Corollaries to Theorems 38 and 39

By combining Theorems 38 and 39 with results of section 1, the following theorems were proved in [Urbanowicz, 1990/1991a]. Their proofs are left to the reader. In the notation of Theorems 38 and 39 we have:

THEOREM 42 *Let d be an odd, negative discriminant, and let $k \geq 3$ be odd. If $\chi_d(2) = 1$ then we have:*

(i) $\text{ord}_2 \, B_{k,\chi_d} \geq 4$,

(ii) $\text{ord}_2 \, B_{k,\chi_d} = 4 \Leftrightarrow 16 \parallel k_2(-4d)$ *(or, equivalently, $4 \parallel h(8d)$)*
     and $k \equiv 3 \ (\text{mod } 4)$,

(iii) $\mathrm{ord}_2 B_{k,\chi_d} = 5 \Leftrightarrow (16 \| k_2(-4d)$ *and* $k \equiv -3 \,(\mathrm{mod}\, 8))$
$$or (32 \| k_2(-4d) \text{ and } k \equiv 3 \,(\mathrm{mod}\, 4)),$$

(iv) $\mathrm{ord}_2 B_{k,\chi_d} \geq 6 \Leftrightarrow k \equiv 1 \,(\mathrm{mod}\, 8)$
$$or (32 \,|\, k_2(-4d) \text{ and } k \equiv -3 \,(\mathrm{mod}\, 8)) \text{ or } 64 \,|\, k_2(-4d).$$

If $|d| = p \equiv -1\,(\mathrm{mod}\, 8)$ *is a prime number we have*:

(i) $\mathrm{ord}_2 B_{k,\chi_d} = 4 \Leftrightarrow p \equiv 7\,(\mathrm{mod}\, 16)$ *and* $k \equiv 3\,(\mathrm{mod}\, 4)$,

(ii) $\mathrm{ord}_2 B_{k,\chi_d} = 5 \Leftrightarrow (p \equiv 7\,(\mathrm{mod}\, 16)$ *and* $k \equiv -3\,(\mathrm{mod}\, 8))$
$$or (\ p \equiv -1\,(\mathrm{mod}\, 16) \text{ and } 8 \| h(8d) \text{ and } k \equiv 3\,(\mathrm{mod}\, 4)),$$

(iii) $\mathrm{ord}_2 B_{k,\chi_d} \geq 6$, *otherwise*.

If $\chi_d(2) = -1$ *then we have*:

(i) $\mathrm{ord}_2 B_{k,\chi_d} \geq 1$,

(ii) $\mathrm{ord}_2 B_{k,\chi_d} = 1 \Leftrightarrow 2 \nmid h(d)$
$$(or\ 4 \| k_2(-4d),\ or\ 2 \| h(8d),\ or\ 4 \| k_2(-8d))$$
$$\Leftrightarrow |d| = p \equiv 3\,(\mathrm{mod}\, 8) \text{ is a prime number},$$

(iii) $\mathrm{ord}_2 B_{k,\chi_d} = 2 \Leftrightarrow 2 \| h(d)$ *(or* $8 \| k_2(-4d)$ *)*,

(iv) $\mathrm{ord}_2 B_{k,\chi_d} = 3 \Leftrightarrow 4 \| h(d)$
$$and\ (k \equiv 1\,(\mathrm{mod}\, 4) \text{ or } (k \equiv 3\,(\mathrm{mod}\, 4) \text{ and } 4 \,|\, h(8d))),$$

(v) $\mathrm{ord}_2 B_{k,\chi_d} \geq 4$, *otherwise*.

THEOREM 43 *Let d be an even, negative discriminant, and let $k \geq 3$ be an odd natural number. Then*

(i) $\mathrm{ord}_2 B_{k,\chi_d} = -1$, *if* $d = -4$,

(ii) $\mathrm{ord}_2 B_{k,\chi_d} = 0$, *if* $d = -8$,

(iii) $\mathrm{ord}_2 B_{k,\chi_d} \geq 1$, *if* $d \neq -4, -8$,

(iv) $\mathrm{ord}_2 B_{k,\chi_d} = \mathrm{ord}_2 h(d)$, *if* $1 \leq \mathrm{ord}_2 h(d) \leq 3$,

(v) $\mathrm{ord}_2 B_{k,\chi_d} \geq 4$, *otherwise*.

REMARK If $d = -4$ we have
$$2(B_{k,\chi_d}/k) \equiv 18 + \frac{13}{k}\,(\mathrm{mod}\, 32),$$

and so
$$E_{k-1} \equiv 14 - \frac{13}{k}\,(\mathrm{mod}\, 32),$$

where E_i denotes the ith Euler number, see [Urbanowicz, 1990/1991a].

2.9   *Corollaries to Theorems 40 and 41*

By combining Theorems 40 and 41 with results of section 1, the following theorem was proved in [Urbanowicz, 1990/1991b]. Its proof is left to the reader.

THEOREM 44 *Let d be a positive discriminant and let $k \geq 4$ be even. Then we have*:

(i) $\operatorname{ord}_2(B_{k,\chi_d}/k) \geq 1$ *if $d \neq 8$, and* $\operatorname{ord}_2(B_{k,\chi_d}/k) = 0$ *if $d = 8$,*

(ii) $\operatorname{ord}_2(B_{k,\chi_d}/k) = \operatorname{ord}_2(k_2(d)) - 1$, *if $2 \leq \operatorname{ord}_2(k_2(d)) \leq 4$,*

and moreover, if d is odd we have:

(iii) $\operatorname{ord}_2(B_{k,\chi_d}/k) = 4 \Leftrightarrow 32 \, \| \, h(d)$
         and ($k \equiv 2 \,(\bmod\, 4)$ or ($k \equiv 0 \,(\bmod\, 4)$ and $16 \, | \, h(-4d)$)),

(iv) $\operatorname{ord}_2(B_{k,\chi_d}/k) \geq 5$, *otherwise.*

Chapter IV

CONGRUENCES AMONG THE VALUES
OF 2-ADIC L-FUNCTIONS

Our purpose in this chapter is to investigate divisibility properties of the values of L-functions attached to quadratic characters at integers, for example, class numbers and the orders of K_2-groups of the integers of quadratic fields. Making use of 2-adic L-functions we extend the linear congruence relations considered in the previous chapters. Extensions of this type have been considered in [Gras, 1989], [Uehara, 1990], [Urbanowicz and Wójcik, 1995/1996] and [Wójcik, 1998]. This chapter will focus on the results given in the latter three papers. We begin with some notation and basic properties of p-adic numbers and functions, especially p-adic L-functions.

1. NOTATION

1.1 p-adic Numbers

Let p be a prime number. As usual we denote the ring of p-adic integers (resp. the field of p-adic numbers) by $\mathbb{Z}_p$ (resp. $\mathbb{Q}_p$). Let $\mathbb{C}_p$ denote the completion of $\overline{\mathbb{Q}}_p$ (the algebraic closure of $\mathbb{Q}_p$, which is not complete). $\mathbb{C}_p$ is algebraically closed. The p-adic absolute value $|\cdot|_p$ (normalized such that $|p|_p = 1/p$) on $\mathbb{Q}_p$ extends uniquely to $\overline{\mathbb{Q}}_p$, and then naturally to $\mathbb{C}_p$ ($\overline{\mathbb{Q}}_p$ is dense in $\mathbb{C}_p$). In what follows the field of all algebraic numbers is regarded as a subfield of $\mathbb{C}_p$ through a fixed embedding. In particular for any Dirichlet character χ we have $\mathbb{Q}(\chi(1), \chi(2), \ldots) \subset \mathbb{C}_p$. For more details see [Koblitz, 1984] or [Washington, 1997].

1.2 Congruences

Let $a, b \in \mathbb{C}_p$ and $\alpha \in \mathbb{Q}$. The notation $a \equiv b \,(\bmod\, p^\alpha)$ means that $|a - b|_p \leq p^{-\alpha}$. For $a, b \in \mathbb{Z}$ and $\alpha \in \mathbb{N}$ these congruences are the usual congruences for integral rational numbers. We say that $a \in \mathbb{C}_p$ is p-integral if $a \equiv 0 \,(\bmod\, p^0)$. For $a \in \mathbb{Q}$, if a is p-integral in the above sense then its denominator is not divisible by p. We say that the p-integral number a is divisible by p^α ($\alpha \geq 1$) if $a \equiv 0 \,(\bmod\, p^\alpha)$. We write $p^\alpha \,|\, a$. If for the p-integral number a we have $a \not\equiv 0 \,(\bmod\, p^\alpha)$, we write $p^\alpha \!\nmid a$ and say that a is not divisible by p^α. We say that $a \,(\in \mathbb{C}_2)$ is even if a is 2-integral and divisible by 2.

1.3   *The p-adic Functions* exp *and* $\log_p$

The p-adic function exp is defined by the p-adic series

$$\exp(x) = \sum_{n=0}^{\infty} \frac{x^n}{n!} \, .$$

This series has radius of convergence $p^{-1/(p-1)} < 1$. The p-adic logarithm is defined by the series

$$\log_p(1+x) = \sum_{n=1}^{\infty} \frac{(-1)^{n+1} x^n}{n} \, .$$

This series has radius of convergence equal to 1. Since for formal power series we have $\exp(xy) = \exp(x) + \exp(y)$ and $\log_p(xy) = \log_p(x) + \log_p(y)$, these identities are true whenever the series converge.

The p-adic logarithm $\log_p$ can be uniquely extended to all of $\mathbb{C}_p^*$ in such a way that $\log_p(xy) = \log_p(x) + \log_p(y)$, $x, y \in \mathbb{C}_p^*$ and $\log_p(p) = 0$. We have $\log_p(x) \in \mathbb{Q}_p(x)$ and $\log_p(x) = 0$ if and only if x is a rational power of p times a root of unity.

For $|x|_p < p^{-1/(p-1)}$ we have the usual identities $\log_p \exp(x) = x$, $\exp \log_p(1+x) = 1 + x$ and moreover

$$|\log_p(1+x)|_p = |x|_p \, .$$

If $|x|_p = p^{-1/(p-1)}$ then

$$|\log_p(1+x)|_p \leq |x|_p \, .$$

1.4   *The Teichmüller Character at p*

Set $q = p$, if $p \geq 3$ and $q = 4$, if $p = 2$. It is well known that for given $a \in \mathbb{Z}_p$, $p \!\nmid a$ there exists a unique $\phi(q)$th root of unity satisfying

$$a \equiv \omega(a) \,(\bmod\, q) \, . \tag{1}$$

Set $\langle a \rangle = \omega(a)^{-1} a$.

If χ is a Dirichlet character then, if we fix (once and for all) an embedding of $\overline{\mathbb{Q}}$ into $\mathbb{C}_p$, we may regard the values of χ as lying in $\mathbb{C}_p$. Then the character χ may be regarded as a p-adic object, more precisely a function $\mathbb{Z}_p \to \mathbb{C}_p$, multiplicative on $\mathbb{Z}_p^*$. Such a function is called a p-adic Dirichlet character. In this sense ω is a (primitive) p-adic Dirichlet character of conductor q and order $\phi(q)$ if we set $\omega(a) = 0$ for $p \mid a$. We call it the Teichmüller character at p. In other words ω can be regarded as a p-adic function $\mathbb{Z}_p \to \mathbb{C}_p$ satisfying the congruence (1). This character may also be regarded as coming (noncanonically) from a complex Dirichlet character generating the group of Dirichlet characters modulo q.

1.5   The Generalized Kummer Congruences

For a p-adic Dirichlet character χ the generalized Bernoulli numbers attached to χ may also be regarded as p-adic numbers from $\mathbb{C}_p$. For example, the generalized Kummer congruence given in section 3.4, Chapter I holds for numbers from $\mathbb{C}_p$.

2. p-ADIC L-FUNCTIONS

2.1   The p-adic L-functions

There are many ways to construct p-adic L-functions. The original construction is due to Kubota and Leopoldt [Kubota and Leopoldt, 1964]. The definition given below is due to Washington [Washington, 1976] (see also [Washington, 1997, §5]). The key point of the construction is the formula giving the values of classical L-functions at negative integers (see the first formula of section 4.5, Chapter I). The p-adic L-functions are defined as functions which agree with these values multiplied by an Euler factor, as is usual in such situations. For other definitions, see for instance [Fresnel, 1967/1968], [Iwasawa, 1969, 1972], [Amice and Fresnel, 1972], [Serre, 1973], [Coates, 1977], [Lang, S., 1990]. See also [Deligne and Ribet, 1980].

2.2   Definition of $L_p(s, \chi)$

Let χ be a Dirichlet character of conductor M and let N be a multiple of M and q. Then there exists a p-adic meromorphic (analytic if $\chi \neq 1$) function $L_p(s, \chi)$ defined on the set $\{s \in \mathbb{C}_p : |s|_p < qp^{-1/(p-1)}\}$ such that

$$L_p(1 - n, \chi) = -(1 - \chi\omega^{-n}(p)p^{n-1})\frac{B_{n,\chi\omega^{-n}}}{n}, \quad n \geq 1. \tag{2}$$

We call $L_p(s, \chi)$ the Kubota-Leopoldt p-adic L-function attached to χ. If $\chi = 1$ then $L_p(s, \chi)$ is analytic except for a pole at $s = 1$ with residue equal to $1 - (1/p)$. Furthermore we have the following explicit formula for $L_p(s, \chi)$

$$L_p(s, \chi) = \frac{1}{s-1}\frac{1}{N} \sum_{\substack{1 \leq a \leq N \\ a \not\equiv 0 \,(\mathrm{mod}\, p)}} \chi(a)\langle a \rangle^{1-s} \sum_{j=0}^{\infty} \binom{1-s}{j}\left(\frac{N}{a}\right)^j B_j$$

(see [Washington, 1976] or [Washington, 1997, Theorem 5.11]). Here ω denotes the Teichmüller character at p, $\langle \cdot \rangle$ is defined in section 1.4 and

$$\binom{X}{n} := \frac{X(X-1)\cdots(X-n+1)}{n!}.$$

2.3   *Classical Dirichlet L-series*

The formula (2) gives the values of p-adic L-functions at negative integers. As usual the values of p-adic and classical complex L-series "differ" by an Euler factor. In fact, we can rewrite (2) in the form

$$L_p(1-n, \chi\omega^n) = -(1 - \chi(p)p^{n-1})\frac{B_{n,\chi}}{n}. \tag{3}$$

Hence by virtue of the first formula of section 4.5 of Chapter I we obtain

$$L_p(1-n, \chi\omega^n) = (1 - \chi(p)p^{n-1})L(1-n, \chi)$$

if $n \geq 1$. In view of (3) the values $L_p(k, \chi\omega^{1-k})$ for $k \leq 0$ can be expressed in terms of Frobenius polynomials (see section 3.3 of Chapter I)

$$L_p(k, \chi\omega^{1-k}) = -(1 - \chi(p)p^{-k})\frac{\tau(\chi, \zeta_M)}{M}\sum_{a=1}^{M-1}\overline{\chi}(a)\zeta_M^a\frac{R_{-k}(\zeta_M^a)}{(1 - \zeta_M^a)^{1-k}},$$

where ζ_M is a primitive Mth root of unity in $\mathbb{C}_p$, $\overline{\chi} = \chi^{-1}$ and as always in the book $\tau(\chi, \zeta_M)$ denotes the Gauss sum defined for ζ_M.

2.4   *Leopoldt's Formula*

If χ is an odd character then n and $\chi\omega^{-n}$ have different parities. Consequently $B_{n,\chi\omega^{-n}} = 0$ which implies that $L_p(s, \chi)$ vanishes identically. Now let χ be an even nontrivial primitive Dirichlet character. Then by the same reasoning it follows that $L_p(s, \chi)$ is not the zero function.

If $k = 1$ we have:

$$L_p(1, \chi) = -\left(1 - \frac{\chi(p)}{p}\right)\frac{\tau(\chi, \zeta_M)}{M}\sum_{a=1}^{M}\overline{\chi}(a)\log_p(1 - \zeta_M^a) \tag{4}$$

(see [Washington, 1997, Theorem 5.18], cf. formula (14) of Chapter I). The above formula evaluates $L_p(k, \chi\omega^{1-k})$ if $k = 1$.

2.5 *Leopoldt's Conjecture*

Let F be a totally real abelian number field of degree n and let X be the group of characters of F. Then we have:

$$\frac{2^{n-1}hR_p}{\sqrt{d}} = \pm \prod_{\substack{\chi \in X \\ \chi \neq 1}} \left(1 - \frac{\chi(p)}{p}\right)^{-1} L_p(1, \chi) \tag{5}$$

(see [Washington, 1997, Theorem 5.24], cf. the second formula of section 6.4, Chapter I). Here $R_p := R_p(F)$ denotes the p-adic regulator of F defined in [Washington, 1997, §5.5], and as usual $h := h(F)$ (resp. $d = d(F)$) denotes the class number (resp. the discriminant) of F. Since both R_p and $\sqrt{d}$ are determined up to sign, the above equality means that we can choose the sign of the right hand side so as to obtain this equality.

If we define the p-adic zeta function of F by the formula

$$\zeta_{F,p}(s) := \prod_{\chi \in X} L_p(s, \chi)$$

then we get the formula

$$\lim_{s \to 1}(s - 1)\zeta_{F,p}(s) = \frac{2^{n-1}hR_p}{\sqrt{d}} \prod_{\chi \in X} \left(1 - \frac{\chi(p)}{p}\right)$$

(see [Washington, 1997, p. 71], cf. (15) of Chapter I). Consequently if $R_p \neq 0$ then $\zeta_{F,p}(s)$ has a simple pole at $s = 1$. It is well known that $R_p \neq 0$ if F is an abelian field over $\mathbb{Q}$. It is a famous conjecture of Leopoldt that this holds for any algebraic number field (for details see [Washington, 1997]).

2.6 *Quadratic Fields*

For $\chi = \chi_d$ we have

$$h(d) = \frac{\sqrt{d}}{2\log_p \varepsilon} \left(1 - \frac{\chi_d(p)}{p}\right)^{-1} L_p(1, \chi_d), \tag{6}$$

if d is positive, which follows from (5). Moreover by (4) and Gauss' evaluation of $\tau(\chi_d)$ we have

$$h(d) = -\frac{1}{2\log_p \varepsilon} \left(1 - \frac{\chi_d(p)}{p}\right)^{-1} \sum_{a=1}^{d} \chi_d(a) \log_p(1 - \zeta^{-a}),$$

where $\zeta = \exp(2\pi i/d)$ (cf. the second formula of section 8.2, Chapter I). If d is negative coupling (3) with (16), Chapter I, we obtain

$$h(d) = \xi^{-1}(1 - \chi_d(p))^{-1} L_p(0, \chi_d \omega). \tag{7}$$

If d is positive making use of (3) and the second formula of section 8.3, Chapter I, we obtain

$$k_2(d) = -2\eta^{-1}(1 - \chi_d(p)p)^{-1}L_p(-1, \chi_d\omega^2). \tag{8}$$

If d is negative then by analogy with (21), Chapter I the p-adic Lichtenbaum conjecture should read

$$k_2(d) = \frac{|d|^{3/2}}{2R_{2,p}(d)}\left(1 - \frac{\chi_d(p)}{p^2}\right)^{-1} L_p(2, \chi_d\omega^{-1}), \tag{9}$$

where by analogy $R_{2,p}(d)$ denotes the second p-adic regulator of the corresponding quadratic field $\mathbb{Q}(\sqrt{d})$. Cf. [Neukirch, 1988], [Schneider, 1988], [Nizioł, 1995], [Kolster, Nguyen Quang Do and Fleckinger, 1996] and [Kolster and Nguyen Quang Do, 1998].

3.   COLEMAN'S RESULTS

3.1   *p-adic Multilogarithms*

Let k be an integer. Let us consider the formal series

$$l_k(s) = \sum_{n=1}^{\infty} \frac{s^n}{n^k}.$$

This series determines an analytic function on the open unit ball in $\mathbb{C}_p$. Using "the action of Frobenius" on some differential equations, Coleman [Coleman, 1982] extended l_k to a locally analytic function $l_{k,p}$ on the set $\mathbb{C}_p - \{1\}$. Using the functions $l_{k,p}$, he found an extension of formulae (4) and (3).

In the complex case l_k can be extended to a multiple-valued analytic function on the set $\mathbb{C} - \{0, 1\}$. Denote by $l_{k,\infty}$ the principal branch of $l_{k,\infty}$ (the unique analytic branch which is a single-valued function on the complement of the ray of real numbers no smaller than the one in $\mathbb{C}$). Then we have the following formula

$$L(k, \chi) = \frac{\tau(\chi, \zeta_M)}{M}\sum_{a=1}^{M-1} \overline{\chi}(a)l_{k,\infty}(\zeta_M^{-a}),$$

where ζ_M is a primitive Mth root of unity in $\mathbb{C}$ and χ is a primitive nonprincipal Dirichlet character of conductor M (cf. formulae of sections 4.5 and 5.1, Chapter I).

3.2   *Coleman's Formulae*

Let χ be a primitive nontrivial p-adic Dirichlet character modulo M and ζ_M be a primitive Mth root of unity in $\mathbb{C}_p$. Then for any integer k we have

$$L_p(k, \chi\omega^{1-k}) = (1 - \chi(p)p^{-k})\frac{\tau(\chi, \zeta_M)}{M}\sum_{a=1}^{M-1} \overline{\chi}(a)l_{k,p}(\zeta_M^{-a}). \tag{10}$$

See [Coleman, 1982]. As usual p-adic and analytic formulae "differ" by an Euler factor. If $k = 1$ then $l_{1,p}(s) = -\log_p(1 - s)$ and we obtain Leopoldt's formula (4).

The functions $l_k = l_{k,p}$ are called multilogarithms. By definition (see [Coleman, 1982])

(i) $l_0(s) = \dfrac{s}{s - 1}$,

(ii) $dl_k(s) = l_{k-1}(s)\dfrac{ds}{s}$,

(iii) $\lim\limits_{s \to 0} l_k(s) = 0$.

For $k \le 0$, Wójcik [Wójcik, 1998] has found an explicit formula for l_k

$$l_k(s) = -\frac{sR_n(s)}{(s - 1)^{n+1}},\qquad (11)$$

where $n = -k$ and the $R_n \in \mathbb{Z}[s]$ are the classical Frobenius polynomials defined by the formula

$$\frac{1 - s}{e^t - s} = \sum_{n=0}^{\infty} \frac{R_n(s)}{(1 - s)^n}\frac{t^n}{n!}$$

(see section 3.3, Chapter I).

In order to prove (11) it suffices to check that the right hand side of the equation satisfies (i), (ii) and (iii). Equations (i) and (iii) are obviously satisfied. As for (ii), we have to prove for $n \ge 0$ that

$$\frac{d}{ds}\left(\frac{sR_n(s)}{(1 - s)^{n+1}}\right) = -\frac{R_{n+1}(s)}{(1 - s)^{n+2}}.\qquad (12)$$

Indeed, by definition of $R_n(s)$ we have

$$\frac{d}{ds}\left(\frac{s}{e^t - s}\right) = \sum_{n=0}^{\infty} \frac{d}{ds}\left(\frac{sR_n(s)}{(1 - s)^{n+1}}\right)\frac{t^n}{n!},$$

and so we obtain

$$\frac{e^t}{(e^t - s)^2} = \sum_{n=0}^{\infty} \frac{d}{ds}\left(\frac{sR_n(s)}{(1 - s)^{n+1}}\right)\frac{t^n}{n!}.\qquad (13)$$

On the other hand we have

$$\frac{d}{dt}\left(\frac{1 - s}{e^t - s}\right) = \frac{-e^t(1 - s)}{(e^t - s)^2}$$

and

$$\frac{d}{dt}\left(\sum_{n=1}^{\infty}\frac{R_n(s)}{(1-s)^n}\frac{t^n}{n!}\right) = \sum_{n=1}^{\infty}\frac{R_n(s)}{(1-s)^n}\frac{t^{n-1}}{(n-1)!} = \sum_{n=0}^{\infty}\frac{R_{n+1}(s)}{(1-s)^{n+1}}\frac{t^n}{n!}\,.$$

Hence we find that

$$\frac{e^t}{(e^t-s)^2} = -\sum_{n=0}^{\infty}\frac{R_{n+1}(s)}{(1-s)^{n+2}}\frac{t^n}{n!}\,.$$

This together with (13) immediately gives (12).

Making use of (11) and Carlitz's formula expressing $B_{m,\chi}/m$ in terms of the Frobenius polynomials (see section 3.3, Chapter I), we obtain for any nontrivial Dirichlet character χ modulo M and $k \leq 0$

$$\frac{B_{1-k,\chi}}{1-k} = -\frac{\tau(\chi,\zeta_M)}{M}\sum_{a=1}^{M-1}\overline{\chi}(a)l_k(\zeta_M^{-a})\,.$$

Coupling the above with (3) we obtain formula (10) for nonpositive k.

3.3   *Nonprimitive Dirichlet Characters*

If the character χ modulo M is induced from a character χ_1 modulo some divisor of M then

$$B_{n,\chi} = B_{n,\chi_1}\prod_{\substack{l\mid M \\ l\,\text{prime}}}\left(1 - \chi_1(l)l^{n-1}\right),$$

where the product is taken over all primes l dividing M. We shall use this formula to find a relation between $L_p(k,\chi)$ and $L_p(k,\chi_1)$.

We have

$$L_p(k,\chi) = \lim_{1-n\to k}L_p(1-n,\chi),$$

where $1 - n \to k$ p-adically as $n \to \infty$. Then by (2) we obtain

$$L_p(k,\chi) = -\lim_{1-n\to k}\left((1-\chi\omega^{-n}(p)p^{n-1})\frac{B_{n,\chi\omega^{-n}}}{n}\right)$$

$$= -\lim_{1-n\to k}\left(\frac{B_{n,\chi}}{n}\right)$$

$$= -\lim_{1-n\to k}\left(\frac{B_{n,\chi_1\omega^{-n}}}{n}\right)\prod_{\substack{l\mid M \\ l\,\text{prime}}}\lim_{1-n\to k}\left(1 - \chi_1\omega^{-n}(l)l^{n-1}\right)$$

$$= L_p(k,\chi_1)\prod_{\substack{l\mid M,\,l\neq p \\ l\,\text{prime}}}\left(1 - \chi_1\omega^{k-1}(l)l^{-k}\right),$$

where the above products are taken over all primes l dividing M (with $l \neq p$ in the second product). The last equality was deduced using

$$\lim_{1-n \to k} l^n = \lim_{1-n \to k} (\omega(l)\langle l \rangle)^n = \lim_{1-n \to k} \langle l \rangle^n = \langle l \rangle^{1-k} = l^{1-k}\omega^{k-1}(l)$$

if $l \neq p$. Thus we have

$$L_p(k, \chi\omega^{1-k}) = L_p(k, \chi_1\omega^{1-k}) \prod_{\substack{l \mid M, l \neq p \\ l \text{ prime}}} (1 - \chi_1(l)l^{-k}).$$

3.4 Basic Properties of Multilogarithms

By definition, for $k = -1, 0, 1$, $l_k(s)$ is given by

$$l_{-1}(s) = \frac{s}{(1-s)^2},$$

$$l_0(s) = \frac{s}{1-s},$$

$$l_1(s) = -\log_p(1-s),$$

where $\log_p$ denotes the p-adic logarithm. The function l_2 is related to the so-called p-adic dilogarithm defined by the formula

$$D(s) = l_2(s) + \frac{1}{2}\log_p(s)\log_p(1-s)$$

(cf. Rogers' dilogarithm, section 5.3, Chapter I).

It is well known (see [Coleman, 1982, Proposition 6.4]) that

$$l_k(s) + (-1)^k l_k(s^{-1}) = -\frac{1}{k!}\log_p^k(s), \qquad (14)$$

where $1/k! = 0$ if $k < 0$ and for any positive integer m the functions l_k satisfy the identity

$$\frac{1}{m}\sum_{\zeta^m=1} l_k(\zeta s) = \frac{l_k(s^m)}{m^k} \qquad (15)$$

provided that $s^m \neq 1$ (see [Coleman, 1982, Proposition 6.1] with z replaced by z^m on the right hand side of the equation of the proposition).

3.5 The Equation $L_p(k, \chi\omega^{1-k}) = 0$

Let χ be a primitive nontrivial Dirichlet character of conductor M. Then it follows from (14) (recall that $\log_p\zeta_M = 0$) that for $k \neq 0$

$$\sum_{a=1}^{M-1} \overline{\chi}(a)l_k(\zeta_M^{-a}) = (1 + (-1)^{k+1}\overline{\chi}(-1)) \sum_{a=1}^{[M/2]} \overline{\chi}(a)l_k(\zeta_M^{-a}),$$

where $[x]$ denotes the integral part of x. Coupling this formula with (10) we obtain

$$L_p(k, \chi\omega^{1-k}) = 0,\tag{16}$$

if $\overline{\chi}(-1) = (-1)^k$ (i.e., if χ and k are of the same parity). If $k = 0$ then by (14) we have

$$l_k(s) + l_k(s^{-1}) = -1$$

and hence

$$\sum_{a=1}^{M-1} \overline{\chi}(a)l_k(\zeta_M^{-a}) = (1 - \overline{\chi}(-1)) \sum_{a=1}^{[M/2]} \overline{\chi}(a)l_k(\zeta_M^{-a}) - \overline{\chi}(-1) \sum_{a=1}^{[M/2]} \overline{\chi}(a).$$

This gives (16) for even nontrivial Dirichlet characters χ and $k = 0$. Recall that the third sum of the above equation is zero for even characters χ and note that by virtue of (10) if $k = 0$ and $\chi(p) = 1$ formula (16) also holds.

We now prove that (16) also holds for χ trivial and k even (i.e., in the case when k has the same parity as χ). Let χ be trivial and let $k \neq 1$. If $k \leq 0$ then, by (3) we have

$$L_p(k, \omega^{1-k}) = -(1 - p^{-k})\frac{B_{1-k,\chi}}{1 - k}.$$

Thus if $k = 0$ formula (16) holds because the Euler factor equals 0. If $k \leq -1$ then (16) holds if $B_{1-k} = 0$, i.e, if k is even.

We now turn to the case when $k \geq 2$. For a prime number p, let E_p be a finitely ramified extension of $\mathbb{Q}_p$ in $\mathbb{C}_p$. If $k \geq 2$ then we have

$$L_p(k, \omega^{1-k}) = (1 - p^{-k}) \lim_{\substack{s \to 1, \\ s \in E_p - \{1\}}} l_{k,p}(s)\tag{17}$$

(see [Coleman, 1982, formula (4), Introduction]). Thus if k is even, formula (14) implies

$$2L_p(k, \omega^{1-k}) = -\frac{1}{k!}(1 - p^{-k}) \lim_{\substack{s \to 1, \\ s \in E_p - \{1\}}} \log_p^k(s) = 0.$$

This completes the proof of (16) if χ is trivial and k is even.

3.6   *Uehara's Functions*

Let p be a prime number. Following [Coleman, 1982] we write

$$l_k^{(p)}(s) = l_k(s) - p^{-k}l_k(s^p),\tag{18}$$

where $l_k = l_{k,p}$. The functions $l_k^{(p)}$ are called p-adic multilogarithms. In particular, in view of the identity (15) (for $m = 2$) we have

$$l_k^{(2)}(s) = \frac{1}{2}(l_k(s) - l_k(-s)).\tag{19}$$

Let f be a natural number and ζ_f be a primitive fth root of unity in $\mathbb{C}_2$. For any Dirichlet character ψ modulo f, any integer k and $s \in \mathbb{C}_2$ we define the functions $\mathcal{L}_{k,\psi}$ by

$$\mathcal{L}_{k,\psi}(s) = (-1)^{k+1} l_k^{(2)}(s) \quad (s \neq \pm 1),$$

if ψ is the trivial character, and

$$\mathcal{L}_{k,\psi}(s) = (-1)^{k+1} \frac{\tau(\overline{\psi}, \zeta_f)}{f} \sum_{a=1}^{f} {}' \psi(a) l_k(\zeta_f^a s) \quad (s \neq \zeta_f^a)$$

otherwise. Throughout the chapter we adopt the notation $\sum_{a=1}^{c} {}'$ to denote a sum taken over integers a prime to c.

In particular, if f is even and ψ is a primitive quadratic character modulo f then by (19) we have

$$\mathcal{L}_{k,\psi}(s) = 2(-1)^{k+1} \frac{\tau(\overline{\psi}, \zeta_f)}{f} \sum_{a=1}^{f/2} {}' \psi(a) l_k^{(2)}(\zeta_f^a s)$$

because

$$\psi\left(\frac{f}{2} + a\right) = -\psi(a).$$

The functions $\mathcal{L}_{k,\psi}$ for $\psi = \chi_{-8}, \chi_{-4}, \chi_1, \chi_8$ and $k \in \{0, 1\}$ are related to the functions $K_e^{\pm}$ ($e = \pm 1, \pm 2$) introduced by Uehara [Uehara, 1990]. Let

$$L^+(s) = \log_2(1 - s), \quad L^-(s) = \frac{1}{1 - s}.$$

Uehara's functions are defined by

$$K_1^{\pm}(s) = -\frac{1}{2}(L^{\pm}(s) - L^{\pm}(-s)),$$

and

$$K_e^{\pm}(s) = -\frac{\tau(\psi, \zeta_f)}{f} \sum_{a=1}^{f} {}' \psi(a) L^{\pm}(\zeta_f^a s),$$

if $f = |4e|$ and $e \in \{-1, \pm 2\}$. It is easy to check that

$$K_1^+(s) = -\frac{1}{2}(2L^+(s) - L^+(s^2)), \quad K_1^-(s) = -(L^-(s) - L^-(s^2)).$$

The functions L^- and L^+ play in the definition of $K_e^{\pm}$ the same role as the functions $l_0(s) = \frac{s}{1-s}$ and $l_1(s) = -\log_2(1-s)$ in the definition of $\mathcal{L}_{k,\psi}$ for

$k = 0, 1$. Making use of the functions $K_e^{\pm}$, Uehara has obtained in a simpler way the p-adic linear congruence relations for class numbers of quadratic fields discovered by Gras [Gras, 1989].

Let $\xi \neq 1$ be a primitive Nth root of unity, where N is an odd natural number. In [Urbanowicz and Wójcik, 1995/1996] (see also [Wójcik, 1998]) the authors have noticed that

$$\mathcal{L}_{1,\psi}(\xi) = K_e^{+}(\xi), \quad \mathcal{L}_{0,\psi}(\xi) = K_e^{-}(\xi),$$

where ψ is the trivial primitive Dirichlet character if $e = 1$, and $\psi = \chi_{4e}$ otherwise. Using Uehara's ideas they have found a linear congruence of the Gras-Uehara type for the values $L_p(k, \chi\omega^{-k})$ if $k \in L \subseteq \{-1, 0, 1, 2\}$ ([Urbanowicz and Wójcik, 1995/1996]) or for any set of consecutive integers ([Wójcik, 1998]). The Gras and Uehara congruence is a special case of their congruence for $L = \{0, 1\}$.

4.   SOME AUXILIARY LEMMAS

4.1   *Preliminary Remarks*

Our purpose in this chapter is to present results of [Urbanowicz and Wójcik, 1995/1996]. In the following, we first prove a sequence of auxiliary lemmas. We then apply the lemmas to give a further generalization of the Hardy-Williams congruence. This congruence and its generalization proved in [Szmidt, Urbanowicz and Zagier, 1995] are special cases of a more general linear congruence between the values of 2-adic L-functions $L_2(k, \chi\omega^{1-k})$, where χ is a primitive quadratic Dirichlet character. We shall consider linear combinations of these values with arbitrary 2-adic integral coefficients (not all even) when k is taken over the integers belonging either to any subset of the set $\{-1, 0, 1, 2\}$ ([Urbanowicz and Wójcik, 1995/1996]) or to any finite set of consecutive integers ([Wójcik, 1998]).

4.2   *Useful Properties of $\mathcal{L}_{k,\psi}$*

The proof of the key lemma of this chapter is based on the concept of [Uehara, 1990, Lemma 3]. We extend Uehara's result to an arbitrary multilogarithm $l_k = l_{k,p}$. In the lemmas of this section we denote, as usual, by ζ_m a primitive mth root of unity in $\mathbb{C}_p$ ($\mathbb{C}_2$ in Lemma 2).

LEMMA 1 ([Urbanowicz and Wójcik, 1995/1996, Lemma 1])   *Let χ be a Dirichlet character modulo $M > 1$ and let N be a multiple of M such that $N/M > 0$ is a rational squarefree integer relatively prime to M. For arbitrary natural number T satisfying $M \mid T \mid N$ we assume that $\zeta_T = \zeta_M \zeta_{T/M}$ and set*

$$S_{k,\chi}(T) = \sum_{a=1}^{T}{}' \chi(a) l_k(\zeta_T^a).$$

Then for any integer k we have

$$S_{k,\chi}(N) = (-1)^{\nu(N/M)} \prod_{\substack{p \mid (N/M) \\ p \text{ prime}}} (1 - \overline{\chi}(p)p^{1-k})S_{k,\chi}(M),$$

where $\nu(t)$ denotes the number of distinct prime factors of t.

PROOF. Let r be a prime number not dividing M. Then for any natural number T satisfying $M \mid T \mid N$ and $r \mid T$ we have

$$S_{k,\chi}(T) = -(1 - \overline{\chi}(r)r^{1-k})S_{k,\chi}(T/r). \tag{20}$$

In fact, it is easy to see that

$$S_{k,\chi}(T) = \sum_{a=1}^{T/r}{}' \chi(a) \sum_{c=0}^{r-1} l_k(\zeta_T^a \zeta_r^c) - \sum_{a=1}^{T/r}{}' \chi(ar)l_k(\zeta_T^{ar}).$$

Coupling the above with identity (15) we obtain

$$S_{k,\chi}(T) = \overline{\chi}(r)r^{1-k} \sum_{a=1}^{T/r}{}' \chi(ar) \sum_{c=0}^{r-1} l_k(\zeta_{T/r}^{ar}\zeta_r^c) - \sum_{a=1}^{T/r}{}' \chi(ar)l_k(\zeta_{T/r}^{ar}).$$

Hence (20) follows at once and the lemma follows by induction on the number of prime factors of N.   ∎

The proof of the main result of the chapter is based on the following property of the 2-adic functions $\mathcal{L}_{k,\psi}$.

LEMMA 2 ([Uehara, 1990], [Urbanowicz and Wójcik, 1995/1996, Lemma 2], [Wójcik, 1998, Lemma 1]) *Given any odd integer M, let χ by a primitive Dirichlet character modulo M. Suppose that N is an odd multiple of M such that $N/M \ (> 0)$ is a rational squarefree integer relatively prime to M. Let ψ be a primitive Dirichlet character being either trivial or of even conductor coprime to N. Denote by ω the Teichmüller character at $p = 2$ and assume that for an arbitrary natural number T satisfying $M \mid T \mid N$ we have $\zeta_T = \zeta_M \zeta_{T/M}$. Then for any integer k we have*

$$\Lambda_{k,\psi} := \Lambda_{k,\psi}(N,\chi) - \frac{\tau(\overline{\chi},\zeta_M)}{M} \sum_{a=1}^{N}{}' \chi(a)\mathcal{L}_{k,\psi}(\zeta_N^a)$$

$$= (-1)^{\nu(N/M)} \prod_{\substack{p \mid (N/M) \\ p \text{ prime}}} (1 - \overline{\chi\psi}(p)p^{1-k})L_2(k, \overline{\chi}\psi\omega^{1-k}),$$

unless $k = 1$ and the characters χ and ψ are trivial, in which case we have

$$\Lambda_{k,\psi} = \sum_{a=1}^{N}{}' \mathcal{L}_{k,\psi}(\zeta_N^a) = \begin{cases} -(\log_2 N)/2, & \text{if } N \text{ is a prime number,} \\ 0, & \text{otherwise.} \end{cases}$$

REMARK In the formulation of Lemma 2 of [Urbanowicz and Wójcik, 1995/1996] there is a small error, which implies the same error in Lemma 1 of [Wójcik, 1998]. The right hand sides of the identities of the lemmas should be multiplied by $(-1)^{k+1}$.

PROOF. We shall distinguish four cases:

(i) either χ or ψ is nontrivial,

(ii) χ, ψ are trivial and $k = 0$ or 1,

(iii) χ, ψ are trivial and $k \geq 2$,

(iv) χ, ψ are trivial and $k \leq -1$.

Case (i). Let ψ be the trivial character. Then χ is a nontrivial character (modulo $M > 1$) and we have

$$(-1)^{k+1}\Lambda_{k,\psi}(N,\chi)$$

$$= \frac{\tau(\overline{\chi},\zeta_M)}{M}(-1)^{k+1}\sum_{a=1}^{N}{}' \chi(a)\mathcal{L}_{k,\psi}(\zeta_N^a) = \frac{\tau(\overline{\chi},\zeta_M)}{M}\sum_{a=1}^{N}{}' \chi(a)l_k^{(2)}(\zeta_N^a)$$

$$= \frac{\tau(\overline{\chi},\zeta_M)}{M}\left(\sum_{a=1}^{N}{}' \chi(a)l_k(\zeta_N^a) - 2^{-k}\overline{\chi}(2)\sum_{a=1}^{N}{}' \chi(2a)l_k(\zeta_N^{2a})\right)$$

$$= \frac{\tau(\overline{\chi},\zeta_M)}{M}(1 - \overline{\chi}(2)2^{-k})\sum_{a=1}^{N}{}' \chi(a)l_k(\zeta_N^a).$$

Thus by Lemma 1 and (10) we obtain

$$(-1)^{r(N/M)+k+1}\Lambda_{k,\psi}(N,\chi)$$

$$= \frac{\tau(\overline{\chi},\zeta_M)}{M}(1 - \overline{\chi}(2)2^{-k}) \prod_{\substack{p\,|\,(N/M) \\ p\,\text{prime}}} (1 - \overline{\chi}(p)p^{1-k})\sum_{b=1}^{M}{}' \chi(b)l_k(\zeta_M^b)$$

$$= (-1)^{k+1} \prod_{\substack{p\,|\,(N/M) \\ p\,\text{prime}}} (1 - \overline{\chi}(p)p^{1-k})L_2(k,\overline{\chi}\omega^{1-k})$$

and the lemma in this case follows.

If ψ is a nontrivial character modulo f then we have

$$(-1)^{k+1}\Lambda_{k,\psi}(N,\chi) = \frac{\tau(\overline{\chi},\zeta_M)\tau(\overline{\psi},\zeta_f)}{Mf} \sum_{a=1}^{N}{}' \chi(a) \sum_{b=1}^{f}{}' \psi(b)l_k(\zeta_N^a\zeta_f^b)$$

$$= \frac{\tau(\overline{\chi\psi},\zeta_{Mf})}{Mf} \sum_{a=1}^{N}{}' \sum_{b=1}^{f}{}' \chi(a)\psi(b)l_k(\zeta_N^a\zeta_f^b)$$

$$= \frac{\tau(\overline{\chi\psi},\zeta_{Mf})}{Mf} \sum_{a=1}^{Nf}{}' (\chi\psi)(a)l_k(\zeta_{Nf}^a).$$

Thus by Lemma 1 and (10) we find that

$$(-1)^{\nu(N/M)+k+1}\Lambda_{k,\psi}(N,\chi)$$

$$= \frac{\tau(\overline{\chi\psi},\zeta_{Mf})}{Mf} \prod_{\substack{p\,|\,(N/M) \\ p\,\text{prime}}} (1 - \overline{\chi\psi}(p)p^{1-k}) \sum_{b=1}^{Mf} (\chi\psi)(b)l_k(\zeta_{Mf}^b)$$

$$= (-1)^{k+1} \prod_{\substack{p\,|\,(N/M) \\ p\,\text{prime}}} (1 - \overline{\chi\psi}(p)p^{1-k})L_2(k,\overline{\chi\psi}\omega^{1-k}).$$

Case (ii). By definition of the functions $\mathcal{L}_{k,\psi}$, we have

$$(-1)^{k+1}\Lambda_{k,\psi} = \sum_{a=1}^{N}{}' l_k^{(2)}(\zeta_N^a) = (1-2^{-k}) \sum_{a=1}^{N}{}' l_k(\zeta_N^a). \tag{21}$$

Thus if $k = 0$ we have $\Lambda_{k,\psi} = 0$. On the other hand it is easy to check (see section 3.5) that $L_2(0,\omega) = 0$. If $k = 1$ then by (21) we have

$$\Lambda_{k,\psi} = -\frac{1}{2} \sum_{a=1}^{N}{}' \log_2(1-\zeta_N^a) = -\frac{1}{2}\log_2\left(\prod_{\substack{1<a<N \\ \gcd(a,N)=1}} (1-\zeta_N^a)\right),$$

and in consequence $\Lambda_{k,\psi} = -\frac{1}{2}\log N$ if N is a prime number, and $\Lambda_{k,\psi} = 0$ otherwise.

Case (iii). Let $N = nr$, where r is a prime number and n is not divisible by r. Coupling (21) with (15) we have

$$(-1)^{k+1}\Lambda_{k,\psi} = \sum_{a=1}^{n}{}' \sum_{c=0}^{r-1} l_k(\zeta_{nr}^a\zeta_r^c) - \sum_{a=1}^{n}{}' l_k(\zeta_{nr}^{ar})$$

$$= \sum_{a=1}^{n}{}' \frac{l_k(\zeta_n^a)}{r^{k-1}} - \sum_{a=1}^{n}{}' l_k(\zeta_n^a) = -(1-r^{1-k})\sum_{a=1}^{n}{}' l_k(\zeta_n^a)$$

(cf. formula (20)). Thus by induction on the number $\nu(N)$ of prime factors of N we obtain

$$\sum_{a=1}^{N}{}' l_k(\zeta_N^a) = (-1)^{\nu(N)-1} \prod_{\substack{p \mid (N/r) \\ p \text{ prime}}} (1 - p^{1-k}) \sum_{a=1}^{r-1} l_k(\zeta_r^a). \qquad (22)$$

On the other hand Corollary 7.1a on p. 201 and formula (4) on p. 173 in [Coleman, 1982] imply that

$$\sum_{a=1}^{r-1} l_k(\zeta_r^a) = -(1 - r^{1-k}) \lim_{s \to 1} l_k(s) = -(1 - r^{1-k})(1 - 2^{-k})^{-1} L_2(k, \omega^{1-k})$$

if $k \geq 2$. Therefore case (iii) follows from (21) and (22).

Case (iv). Let r be a prime number and $\zeta_r \neq 1$ be a rth root of unity. Let $k \leq -1$. Making use of formula (11) we shall prove that

$$\sum_{a=1}^{r-1} l_k(\zeta_r^a) = -(1 - r^{1-k}) \frac{B_{1-k}}{1 - k}. \qquad (23)$$

Indeed, appealing to identity (11) we have

$$\frac{s}{e^t - s} = \sum_{n=0}^{\infty} (-1)^n l_{-n}(s) \frac{t^n}{n!},$$

which implies

$$t \sum_{a=1}^{r-1} \frac{\zeta_r^a}{e^t - \zeta_r^a} = \sum_{a=1}^{r-1} \sum_{n=0}^{\infty} (-1)^n l_{-n}(\zeta_r^a) \frac{t^{n+1}}{n!}$$

$$= \sum_{n=1}^{\infty} \sum_{a=1}^{r-1} (-1)^{n-1} l_{1-n}(\zeta_r^a) n \frac{t^n}{n!}. \qquad (24)$$

On the other hand, we have formally

$$\sum_{a=1}^{r-1} \frac{\zeta_r^a}{e^t - \zeta_r^a} = -\sum_{n=0}^{\infty} \left(\sum_{a=1}^{r-1} \zeta_r^{-an} \right) e^{tn} = \sum_{n \geq 0, r \nmid n} e^{tn} - (r-1) \sum_{n \geq 0, r \mid N} e^{tn}$$

$$= \sum_{n=0}^{\infty} e^{tn} - r \sum_{n=0}^{\infty} e^{rtn} = \frac{1}{1 - e^t} - \frac{r}{1 - e^{rt}},$$

and in consequence by the definition of Bernoulli numbers we obtain

$$t \sum_{a=1}^{r-1} \frac{\zeta_r^a}{e^t - \zeta_r^a} = -\sum_{n=0}^{\infty} B_n(1 - r^n) \frac{t^n}{n!}.$$

Coupling the above with (24) we deduce that

$$\sum_{a=1}^{r-1} l_{1-n}(\zeta_r^a) = (-1)^n (1 - r^n)\frac{B_n}{n}.$$

Hence (23) follows at once.

Thus if $k \leq -1$, by virtue of (21), (22), (23) and (3), we have

$$\sum_{a=1}^{N}{}' \mathcal{L}_{k,\psi}(\zeta_N^a) = (1 - 2^{-k})(-1)^{\nu(N)-1} \prod_{\substack{p \mid N \\ p \, \text{prime}}} (1 - p^{1-k})\frac{B_{1-k}}{1-k}$$

$$= (-1)^{\nu(N)} \prod_{\substack{p \mid N \\ p \, \text{prime}}} (1 - q^{1-k}) L_2(k, \omega^{1-k}). \quad \blacksquare$$

4.3   Some Binomial Coefficient Identities

In the lemmas presented in Section 4.5 (Lemmas 5 and 6) we require two identities involving binomial coefficients (Lemma 3) and their extension (Lemma 4). The proofs of these two identities given below are due to A. Granville.

LEMMA 3 ([Urbanowicz and Wójcik, 1995/1996, Lemma 3]) *Let $n \geq 0$ be an integer. Set $\gamma_n = -1$, if $n \equiv 1, 2 \,(\text{mod}\, 4)$, and $\gamma_n = 1$, otherwise. Then we have:*

(i) $\displaystyle\sum_{k=0}^{n} \binom{n+k}{n-k} \frac{4^{2k}(-1)^k}{(2k+1)^2} \binom{2k}{k}^{-1} = \frac{1}{(2n+1)^2},$

(ii) $\displaystyle\sum_{k=0}^{n} \binom{n+k}{n-k} \frac{4^{2k}(-1)^k}{(2k+1)^2} \binom{2k}{k}^{-1} \sum_{l=0}^{k} \binom{2l}{l} 2^{-3l} = \frac{\gamma_n}{(2n+1)^2}.$

PROOF. Set

$$\lambda_k = (-1)^k 2^{4k} \binom{n+k}{n-k} \binom{2k}{k}^{-1}.$$

The main idea of the proof is to note the following identity

$$\lambda_k - \lambda_{k+1} = \left(\frac{2n+1}{2k+1}\right)^2 \lambda_k$$

for all $k \geq 0$. In this notation the left hand side of the equation (i) of the lemma equals

$$\sum_{k=0}^{n} \frac{\lambda_k}{(2k+1)^2} = \frac{1}{(2n+1)^2} \sum_{k=0}^{n}(\lambda_k - \lambda_{k+1}) = \frac{\lambda_0}{(2n+1)^2} = \frac{1}{(2n+1)^2}.$$

The identity (ii) is a little more subtle. Multiplying the left hand side of (ii) by $(2n+1)^2$ we obtain

$$(2n+1)^2 \sum_{k=0}^{n} \frac{\lambda_k}{(2k+1)^2} \sum_{l=0}^{k} \binom{2l}{l} 2^{-3l} = \sum_{k=0}^{n} (\lambda_k - \lambda_{k+1}) \sum_{l=0}^{k} \binom{2l}{l} 2^{-3l}$$

$$= \sum_{k=0}^{n} \lambda_k \binom{2k}{k} 2^{-3k} = \sum_{k=0}^{n} \binom{n+k}{n-k} (-2)^k.$$

To evaluate the last sum, note that $\binom{n+k}{n-k}$ is the coefficient of t^n in

$$\frac{t^k}{(1-t)^{2k+1}}.$$

Thus our sum is the coefficient of t^n in

$$\sum_{k \geq 0} \frac{(-2t)^k}{(1-t)^{2k+1}} = \frac{1}{1-t} \frac{1}{1-((-2t)/(1-t)^2)} = \frac{1-t}{1+t^2} = \frac{1-t-t^2+t^3}{1-t^4}$$

and so equals -1, if $n \equiv 1, 2 \,(\mathrm{mod}\, 4)$ and 1, otherwise.   ∎

Wójcik [Wójcik, 1998] has extended Granville's ideas to obtain other useful binomial identities. Making use of these identities Wójcik has found some useful 2-adic expansions of the numbers $\mathcal{L}_{k,\psi}(\xi)$ for all $k \in \mathbb{Z}$. Applying these expansions he has obtained a further generalization of the congruences of the Hardy-Williams type which seems to be at present the most general congruence of the Gras-Uehara type.

4.4   The numbers $W_{k,e}(n)$

Let $\gamma_{n,e} = -1$ if $n \equiv 1, 2 \,(\mathrm{mod}\, 4)$ and $e \in \mathcal{T}_8 - \mathcal{T}_4$, and $\gamma_{n,e} = 1$ otherwise. Here as before we denote by $\mathcal{T}_d$ the set of all fundamental discriminants dividing d. For any integers n, k with $n \geq 0$, and $e \in \mathcal{T}_8$ set

$$W_{k,e}(n) = \sum_{l=0}^{n} (-1)^{l(k+1)} (2l+1)^{1-k} \gamma_{l,e} \binom{2n+1}{n-l}.$$

It is easily seen that the numbers $W_{k,e}(n)$ are 2-integral rational numbers. Moreover an easy computation shows that $W_{k,e}(0) = 1$, and if $n \geq 1$

$$W_{k,e}(n) = \begin{cases} 0, & \text{if } k = 0 \text{ and } e \in \mathcal{T}_4, \\ (2n+1)2^n, & \text{if } k = 0 \text{ and } e \notin \mathcal{T}_4, \\ 4^n, & \text{if } k = 1 \text{ and } e \in \mathcal{T}_4, \\ 2^n, & \text{if } k = 1 \text{ and } e \notin \mathcal{T}_4. \end{cases} \tag{25}$$

Furthermore, if $n \geq 1$ and $e \in \mathcal{T}_8$, by an easy calculation, we have

$$W_{k-2,e}(n) = (2n + 1)^2 W_{k,e}(n) - 8n(2n + 1)W_{k,e}(n - 1), \qquad (26)$$

and hence, by a simple induction, we can deduce that

$$W_{k,e}(n) = \frac{2^{4n}}{2n + 1}\binom{2n}{n}^{-1} \sum_{l=0}^{n} \frac{1}{2l + 1}\binom{2l}{l} 2^{-4l} W_{k-2,e}(l). \qquad (27)$$

Moreover from (26) and (27) we see that

$$\mathrm{ord}_2(W_{k,e}(n)) \geq n. \qquad (28)$$

By (25) it is obvious that (28) holds for $k = 0$ and 1. If $k \leq 0$ (resp. $k \geq 1$) formula (28) follows from (26) (resp. (27)) by a simple induction on k. If $k \leq 0$ the proof is straightforward and if $k \geq 1$ it suffices to notice that

$$\mathrm{ord}_2(W_{k,e}(n)) = \mathrm{ord}_2\left(\frac{2^{4n}}{2n + 1}\binom{2n}{n}^{-1} \sum_{l=0}^{n} \frac{2^{4l}}{2l + 1}\binom{2l}{l} W_{k-2,e}(l) \right)$$

$$= \mathrm{ord}_2\left(\sum_{l=0}^{n} \frac{n!(2l + 1)!!}{l!(2l + 1)(2n + 1)!!} 2^{3(n-l)} W_{k-2,e}(l) \right),$$

where $t!!$ denotes the product of all odd integers $\leq t$.

LEMMA 4 ([Wójcik, 1998, Lemma 2]) *Let n, m be integers with $n \geq 0$. Assume that $e \in \mathcal{T}_8$. Then in the notation of Lemma 3 we have:*

$$\sum_{k=0}^{n}\binom{n + k}{n - k}\frac{(-1)^k}{2k + 1}W_{m,e}(k) = \frac{(-1)^{nm}\gamma_n}{(2n + 1)^m}.$$

PROOF. First we shall prove the following identity

$$(2n + 1)^2 \sum_{k=0}^{n}\binom{n + k}{n - k}\frac{(-1)^k}{2k + 1}W_{m,e}(k)$$

$$= \sum_{k=0}^{n}\binom{n + k}{n - k}\frac{(-1)^k}{2k + 1}W_{m-2,e}(k). \qquad (29)$$

We set

$$\lambda_k = (-1)^k 2^{4k}\binom{n + k}{n - k}\binom{2k}{k}^{-1}.$$

Then for all $k \geq 0$ we have

$$\lambda_k - \lambda_{k+1} = \left(\frac{2n+1}{2k+1}\right)^2 \lambda_k$$

and so we obtain

$$(2n+1)^2 \sum_{k=0}^{n} \binom{n+k}{n-k} \frac{(-1)^k}{2k+1} W_{m,e}(k)$$

$$= (2n+1)^2 \sum_{k=0}^{n} \frac{\lambda_k}{(2k+1)^2} \sum_{l=0}^{k} \frac{1}{2l+1} \binom{2l}{l} 2^{-4l} W_{m-2,e}(l)$$

$$= \sum_{k=0}^{n} (\lambda_k - \lambda_{k+1}) \sum_{l=0}^{k} \frac{1}{2l+1} \binom{2l}{l} 2^{-4l} W_{m-2,e}(l)$$

$$= \sum_{k=0}^{n} \frac{\lambda_k}{2k+1} \binom{2k}{k} 2^{-4k} W_{m-2,e}(k) = \sum_{k=0}^{n} \binom{n+k}{n-k} \frac{(-1)^k}{2k+1} W_{m-2,e}(k),$$

which gives (29). Now the lemma follows from (29) by induction on m in virtue of the identities

$$\sum_{k=0}^{n} \binom{n+k}{n-k} (-2)^k = \gamma_n, \qquad \sum_{k=0}^{n} \binom{n+k}{n-k} \frac{(-1)^k W_{0,1}(k)}{2k+1} = 1,$$

$$\sum_{k=0}^{n} \binom{n+k}{n-k} \frac{(-4)^k}{2k+1} = \frac{(-1)^n}{2n+1}, \qquad \sum_{k=0}^{n} \binom{n+k}{n-k} \frac{(-2)^k}{2k+1} = \frac{(-1)^n \gamma_n}{2n+1}.$$

For the first identity, see the proof of Lemma 3 (ii). The identity with the numbers $W_{0,1}(k)$ follows immediately from the defintion of these numbers. The two remaining identities follow from the obvious formula

$$\frac{2n+1}{2k+1} \binom{n+k}{n-k} = \binom{n+k+1}{n-k} + \binom{n+k}{n-k-1}.$$

It suffices to notice that $\binom{n+k+1}{n-k}$ (resp. $\binom{n+k}{n-k-1}$) is the coefficient of t^n (resp. t^{n+1}) in

$$\frac{t^k}{(1-t)^{2k+2}}.$$

The rest of the proof is the same as the proof of Lemma 3 (ii).   ∎

4.5   *2-adic Expansions of $\mathcal{L}_{k,\psi}(\xi)$*

Next we extend Lemma 6, [Uehara, 1990]. For $\psi = \chi_e$ set $\mathcal{L}_{k,\psi} = \mathcal{L}_{k,e}$. In the lemmas of this section ξ ($\neq 1$) denotes a primitive Nth root of unity, where N is an odd natural number.

LEMMA 5 ([Urbanowicz and Wójcik, 1995/1996, Lemma 4]) *For any $e \in \mathcal{T}_8$ write $\alpha = \operatorname{sgn} e$ and set*

$$w_\alpha = \frac{\alpha\xi}{1 + \alpha\xi^2}.$$

Then we have

$$\mathcal{L}_{-1,e}(\xi) = \sum_{k=0}^{\infty}(4\alpha)^k w_\alpha^{2k+1}, \quad \mathcal{L}_{0,e}(\xi) = w_{-\alpha},$$

$$\mathcal{L}_{1,e}(\xi) = \sum_{k=0}^{\infty}\frac{(4\alpha)^k \omega_\alpha^{2k+1}}{2k+1}, \quad \mathcal{L}_{2,e}(\xi) = \sum_{k=0}^{\infty}\frac{(-16\alpha)^k \omega_{-\alpha}^{2k+1}}{(2k+1)^2}\binom{2k}{k}^{-1},$$

if $e \in \mathcal{T}_4$, and

$$\mathcal{L}_{-1,e}(\xi) = -\sum_{k=0}^{\infty}(2\alpha)^k(2k-1)\omega_\alpha^{2k+1},$$

$$\mathcal{L}_{0,e}(\xi) = \sum_{k=0}^{\infty}(-2\alpha)^k \omega_{-\alpha}^{2k+1},$$

$$\mathcal{L}_{1,e}(\xi) = \sum_{k=0}^{\infty}\frac{(2\alpha)^k \omega_\alpha^{2k+1}}{2k+1},$$

$$\mathcal{L}_{2,e}(\xi) = \sum_{k=0}^{\infty}\frac{(-16\alpha)^k \omega_{-\alpha}^{2k+1}}{(2k+1)^2}\binom{2k}{k}^{-1}\sum_{l=0}^{k}\binom{2l}{l}2^{-3l},$$

if $e \in \mathcal{T}_8 - \mathcal{T}_4$.

PROOF. We first note that the w_α are 2-adic integers. Let $\beta = 1$ (resp. i) if $e = 1$ (resp. -4). As for the expansions of $\mathcal{L}_{\nu,e}(\xi)$ for $\nu = 0, 1$, following [Uehara, 1990] we have for $e \in \mathcal{T}_4$

$$\mathcal{L}_{0,e}(\xi) = -\frac{\beta}{2}(l_0(\beta\xi) - l_0(-\beta\xi)) = -\frac{\beta^2\xi}{2}(L^-(\beta\xi) + L^-(-\beta\xi))$$

$$= -\frac{\beta}{2}(L^-(\beta\xi) - L^-(-\beta\xi)) = w_{-\alpha}$$

and

$$\mathcal{L}_{1,e}(\xi) = \frac{\beta}{2}(l_1(\beta\xi) - l_1(-\beta\xi)) = -\frac{\beta}{2}(L^+(\beta\xi) - L^+(-\beta\xi))$$

$$= -\frac{\beta}{4}(\log_2(1 - 2\beta\xi + \alpha\xi^2) - \log_2(1 + 2\beta\xi + \alpha\xi^2))$$

$$= -\frac{\beta}{4}\log_2\left(\frac{1 - 2\alpha\beta w_\alpha}{1 + 2\alpha\beta w_\alpha}\right) = \sum_{k=0}^{\infty}\frac{(4\alpha)^k w_\alpha^{2k+1}}{2k+1}.$$

If $e \in \mathcal{T}_8 - \mathcal{T}_4$ then putting $\eta = (1+i)/\sqrt{2}$ (so that $\eta^2 = i$) we have by definition

$$\mathcal{L}_{0,e}(\xi) = -\frac{\sqrt{2\alpha}}{4}(l_0(\eta\xi) - \alpha l_0(-\eta^{-1}\xi) - l_0(-\eta\xi) + \alpha l_0(\eta^{-1}\xi))$$

$$= -\frac{\xi\sqrt{2\alpha}}{4}(\eta(L^-(\eta\xi) + L^-(-\eta\xi)) + \alpha\eta^{-1}(L^-(\eta^{-1}\xi) + L^-(-\eta^{-1}\xi)))$$

$$= -\frac{\sqrt{2\alpha}}{4}((L^-(\eta\xi) - L^-(-\eta\xi)) + \alpha(L^-(\eta^{-1}\xi) - L^-(-\eta^{-1}\xi)))$$

$$= -\frac{\sqrt{2\alpha}}{4}\left(\frac{2\eta\xi}{1 - i\xi^2} + \alpha\frac{2\eta^{-1}\xi}{1 + i\xi^2}\right)$$

$$= \frac{\alpha\xi(1 - \alpha\xi^2)}{1 + \xi^4} = \frac{w_{-\alpha}}{1 + 2\alpha w_{-\alpha}^2} = \sum_{k=0}^{\infty}(-2\alpha)^k w_{-\alpha}^{2k+1}$$

and

$$\mathcal{L}_{1,e}(\xi) = \frac{\sqrt{2\alpha}}{4}(l_1(\eta\xi) - \alpha l_1(-\eta^{-1}\xi) - l_1(-\eta\xi) + \alpha l_1(\eta^{-1}\xi))$$

$$= -\frac{\sqrt{2\alpha}}{4}(L^+(\eta\xi) + \alpha L^+(\eta^{-1}\xi) - L^+(-\eta\xi) - \alpha L^+(-\eta^{-1}\xi))$$

$$= -\frac{\sqrt{2\alpha}}{4}(\log_2(1 - \sqrt{2\alpha}\xi + \alpha\xi^2) - \log_2(1 + \sqrt{2\alpha}\xi + \alpha\xi^2))$$

$$= -\frac{\sqrt{2\alpha}}{4}\log_2\left(\frac{1 - \sqrt{2\alpha}\alpha w_\alpha}{1 + \sqrt{2\alpha}\alpha w_\alpha}\right) = \sum_{k=0}^{\infty}\frac{(2\alpha)^k w_\alpha^{2k+1}}{2k+1}.$$

Let us consider the case of $\nu = -1$. Then we have

$$\mathcal{L}_{-1,e}(\xi) = -\frac{\alpha\xi(1 + \alpha\xi^2)}{(1 - \alpha\xi^2)^2},$$

if $e \in \mathcal{T}_4$. In this case it suffices to use the 2-adic expansion

$$\sum_{\substack{n \geq 1 \\ n \text{ odd}}} x^n = \frac{x}{1 - x^2} \tag{30}$$

with $x = 2w_\alpha \sqrt{\alpha}$. Furthermore, if $e \notin \mathcal{T}_4$ we have

$$\mathcal{L}_{-1,e}(\xi) = -\frac{\alpha\xi(\xi^2 + \alpha)(\xi^4 - 4\alpha\xi^2 + 1)}{(1 + \xi^4)^2}.$$

In this case it is sufficient to apply (30) together with the 2-adic series

$$\sum_{\substack{n \geq 1 \\ n \text{ odd}}} nx^n = \frac{x(1 + x^2)}{(1 - x^2)^2}$$

with $x = w_\alpha \sqrt{2\alpha}$ in both the formulae of the lemma. Indeed we have

$$\sqrt{2\alpha} \sum_{k=0}^{\infty} (2\alpha)^k (2k - 1)w_\alpha^{2k+1} = \sum_{k=0}^{\infty} (2k - 1)(w_\alpha\sqrt{2\alpha})^{2k+1}$$

$$= \sum_{\substack{n \geq 1 \\ n \text{ odd}}} n(w_\alpha\sqrt{2\alpha})^n - 2 \sum_{\substack{n \geq 1 \\ n \text{ odd}}} (w_\alpha\sqrt{2\alpha})^n.$$

It remains to prove the lemma when $\nu = 2$. We consider the 2-adic series

$$G_e(x) := \sum_{k=0}^{\infty} \frac{(-16\alpha)^k \nu_{k,e}}{(2k + 1)^2} \left(\frac{-\alpha x}{1 - \alpha x^2}\right)^{2k+1},$$

where

$$\nu_{k,e} = \begin{cases} \displaystyle \binom{2k}{k}^{-1} \sum_{l=0}^{k} \binom{2l}{l} 2^{-3l}, & \text{if } e \notin \mathcal{T}_4, \\[4mm] \displaystyle \binom{2k}{k}^{-1}, & \text{if } e \in \mathcal{T}_4. \end{cases}$$

Since

$$\mathrm{ord}_2 \binom{2k}{k} = s_2(k),$$

where $s_2(t)$ denotes the sum of the digits in the 2-adic expansion of t, we have

$$\mathrm{ord}_2 \left(2^{3k} \binom{2k}{k}^{-1}\right) = 3k - s_2(k) \geq 2k. \tag{31}$$

Moreover

$$\text{ord}_2\left(2^{3k}\binom{2k}{k}^{-1}\sum_{l=0}^{k}\binom{2l}{l}2^{-3l}\right)$$

$$= \text{ord}_2\left(\sum_{l=0}^{k}\frac{k!(2l-1)!!}{l!(2k-1)!!}2^{2(k-l)}\right) = 0,\tag{32}$$

where $t!!$ denotes the product of all odd integers $\leq t$ and we set $(-1)!! = 1$. Thus the series $G_e(x)$ determines an analytic function on the open unit ball in $\mathbb{C}_2$. Furthermore, setting $\gamma^2 = \alpha$ we have

$$G_e(x) = \sum_{k=0}^{\infty}\frac{(-16\alpha)^k\nu_{k,e}}{(2k+1)^2}\left(-\alpha x\sum_{l=0}^{\infty}(\gamma x)^{2l}\right)^{2k+1}$$

$$= -\gamma\sum_{k=0}^{\infty}\frac{(-16)^k\nu_{k,e}}{(2k+1)^2}\left(\sum_{l=0}^{\infty}(\gamma x)^{2l+1}\right)^{2k+1}$$

$$= -\gamma\sum_{k=0}^{\infty}\frac{(-16)^k\nu_{k,e}}{(2k+1)^2}\sum_{l=0}^{\infty}\binom{2k+l}{l}(\gamma x)^{2(k+l)+1},$$

where x belongs to the open unit ball in $\mathbb{C}_2$. Hence, by virtue of the well known identity

$$\sum_{k=0}^{\infty}\sum_{l=0}^{\infty}f(k,l) = \sum_{l=0}^{\infty}\sum_{k=0}^{l}f(k,l-k),\tag{33}$$

we obtain

$$G_e(x) = -\gamma\sum_{l=0}^{\infty}\sum_{k=0}^{l}\frac{(-16)^k\nu_{k,e}}{(2k+1)^2}\binom{l+k}{l-k}(\gamma x)^{2l+1}$$

$$= -\gamma\sum_{l=0}^{\infty}(\gamma x)^{2l+1}\sum_{k=0}^{l}\binom{l+k}{l-k}\frac{(-16)^k\nu_{k,e}}{(2k+1)^2}.$$

Thus appealing to Lemma 3, on the open unit ball in $\mathbb{C}_2$ we have

$$G_e(x) = \mathcal{L}_{2,e}(x).$$

Consequently Lemma 5 for $\nu = 2$ follows from [Coleman, 1982, Theorem 5.11] and the uniqueness principle (see [Coleman, 1982, p. 176]). ∎

REMARK Uehara in a letter to the first author has observed that the formulae for $\mathcal{L}_{-1,e}(\xi)$ and $\mathcal{L}_{2,e}(\xi)$ given in the above lemma can be deduced easily from his formulae for $\mathcal{L}_{0,e}(\xi)$, $\mathcal{L}_{1,e}(\xi)$, and differential properties of Coleman's multilogarithms. The details of the proof are left to the reader as an exercise.

Wójcik has extended Lemma 5 by proving the following result.

LEMMA 6 ([Wójcik, 1998, Lemma 3]) *For any* $e \in T_8$ *and* $m \in \mathbb{Z}$ *write* $\alpha = (-1)^{m+1} \operatorname{sgn} e$ *and let*

$$w_\alpha = \frac{\alpha\xi}{1 + \alpha\xi^2}.$$

Then we have

$$\mathcal{L}_{m,e}(\xi) = \sum_{k=0}^{\infty} \frac{\alpha^k W_{m,e}(k)}{2k+1} w_\alpha^{2k+1}.$$

PROOF. First we observe that the 2-adic series on the right hand side of the above equation converges, which follows immediately from (28).

Now write $\gamma^2 = \alpha$. We proceed in the same way as in the proof of Lemma 5. For x belonging to the open unit ball in $\mathbb{C}_2$ we have (recall formula (33))

$$\sum_{k=0}^{\infty} \frac{\alpha^k W_{m,e}(k)}{2k+1} \left(\frac{\alpha x}{1 + \alpha x^2} \right)^{2k+1}$$

$$= -i\gamma \sum_{k=0}^{\infty} \frac{(-1)^k W_{m,e}(k)}{2k+1} \left(\sum_{l=0}^{\infty} (i\gamma x)^{2l+1} \right)^{2k+1}$$

$$= -i\gamma \sum_{k=0}^{\infty} \frac{(-1)^k W_{m,e}(k)}{2k+1} \sum_{l=0}^{\infty} \binom{2k+l}{l} (i\gamma x)^{2(k+l)+1}$$

$$= -i\gamma \sum_{l=0}^{\infty} \sum_{k=0}^{l} \frac{(-1)^k W_{m,e}(k)}{2k+1} \binom{l+k}{l-k} (i\gamma x)^{2l+1}$$

$$= -i\gamma \sum_{l=0}^{\infty} (i\gamma x)^{2l+1} \sum_{k=0}^{l} \binom{l+k}{l-k} \frac{(-1)^k W_{m,e}(k)}{2k+1}.$$

The lemma now follows immediately from Lemma 4, [Coleman, 1982, Theorem 5.11] and the uniqueness principle (see [Coleman, 1982, p. 176]). ∎

4.6 A Useful Numerical Lemma

Let K be a finite nonempty subset of the rational integers and let d be a fundamental discriminant. We consider a finite set of 2-adic integers

$$x = \{x_{k,e}\}_{(k,e) \in K \times T_d} \subseteq \mathbb{C}_2.$$

For any $L \subseteq K$ the x is said to be defined on L, if $x_{k,e} = 0$ for $k \notin L$.

Given 2-adic integers $a_{k,e}(n)\,(\in \mathbb{C}_2)$ with $k \in K$, $e \in \mathcal{T}_d$, $n \geq 0$ we consider a sequence of linear combinations of $x_{k,e}$ of the form

$$y_n(x) = \sum_{(k,e)\in K\times\mathcal{T}_d} a_{k,e}(n)x_{k,e}, \quad n \geq 0. \tag{34}$$

For any $L \subseteq K$ the sequence $(y_n)_{n\geq 0}$ of the form (34) is said to be defined on L, if the sum is taken over $k \in L$, $e \in \mathcal{T}_d$. The sequence $(y_n)_{n\geq 0}$ defined on K can be considered as a sequence defined on L if we put $x_{k,e} = 0$ for $k \notin L$.

Let $y = (y_n)_{n\geq 0}$ be a sequence of the form (34). A nonnegative integer $c = c(y)$ is called the exponent of the sequence y if it satisfies the following two conditions:

(i) there exist 2-adic integers $x_{k,e}$ not all even satisfying

$$y_n(x) \equiv 0\,(\mathrm{mod}\,2^c), \quad n \geq 0,$$

(ii) if for some 2-adic integers $x_{k,e}$ we have

$$y_n(x) \equiv 0\,(\mathrm{mod}\,2^{c+1}), \quad n \geq 0,$$

then all the numbers $x_{k,e}$ are even.

Let $K = \{-1,0,1,2\}$. We consider a sequence $z = (z_n)_{n\geq 0}$, where $z_n = z_n(x)$ are defined for $x = \{x_{k,e}\}_{(k,e)\in K\times\mathcal{T}_8}$ by

$$z_0 = \sum_{(k,e)\in K\times\mathcal{T}_8} x_{k,e}, \qquad z_1 = 2\sum_{\substack{(k,e)\in K\times\mathcal{T}_8,\\ \mathrm{sgn}\,e=(-1)^k}} x_{k,e},$$

$$\begin{aligned}
z_{2l+\varrho} = 2^{l+\varrho}\Big(&2^l(2l+1)^2((1-\varrho)x_{-1,1} + x_{-1,-4})\\
&- (2l-1)(2l+1)^2((1-\varrho)x_{-1,8} + x_{-1,-8})\\
&+ (2l+1)^2((1-\varrho)x_{0,-8} + x_{0,8})\\
&+ 2^l(2l+1)((1-\varrho)x_{1,1} + x_{1,-4}) + (2l+1)((1-\varrho)x_{1,8} + x_{1,-8})\\
&+ \nu_{l,1}2^{3l}((1-\varrho)x_{2,-4} + x_{2,1}) + \nu_{l,8}2^{3l}((1-\varrho)x_{2,-8} + x_{2,8})\Big),
\end{aligned}$$

if $l \geq 1$, $\varrho \in \{0,1\}$ and the numbers $\nu_{l,e}$ ($e \in \mathcal{T}_8$) are defined in the proof of Lemma 5.

REMARKS By definition we have

$$z_3 = 4(18x_{-1,-4} - 9x_{-1,-8} + 9x_{0,8} + 6x_{1,-4} + 3x_{1,-8} + 5x_{2,8} + 4x_{2,1}). \quad (35)$$

We shall use the above formulae in the proof of Lemma 7. Throughout the proof of the lemma, we set

$$t_l = z_{2l+3} - \frac{8(l+1)}{2l+1} z_{2l+1},$$

where $l \geq 1$. We have

$$t_l = 2^{l+2}(x_{2,8} - (8l^3 - 12l^2 - 34l - 13)x_{-1,-8} \quad (36)$$

$$+ (5 - 4l^2)x_{0,8} - (2l+1)x_{1,-8}) + 2^{2l+3}((6l+7)x_{-1,-4} + x_{1,-4}),$$

which is clear from

$$\nu_{l+1,1} 2^{3(l+1)} - \frac{4(l+1)}{2l+1} \nu_{l,1} 2^{3l} = 0$$

and

$$\nu_{l+1,8} 2^{3(l+1)} - \frac{4(l+1)}{2l+1} \nu_{l,8} 2^{3l} = 1.$$

The sequence $z = z(x)$ is of the form (34) since $z_{2l+\varrho}$ are linear combinations of the $x_{k,e}$ ($k \in K$, $e \in \mathcal{T}_8$) with 2-integral coefficients (recall (31) and (32)). Note that $z_{2l+\varrho}$ are divisible by $2^{l+\varrho}$. We compute the exponent $c(z)$ of this sequence.

LEMMA 7 ([Urbanowicz and Wójcik, 1995/1996, Lemma 5]) *Let K equal $\{-1, 0, 1, 2\}$ and let L be a nonempty subset of K. Write $c(L) = c(z)$, where $z = (z_n)_{n \geq 0}$ is the sequence given above, defined on L. Then we have*

$$c(L) = 12, \ 9, \ 5, \ resp. \ 2,$$

if $\mathrm{card}(L) = 4, \ 3, \ 2, \ resp. \ 1$, *unless* $L = \{-1, 1\}$ *or* $\{0, 2\}$, *in which cases*

$$c(L) = 6.$$

PROOF. We briefly sketch the proof of the lemma. We consider four cases:

1. $\mathrm{card}(L) = 4$.

2. $\mathrm{card}(L) = 3$.

3. $\mathrm{card}(L) = 2$.

4. $\mathrm{card}(L) = 1$.

Set $r = 12$, 9, 5, resp. 2 if $\mathrm{card}(L) = 4$, 3, 2, resp. 1, unless $L = \{-1, 1\}$ or $\{0, 2\}$, in which cases set $r = 6$. In each of these cases the proof is divided into 2 steps.

(i) We find 2-adic integers $x_{k,e}$ defined on L such that the numbers $z_n(x)$ ($n \geq 0$) are divisible by 2^r (which proves that $c(L) \geq r$).

(ii) Assuming that

$$z_n(x) \equiv 0 \,(\mathrm{mod}\, 2^{r+1}), \quad n \geq 0 \tag{37}$$

for some 2-adic integers $x_{k,e}$, we prove that all the numbers $x_{k,e}$ must be even (which proves that $c(L) \leq r$).

Step (ii) falls naturally into two parts. Making use of the congruences (37) with n odd we first prove that the $x_{k,e}$ are even when $\mathrm{sgn}\, e = (-1)^k$. We observe next that for n even

$$2z_n = z_{n+1} + \tilde{z}_{n+1},$$

where the $\tilde{z}_{n+1}$ comes from z_{n+1} by replacing $x_{k,-4}$ (resp. $x_{k,1}$, $x_{k,-8}$ or $x_{k,8}$) by $x_{k,1}$ (resp. $x_{k,-4}$, $x_{k,8}$ or $x_{k,-8}$). Moreover, coupling the above equation with (37) we deduce that

$$\tilde{z}_{n+1}(x) \equiv 0 \,(\mathrm{mod}\, 2^{r+1}).$$

Thus by the same reasoning as in the case of $\mathrm{sgn}\, e = (-1)^k$, applied to the numbers $\tilde{z}_{2l+1}(x)$ we obtain that the $x_{k,e}$ are also even if $\mathrm{sgn}\, e \neq (-1)^k$, which completes the proof. Making use of the congruences (37) with n odd we will therefore just focus on the case $\mathrm{sgn}\, e = (-1)^k$.

We are now ready to prove the lemma. We omit some less important details because of limitation of space. For a thorough treatment we refer the reader to [Urbanowicz and Wójcik, 1995/1996].

If $l \geq 1$ by definition we have $z_{2l+1}(x) \equiv 0 \,(\mathrm{mod}\, 2^{l+1})$ for arbitrary 2-adic integers $x_{k,e}$. Moreover, in the notation of Remark before Lemma 7, we have

$$t_l(x) \equiv 0 \,(\mathrm{mod}\, 2^{l+2})$$

since $2l + 3 \geq l + 4$. Furthermore, the congruence $t_l \equiv 0 \,(\mathrm{mod}\, 2^{r+1})$ implies that

$$c_l := x_{2,8} + (8l^3 - 12l^2 - 34l - 13)x_{-1,-8} + (5 - 4l^2)x_{0,8}$$

$$- (2l + 1)x_{1,-8} + 2^{l+1}((6l + 7)x_{-1,-4} + x_{1,-4}) \equiv 0 \,(\mathrm{mod}\, 2^{r-l-1}). \tag{38}$$

Following the above observations, we consider four cases. The proof of the lemma in each of the cases falls into two steps (i) and (ii).

Case 1. We prove that $c(L) = 12$.

(i) We define $x = \{x_{k,e}\}$ by

$$x_{-1,-4} = x_{-1,-8} = 1,$$
$$x_{0,1} = x_{0,8} = -3,$$
$$x_{1,-4} = x_{1,-8} = 61,$$
$$x_{2,1} = x_{2,8} = 63,$$

and

$$x_{k,1} = -x_{k,-4}, \quad x_{k,8} = -x_{k,-8}.$$

We must prove that $z_n(x) \equiv 0 \,(\bmod\, 2^{12})$ if $n \geq 0$.

It is easily seen that $z_1 = z_3 = 0$ and $z_{2l} = 0$, if $l \geq 0$. Therefore it suffices to show that $z_{2l+1}(x) \equiv 0 \,(\bmod\, 2^{12})$ for $2 \leq l \leq 10$, or equivalently, that t_l is divisible by 2^{12} for $1 \leq l \leq 9$. We show the latter assertion using an easy computation (note that $t_l = 0$ if $l \leq 5$).

(ii) Assuming on the contrary that for some 2-adic integers $x_{k,e}$ not all even and $n \geq 0$ we have $z_n(x) \equiv 0 \,(\bmod\, 2^{13})$ (and in consequence $t_n \equiv 0 \,(\bmod\, 2^{13})$), by (36) we obtain

$$-2^7 t_1 + 2^7 t_2 - 2^3 t_3 - 2^2 \cdot 7 t_4 + 2 \cdot 5 t_5 - t_6 \equiv 2^{12} x_{0,8} \equiv 0 \,(\bmod\, 2^{13}),$$

and consequently $x_{0,8}$ must be even. This is the key point of the proof of (ii). The rest follows from congruence (38). By (38) we have

$$c_7 + c_8 \equiv 0 \,(\bmod\, 8), \quad c_6 - c_4 \equiv 0 \,(\bmod\, 32), \quad c_6 - c_7 \equiv 0 \,(\bmod\, 8).$$

Hence we deduce that

$$x_{2,8} \equiv 0 \,(\bmod\, 2), \quad 3x_{-1,-8} - x_{1,-8} \equiv 0 \,(\bmod\, 8),$$
$$-x_{-1,-8} + x_{1,-8} \equiv 0 \,(\bmod\, 4)$$

because $x_{0,8}$ is even.

Making use of the above congruence we obtain $x_{-1,-8} \equiv x_{1,-8} \equiv 0$ $(\bmod\, 4)$. Consequently, by (38), we deduce that

$$2c_1 - c_2 + c_6 - 2c_l \equiv 16x_{-1,-4} \equiv 0 \,(\bmod\, 32),$$

and

$$c_1 + c_6 - c_7 - c_8 \equiv 4x_{1,-4} \,(\bmod\, 8).$$

Therefore $x_{-1,-4}$ and $x_{1,-4}$ must be even.

In order to prove that $x_{2,1}$ is even we shall use the congruence $z_3 \equiv 0$ (mod 32). By (35) and (38), we obtain

$$(z_3/4) + c_8 - c_6 \equiv 2(x_{-1,-4} - x_{1,-4}) + 4x_{2,1} \equiv 0 \,(\mathrm{mod}\, 8)$$

and

$$c_1 - c_7 \equiv -4(x_{-1,-8} + x_{1,-8}) + 4(x_{-1,-4} + x_{1,-4}) \equiv 0 \,(\mathrm{mod}\, 16).$$

Hence we deduce that

$$x_{-1,-4} + x_{1,-4} \equiv 0 \,(\mathrm{mod}\, 4)$$

because $-x_{-1,-8} + x_{1,-8} \equiv 0 \,(\mathrm{mod}\, 4)$. Therefore $x_{2,1}$ must be even.

In order to obtain the required contradiction it remains to show that $x_{0,1}$ is even. But this follows easily from the congruence $z_1 \equiv 0 \,(\mathrm{mod}\, 4)$.

Case 2. We prove that $c(L) = 9$.

(i) We define $x = \{x_{k,e}\}$ by

$$x_{-1,-4} = -x_{-1,-8} = a_{-1},$$
$$x_{0,1} = -x_{0,8} = a_0,$$
$$x_{1,-4} = -x_{1,-8} = a_1,$$
$$x_{2,1} = -x_{2,8} = a_2,$$

and

$$x_{k,1} = -x_{k,-4}, \quad x_{k,8} = -x_{k,-8} \quad (k \in L),$$

where $a_k := a_k(L)$ are defined as follows

$$a_{-1} = 1, \quad a_0 = -2, \quad a_1 = -15, \quad \text{if } L = \{-1,0,1\},$$
$$a_{-1} = 2, \quad a_0 = -19, \quad a_2 = 225, \quad \text{if } L = \{-1,0,2\},$$
$$a_{-1} = 1, \quad a_1 = -19, \quad a_2 = -30, \quad \text{if } L = \{-1,1,2\},$$
$$a_0 = -1, \quad a_1 = 2, \quad a_2 = 15, \quad \text{if } L = \{0,1,2\},$$

and $a_k = 0$ if $k \notin L$. We should prove that $z_n(x) \equiv 0 \,(\mathrm{mod}\, 2^9)$ for $n \geq 0$.

In all the cases we have $z_1 = z_3 = z_{2l} = 0, l \geq 0$ (note that, in view of (35), we have

$$z_3 = 4(27a_{-1} - 9a_0 + 3a_1 - a_2)).$$

Furthermore, by (36) it follows easily that $t_l \equiv 0 \,(\mathrm{mod}\, 2^9)$ if $1 \leq l \leq 6$, which implies that $z_{2l+1} \equiv 0 \,(\mathrm{mod}\, 2^9)$ for $2 \leq l \leq 7$ (note that $t_l = 0$ if $1 \leq l \leq 3$).

(ii) Suppose on the contrary that $z_n(x) \equiv 0\,(\mathrm{mod}\,2^{10})$ $(n \geq 0)$ for some 2-integers $x_{k,e}$ not all even. We proceed by considering subcases:

2a. $2 \notin L$.   2b. $1 \notin L$.   2c. $0 \notin L$.   2d. $-1 \notin L$.

Subcase 2a. By virtue of (38) we have

$$c_1 - c_5 \equiv 0\,(\mathrm{mod}\,8), \text{ resp. } c_2 - c_4 \equiv 0\,(\mathrm{mod}\,16).$$

Consequently, we obtain

$$x_{-1,-4} + x_{1,-4} \equiv 0\,(\mathrm{mod}\,2), \text{ resp. } x_{-1,-8} + x_{1,-8} \equiv 0\,(\mathrm{mod}\,4).$$

Therefore we have

$$c_6 + x_{-1,-8} + x_{1,-8} \equiv x_{0,8} \equiv 0\,(\mathrm{mod}\,4), \quad c_4 + c_5 \equiv 4x_{1,-8} \equiv 0\,(\mathrm{mod}\,8).$$

On the other hand, by (35) we find that

$$(z_3/4) - c_2 \equiv 2(x_{-1,-4} - x_{1,-4}) \equiv 0\,(\mathrm{mod}\,8),$$

and in consequence

$$c_1 + c_3 - 2c_2 \equiv 4(x_{-1,-4} + x_{1,-4}) \equiv 8x_{-1,-4} \equiv 0\,(\mathrm{mod}\,16).$$

Consequently $x_{-1,-4}$ and $x_{1,-4}$ must be even.

Finally since $z_1/2$ is even, $x_{0,1}$ must be also even. This gives a contradiction.

Subcase 2b. By (38) we have

$$c_1 - c_5 \equiv 4x_{-1,-4} \equiv 0\,(\mathrm{mod}\,8), \text{ resp. } c_2 - c_4 \equiv 4x_{-1,-8} \equiv 0\,(\mathrm{mod}\,16),$$

and so $x_{-1,-4} \equiv 0\,(\mathrm{mod}\,2)$, resp. $x_{-1,-8} \equiv 0\,(\mathrm{mod}\,4)$. Consequently, we obtain

$$c_4 - c_5 \equiv 4x_{0,8} \equiv 0\,(\mathrm{mod}\,8).$$

Hence we find that

$$c_4 + c_3 \equiv 6x_{0,8} + 2x_{2,8} \equiv 0\,(\mathrm{mod}\,16).$$

which gives

$$c_1 + c_4 \equiv 4x_{-1,-4} \equiv 0\,(\mathrm{mod}\,16), \quad (z_3/4) - c_1 \equiv 4x_{2,1} \equiv 0\,(\mathrm{mod}\,8)$$

(recall that $x_{-1,-8}$ is even). Hence $x_{2,1}$ is even and, by virtue of $(z_1/2) \equiv 0$ $(\mathrm{mod}\,2)$, $x_{0,1}$ must be also even, a contradiction.

Subcase 2c. By (38) we have

$$c_4 + c_5 \equiv 4x_{1,-8} + 2x_{2,8} \equiv 0 \,(\,\mathrm{mod}\,8\,),$$
$$c_1 - c_5 \equiv 4(x_{-1,-4} + x_{1,-4}) \equiv 0 \,(\,\mathrm{mod}\,8\,).$$

Thus $x_{2,8}$ and $x_{2,1}$ must be even (recall that c_6, $x_{2,8}$ and $z_1/2$ are even). Furthermore, by (38) we obtain

$$c_4 - c_2 \equiv -4(x_{-1,-8} + x_{1,-8}) \equiv 0 \,(\,\mathrm{mod}\,16\,),$$
$$c_2 + c_3 \equiv 4x_{1,-8} \equiv 0 \,(\,\mathrm{mod}\,8\,).$$

Consequently, $x_{-1,-8} + x_{1,-8} \equiv 0 \,(\,\mathrm{mod}\,4\,)$ and $x_{2,8}$, $x_{1,-8}$ must be even (recall that c_6 is divisible by 4). Moreover, since c_6 and $x_{2,8}$ are even, $x_{-1,-8}$ must be also even. Thus we have

$$(z_3/4) - c_2 \equiv 2(x_{-1,-4} - x_{1,-4}) \equiv 0 \,(\,\mathrm{mod}\,8\,),$$

and

$$c_1 - c_3 \equiv 4(x_{-1,-4} + x_{1,-4}) \equiv 0 \,(\,\mathrm{mod}\,16\,),$$

and in consequence $x_{-1,-4}$ and $x_{1,-4}$ are even. Therefore, by the same argument as in the previous cases all the $x_{k,e}$ must be even, a contradiction.

Subcase 2d. By (38) we have

$$c_1 - c_5 \equiv 4x_{1,-4} \equiv 0 \,(\,\mathrm{mod}\,8\,), \quad c_2 - c_4 \equiv 4x_{1,-8} \equiv 0 \,(\,\mathrm{mod}\,16\,),$$
$$c_3 - c_4 \equiv 4x_{0,8} \equiv 0 \,(\,\mathrm{mod}\,8\,),$$

and consequently $x_{1,-4} \equiv 0 \,(\,\mathrm{mod}\,2\,)$, $x_{1,-8} \equiv 0 \,(\,\mathrm{mod}\,4\,)$, $x_{0,8} \equiv 0 \,(\,\mathrm{mod}\,2\,)$ (and thus $x_{2,8}$ must be even because c_6 is even).
On the other hand, we have

$$c_1 - c_3 \equiv 4x_{1,-4} \equiv 0 \,(\,\mathrm{mod}\,16\,), \quad (z_3/4) - c_1 \equiv 4x_{2,1} \equiv 0 \,(\,\mathrm{mod}\,8\,),$$

and consequently $x_{1,-4}$ is divisible by 4 and $x_{2,1}$ is even (recall that $x_{1,-8} \equiv x_{1,-8} \equiv x_{2,1} \equiv 0 \,(\,\mathrm{mod}\,4\,)$ and $x_{2,8} \equiv 0 \,(\,\mathrm{mod}\,2\,)$).
The rest of the proof may be handled in the same way as in the previous cases and again we obtain a contradiction.

Case 3. We prove that $c(L) = c_0$, where $c_0 = 5$ unless $L = \{-1, 1\}$ or $L = \{0, 2\}$, in which cases $c_0 = 6$.

We proceed by considering subcases:

3a. $L = \{-1, 1\}$.   3b. $L = \{0, 2\}$.   3c. $L = \{-1, 0\}$.
3d. $L = \{-1, 2\}$.   3e. $L = \{1, 2\}$.   3f. $L = \{0, 1\}$.

(i) We define $x = \{x_{k,e}\}$ by

$$x_{-1,-4} = x_{-1,-8} = b_{-1},$$
$$x_{0,1} = x_{0,8} = b_0,$$
$$x_{1,-4} = x_{1,-8} = b_1,$$
$$x_{2,1} = x_{2,8} = b_2,$$

and

$$x_{k,1} = -x_{k,-4}, \quad x_{k,8} = -x_{k,-8}, \quad k \in L,$$

where b_t, $t \in K$ are defined for $L = \{k_1, k_2\}$, $k_1 < k_2$ as follows:

$$b_{k_1} = -b_{k_2} = 1,$$

and $b_k = 0$ if $k \notin L$. We should prove that $z_n \equiv 0 \,(\bmod\, 2^{c_0})$, $n \geq 0$.
In all the cases we have $z_1 = z_3 = z_{2l} = 0$, $l \geq 0$ (note that by (35) we have

$$z_3 = 3(b_{-1} + b_0 + b_1 + b_2)).$$

Furthermore, by virtue of (36) it follows easily that in the first two subcases we have $t_1 = 0$ and $2^6 \,|\, t_l$, if $l \geq 2$. In the remaining subcases it is easily seen that $2^5 \,|\, t_l$, if $l \geq 1$.

(ii) Suppose on the contrary that $z_n \equiv 0 \,(\bmod\, 2^{c_0+1})$ $(n \geq 0)$ for some 2-adic integers $x_{k,e}$ not all even. We proceed by considering subcases.

Subcase 3a. By (35) and (38) we obtain

$$(z_3/4) - c_2 \equiv 2(x_{-1,-4} - x_{1,-4}) \equiv 0 \,(\bmod\, 8),$$

that is, $x_{-1,-4} \equiv x_{1,-4} \,(\bmod\, 4)$. This implies that

$$c_1 + c_2 \equiv 4x_{-1,-8} \equiv 0 \,(\bmod\, 8).$$

Hence we see that $x_{1,-8}$ is even since $z_1/2$ is even. Finally, we have

$$(z_1/2) - c_3(x_{-1,-4} - x_{1,-4}) \equiv 2x_{1,-4} \,(\bmod\, 4),$$

which gives the required contradiction.

Subcase 3b. By (38) we have

$$c_2 - c_1 = 4x_{0,8} \equiv 0 \,(\bmod\, 8),$$

and hence $x_{2,8}$ is even because c_1 is even. Moreover, by virtue of (35) we find that $(z_3/4) \equiv 4x_{2,1} \equiv 0 \,(\bmod\, 8)$ (note that $c_1 \equiv x_{0,8} + x_{2,8} \equiv 0 \,(\bmod\, 8)$). We obtain the required contradiction. ∎

Subcase 3c. By (35) and (38) we have

$$(z_3/4) - c_2 \equiv 2x_{-1,-4} \equiv 0\,(\mathrm{mod}\,4), \quad c_1 + c_2 \equiv 2x_{0,8} \equiv 0\,(\mathrm{mod}\,4),$$

and in consequence $x_{-1,-4} \equiv x_{0,8} \equiv 0\,(\mathrm{mod}\,2)$. Furthermore $x_{-1,-8}$ must be even because c_2 is even. We obtain the required contradiction.

Subcase 3d. By (38) we have

$$c_1 - c_2 \equiv 2x_{-1,-8} \equiv 0\,(\mathrm{mod}\,4),$$

and consequently $x_{2,8}$ must be even because c_2 is even. Thus by (35) we obtain

$$(z_3/4) - c_2 \equiv 2x_{-1,-4} \equiv 0\,(\mathrm{mod}\,4),$$

which gives the required contradiction.

Subcase 3e. By (38) and (35) we have

$$c_1 + c_2 \equiv 2x_{2,8} \equiv 0\,(\mathrm{mod}\,4), \quad (z_3/4) - c_2 \equiv 2x_{1,-4} \equiv 0\,(\mathrm{mod}\,4),$$

and so $x_{2,8} \equiv x_{1,-8} \equiv x_{1,-4} \equiv 0\,(\mathrm{mod}\,2)$ (recall that c_2 is even).

In order to obtain the required contradiction it remains to prove that $x_{2,1}$ is even. But this follows easily because $z_1/2$ is even.

Subcase 3f. The case $L = \{0, 1\}$ was considered in [Gras, 1989] and [Uehara, 1990]. By (38) and (35), we have

$$c_1 + c_2 \equiv 2x_{0,8} \equiv 0\,(\mathrm{mod}\,4), \quad (z_3/4) - c_2 \equiv 2x_{1,-4} \equiv 0\,(\mathrm{mod}\,4),$$

and hence $x_{0,8} \equiv x_{1,-8} \equiv x_{1,-4} \equiv 0\,(\mathrm{mod}\,2)$ (recall that c_2 is even).

We obtain the required contradiction, which completes the proof of the lemma in case 3.

Case 4. We prove that $c(L) = 2$. Let $L = \{k\}$.

(i) We define $x = \{x_{k,e}\}$ by

$$x_{k,1} = -x_{k,8} = 1,$$

if k is even, and

$$x_{k,-4} = -x_{k,-8} = 1,$$

if k is odd, and

$$x_{k,1} = -x_{k,-4}, \quad x_{k,8} = -x_{k,-8},$$

and $x_{l,e} = 0$, if $l \in K, l \neq k$.

It is evident that for an arbitrary $k \in K$ we have $z_1 = z_{2l} = 0\ (l \geq 0)$ and that $z_{2l+1}\ (l \geq 1)$ is divisible by 4.

(ii) Assume on the contrary that $z_n \equiv 0 \pmod 8$ $(n \geq 0)$ for some 2-adic integers $x_{k,e}$ not all even. Then coupling (35) with the congruence $(z_1/2) \equiv 0 \pmod 4$, we obtain

$$x_{k,1} + x_{k,8} \equiv x_{k,-4} + x_{k,-8} \equiv 0 \pmod 4,$$

and since $(z_3/4)$ is even, all the $x_{k,e}$ with $\operatorname{sgn} e = (-1)^k$ are even. Consequently by the same reasoning as previously all the $x_{k,e}$ $(k \in L, e \in \mathcal{T}_8)$ are even and we obtain the required contradiction. Lemma 7 is proved. ∎

4.7 Another Sequence of the Type (34)

Wójcik [Wójcik, 1998] has considered another sequence, $u = (u_n)_{n \geq 0}$, of the type (34) which is defined, for any finite subset K of rational integers, by

$$u_{2l+\varrho} = 2^\varrho \sum_{k,e} (-1)^{l(k+1)} (2l+1)^{1-k} \gamma_{l,e} x_{k,e},$$

where $l \geq 0$, $\varrho \in \{0, 1\}$ and the sum is taken over all $k \in K$, $e \in \mathcal{T}_8$ if $\varrho = 0$, and over $k \in K$, $e \in \mathcal{T}_8$ with $\operatorname{sgn} e = (-1)^k$ if $\varrho = 1$. Recall that $\gamma_{l,e}$ is defined by

$$\gamma_{l,e} = \begin{cases} -1, & \text{if } l \equiv 1, 2 \pmod 4 \text{ and } e \in \mathcal{T}_8 - \mathcal{T}_4, \\ 1, & \text{otherwise} \end{cases}$$

(see section 4.4).

We now prove the key lemma of the paper [Wójcik, 1998] (Lemma 9). The proof of this lemma is based on the following well known elementary fact

LEMMA 8 [Washington, 1997, Lemmas 5.19 and 5.21], [Wójcik, 1998, Lemma 4] *For integers b, $h \geq 0$, we have*

$$\sum_{a=0}^{b} a^h (-1)^a \binom{b}{a} = \begin{cases} 0, & \text{if } h < b, \\ (-1)^b b!, & \text{if } h = b, \\ b! \times (\text{integer}), & \text{if } h > b. \end{cases}$$

PROOF. The first and the third identities are proved in [Washington, 1997]. As for the second equation, it follows in the same manner as the third (for details, see the proof of Lemma 5.21, [Washington, 1997]). ∎

LEMMA 9 ([Wójcik, 1998, Lemma 5]) *Let $m \geq 1$ be an integer and let $K = \{-m + 2, -m + 3, \ldots, 1\}$. Then in the notation of section 4.6 we have*

$$c(u_n) = 3m - 1 + \operatorname{ord}_2 ((m-1)!).$$

PROOF. In what follows, we denote by r the right hand side of the above equation. The proof of Lemma 9 is divided into the same two steps as the proof of Lemma 7.

(i) We find 2-adic integers $x_{k,e}$ ($k \in K, e \in \mathcal{T}_8$) such that the numbers $u_n(x)$ ($n \geq 0$) are divisible by 2^r (which proves that $c(u_n) \geq r$).

(ii) Assuming that the infinite system of congruences

$$u_n(x) \equiv 0 \,(\mathrm{mod}\, 2^\rho), \quad n \geq 0 \qquad (39)$$

with $\rho = r + 1$ has a solution $x = \{x_{k,e}\}$, we prove that all the numbers $x_{k,e}$ must be even (which proves that $c(u_n) \leq r$).

We shall deduce that all the numbers $x_{k,e}$ are even from (39) with n odd, $1 \leq n \leq 4m - 1$. Analysis similar to that in the proof of Lemma 7 shows that it suffices to focus on the case $\mathrm{sgn}\, e = (-1)^k$. Namely, as previously, we observe that for n even

$$2u_n = u_{n+1} + \tilde{u}_{n+1},$$

where the $\tilde{u}_{n+1}$ comes from u_{n+1} by replacing $x_{k,-4}$ (resp. $x_{k,1}$, $x_{k,-8}$ or $x_{k,8}$) by $x_{k,1}$ (resp. $x_{k,-4}$, $x_{k,8}$ or $x_{k,-8}$). Thus, coupling the above equation with (39) for $\rho = r + 1$ we deduce that

$$\tilde{z}_{n+1}(x) \equiv 0 \,(\mathrm{mod}\, 2^{r+1}).$$

Suppose now that $x_{k,e}$ are even if $k \in K$, $e \in \mathcal{T}_8$ and $\mathrm{sgn}\, e = (-1)^k$. Then by the same reasoning as in the case of $\mathrm{sgn}\, e = (-1)^k$, applied to the numbers $\tilde{u}_{n+1}(x)$ we obtain that the $x_{k,e}$ are also even if $\mathrm{sgn}\, e \neq (-1)^k$.

Now it is convenient to start with step (ii).

Substituting in (39) $n = 2l + 1$ and

$$x_k = x_{2-k,e} + x_{2-k,e'} \quad \text{and} \quad y_k = x_{2-k,e} - x_{2-k,e'},$$

where $k = 1, 2, \ldots, m$ and $e \in \mathcal{T}_4$, $e' \notin \mathcal{T}_4$, we can rewrite the system (39) with $0 \leq l \leq 2m - 1$ in the form of two subsystems of congruences modulo 2^ρ with $\rho = r$

$$\left. \begin{aligned} &\sum_{j=0}^{m} \alpha_{ij} x_j \equiv 0 \,(\mathrm{mod}\, 2^\rho) \\ &\alpha_{ij} = (2(-1)^{i+1}(2i - 1) - 1)^{j-1} \\ &i, j = 1, 2, \ldots, m \end{aligned} \right\} \qquad (40)$$

and

$$\left.\begin{array}{c} \sum_{j=1}^{m} \beta_{ij} y_j \equiv 0 \, (\bmod \, 2^{\rho}) \\[2mm] \beta_{ij} = (2(-1)^i(2i-1)-1)^{j-1} \\[2mm] i,j = 1,2,\ldots,m \end{array}\right\} \qquad (41)$$

where in the former subsystem we consider congruences of (39) with $l \equiv 0$ or $3 \, (\bmod \, 4)$ and in the latter one those with $l \equiv 1$ or $2 \, (\bmod \, 4)$.

We shall rewrite the above systems in the equivalent triangular forms applying Lemma 8, more precisely the identity

$$\sum_{a=0}^{b} (8(c-a)+f)^h (-1)^a \binom{b}{a} = \begin{cases} 0, & \text{if } h < b, \\ 8^b b!, & \text{if } h = b, \qquad (42) \\ 8^b b! \times (\text{integer}), & \text{if } h > b, \end{cases}$$

which follows from the lemma for any $c \in \mathbb{Z}$ after an easy computation. We shall use formula (42) with $0 \le b, \, h \le m-1, \, f = 1$ or -3 and $c = [b/2]$.

We first observe that for $0 \le b \le m-1$ there are two one-one correspondences θ_1, resp. θ_2 between integers $k \in [1, b+1]$ and $a \in [0, b]$ such that

$$2(-1)^{k+1}(2k-1)-1 = 8([b/2]-a)+1,$$
$$\text{resp. } 2(-1)^k(2k-1)-1 = 8([b/2]-a)-3.$$

These correspondences are of the form

$$\theta_1(k) = [b/2] + (-1)^k[k/2], \quad \text{resp. } \theta_2(k) = [b/2] + (-1)^{k+1}[k/2].$$

Moreover we have

$$\theta_1^{-1}(a) = 2a - 2[b/2], \quad \text{resp. } \theta_2^{-1}(a) = 2a - 2[b/2] + 1,$$

if $a \ge [b/2]$ and

$$\theta_1^{-1}(a) = 2[b/2] - 2a + 1, \quad \text{resp. } \theta_2^{-1}(a) = 2[b/2] - 2a,$$

if $a < [b/2]$.

Making use of these correspondences and formula (42) we can rewrite the systems (40) and (41) in the equivalent triangular forms

$$\left.\begin{array}{c} \lambda_{11}x_1 + \lambda_{12}x_2 + \lambda_{13}x_3 + \ldots + \lambda_{1,m-1}x_{m-1} + \lambda_{1,m}x_m \equiv 0 \\[1mm] \lambda_{22}x_2 + \lambda_{23}x_3 + \ldots + \lambda_{2,m-1}x_{m-1} + \lambda_{2,m}x_m \equiv 0 \\[1mm] \lambda_{33}x_3 + \ldots + \lambda_{3,m-1}x_{m-1} + \lambda_{3,m}x_m \equiv 0 \\[1mm] \cdots \qquad \cdots \qquad \cdots \\[1mm] \lambda_{m-1,m-1}x_{m-1} + \lambda_{m-1,m}x_m \equiv 0 \\[1mm] \lambda_{mm}x_m \equiv 0 \end{array}\right\} \qquad (43)$$

and

$$
\left.\begin{array}{r}
\mu_{11}y_1 + \mu_{12}y_2 + \mu_{13}y_3 + \cdots + \mu_{1,m-1}y_{m-1} + \mu_{1,m}y_m \equiv 0 \\[2mm]
\mu_{22}y_2 + \mu_{23}y_3 + \cdots + \mu_{2,m-1}y_{m-1} + \mu_{2,m}y_m \equiv 0 \\[2mm]
\mu_{33}y_3 + \cdots + \mu_{3,m-1}y_{m-1} + \mu_{3,m}y_m \equiv 0 \\[2mm]
\cdots \qquad \cdots \qquad \cdots \\[2mm]
\mu_{m-1,m-1}y_{m-1} + \mu_{m-1,m}y_m \equiv 0 \\[2mm]
\mu_{mm}\, y_m \equiv 0
\end{array}\right\}, \quad (44)
$$

where all the congruences are modulo 2^ρ (recall that $\rho = r$) and

$$
\lambda_{1j} = 1, \quad \mu_{1j} = (-3)^{j-1},
$$
$$
\lambda_{mm} = \mu_{mm} = 8^{m-1} \times (m-1)!,
$$
$$
\lambda_{ij}, \mu_{ij} = 8^{i-1} \times (i-1)! \times (\text{integer}),
$$

if $2 \le i \le m-1, 1 \le j \le m$.

To obtain the ith congruence of (43) (resp. (44)) we multiply the kth congruence of (40) (resp. (41)) for $1 \le k \le i$ through by $(-1)^a \binom{b}{a}$, where $b = i-1$ and $a = [b/2] + (-1)^k[k/2]$ (resp. $a = [b/2] + (-1)^{k+1}[k/2]$) and next, making use of identity (42), add up the first i congruences of each of the systems.

We observe that

$$
x_{k,e} = \frac{x_{2-k} + y_{2-k}}{2} \quad \text{and} \quad x_{k,e'} = \frac{x_{2-k} - y_{2-k}}{2},
$$

if $k \in K$, $e \in \mathcal{T}_4$ and $e' \notin \mathcal{T}_4$. Therefore to prove that $x_{k,e}$ and $x_{k,e'}$ are even it suffices to prove that x_k and y_k are divisible by 4 (recall that $\operatorname{sgn} e = \operatorname{sgn} e' = (-1)^k$).

By the last congruence of the system (43) (resp. (44)) (recall that $\rho = r$) we deduce that x_m (resp. y_m) must be divisible by 4. Further we proceed by induction. Assuming that $x_m, x_{m-1}, \ldots, x_t$ (resp. $y_m, y_{m-1}, \ldots, y_t$) are divisible by 4, we consider the $(t-1)$th congruence of (43) (resp. (44)). This congruence is of the form

$$
8^{t-2}(t-2)!x_{t-1} \equiv 4 \times 8^{t-2}(t-2)! \times (\text{integer}) \ (\operatorname{mod} 2^r)
$$

(resp. $8^{t-2}(t-2)!y_{t-1} \equiv 4 \times 8^{t-2}(t-2)! \times (\text{integer}) \ (\operatorname{mod} 2^r)$).

Consequently x_{t-1} and y_{t-1} are also divisible by 4 as required.

We now turn to step (i).

In order to prove that $c(u_n) \geq r$, we should find 2-adic integers $x_{k,e}$ not all even satisfying (39) with $\rho = r$.

We begin the construction by putting

$$x_{k,1} = -x_{k,-4} \text{ and } x_{k,8} = -x_{k,-8},$$

which implies immediately that $u_n(x) = 0$ if n is even. We next find 2-adic integers $x_{k,-4}$, $x_{k,-8}$ ($k \in K$) not all even satisfying (39) with $1 \leq n \leq 4m-1$ (n odd) and prove that the $x_{k,e}$ satisfy (39) with n odd, not smaller than $4m$.

As previously, by the same substituting we can rewrite (39) with $1 \leq n \leq 4m - 1$ (n odd) in the form of two subsystems (40) and (41) with $\rho = r - 1$, or in the equivalent triangular forms (43) and (44) with the same ρ.

Making use of (43) and (44) we construct, by induction, 2-adic integers $x_{k,-4}$, $x_{k,-8}$ ($k \in K$) not all not all even satisfying (39) with $1 \leq t \leq 4m - 1$ (t odd).

Appealing to the last congruence of (43) (resp. (44)) we start the construction by putting

$$x_m = 2 \text{ (resp. } y_m = 0).$$

This gives

$$x_{-m+2,-4} = x_{-m+2,-8} = 1$$

and not all $x_{k,e}$ are even as required.

Assuming that

$$x_m = 2, x_{m-1}, \ldots, x_t \text{ (resp. } y_m = 0, y_{m-1}, \ldots, y_t)$$

are already defined we shall define x_{t-1} (resp. y_{t-1}).

For this purpose it suffices to use the $(t-1)$th congruence of (43) (resp. (44)) (recall that $\rho = r - 1$), namely the congruence

$$8^{t-2}(t-2)! x_{t-1} \equiv 8^{t-2}(t-2)! \times \text{(integer)} \times x_t$$

$$+ \ldots + 8^{t-2}(t-2)! \times \text{(integer)} \times x_{m-1}$$

$$+ 2 \times 8^{t-2}(t-2)! \times \text{(integer)} \, (\text{mod } 2^{r-1})$$

(resp. $8^{t-2}(t-2)! y_{t-1} \equiv 8^{t-2}(t-2)! \times \text{(integer)} \times y_t$

$$+ \ldots + 8^{t-2}(t-2)! \times \text{(integer)} \times y_{m-1} \, (\text{mod } 2^{r-1})).$$

We define x_{t-1} (resp. y_{t-1}) by the congruence

$$x_{t-1} \equiv \text{(integer)} \times x_t + \ldots + \text{(integer)} \times x_{m-1} + 2 \times \text{(integer)} \, (\text{mod } 2^{\mu})$$

$$(\text{resp. } y_{t-1} \equiv \text{(integer)} \times y_t + \ldots + \text{(integer)} \times y_{m-1} \, (\text{mod } 2^{\mu})),$$

where

$$\mu = 3(m - t) + \mathrm{ord}_2\left(\frac{(m-1)!}{(t-2)!}\right) + 4\,.$$

Obviously the solutions x_k and y_k ($k \in K$) determine the required $x_{k,e}$. We need only prove that the determined $x_{k,e}$ satisfy (39) with $n \geq 4m$ (n odd). To do this, we proceed by induction on n, $n \geq 4m$ using formula (42) with $b \geq m$. The details of the induction are left to the reader. ∎

REMARK Analysis similar to that in the proof of Lemma 9 shows that the equation of the lemma holds for an arbitrary set K consisting of m consecutive integers.

4.8   Linear Combinations of $\mathcal{L}_{k,e}(\xi)$

Let $\xi \neq 1$ be a primitive Nth root of unity, where N is an odd natural number. Given 2-adic integers $x_{k,e}$ not all even, defined on a nonempty subset L of K, our purpose is to evaluate the linear combinations

$$\sum_{(k,e)\in K\times T_8} x_{k,e}\mathcal{L}_{k,e}(\xi)$$

modulo powers of 2. Combining the obtained congruences with Lemma 2 we shall derive some new congruences for linear combinations of the values of 2-adic L-functions $L_2(k, \chi\omega^{1-k})$ with arbitrary 2-adic integral coefficients, where χ is a primitive quadratic Dirichlet character.

LEMMA 10 ([Urbanowicz and Wójcik, 1995/1996, Lemma 5]) *Set* $K = \{-1, 0, 1, 2\}$. *Let* $x_{k,e}$ ($k \in K$, $e \in T_8$) *be 2-adic integers not all even defined on a nonempty subset* L *of* K. *Then in the notation of section 4.6 we have*

$$\Lambda := \sum_{(k,e)\in L\times T_8} x_{k,e}\mathcal{L}_{k,e}(\xi) \equiv 0\,(\mathrm{mod}\,2^\lambda),$$

where 2^λ *is the greatest common divisor of*

$$2^{c(L)}\text{ and }z_n\,,\ 0 \leq n \leq \max(2c(L) - 4, 2)\,,$$

and

$$c(L) = 12,\ 9,\ 5,\ resp.\ 2,$$

if $\mathrm{card}(L) = 4,\ 3,\ 2,\ resp.\ 1,\ unless\ L = \{-1, 1\}\ or\ \{0, 2\},\ in\ which\ cases$

$$c(L) = 6.$$

PROOF. The formula for $c(L)$ is the hypothesis of Lemma 7. In order to prove the congruence of Lemma 10 we shall apply Lemma 5. We have

$$\Lambda = (x_{-1,1} + x_{-1,8} + x_{0,-4} + x_{0,-8} + x_{1,1} + x_{1,8} + x_{2,-4} + x_{2,-8})w_1$$
$$+ (x_{-1,-4} + x_{-1,-8} + x_{0,1} + x_{0,8} + x_{1,-4} + x_{1,-8} + x_{2,1} + x_{2,8})w_{-1}$$
$$+ \sum_{k=1}^{\infty} \Big(4^k x_{-1,1} - 2^k(2k-1)x_{-1,8} + 2^k x_{0,-8}$$
$$+ \frac{4^k}{2k+1}x_{1,1} + \frac{2^k}{2k+1}x_{1,8} + \frac{16^k}{(2k+1)^2}\binom{2k}{k}^{-1}x_{2,-4}$$
$$+ \frac{16^k}{(2k+1)^2}\binom{2k}{k}^{-1}\sum_{l=0}^{k}\binom{2l}{l}2^{-3l}x_{2,-8}\Big)w_1^{2k+1}$$
$$+ \sum_{k=1}^{\infty}(-1)^k \Big(4^k x_{-1,-4} - 2^k(2k-1)x_{-1,-8} + 2^k x_{0,8}$$
$$+ \frac{4^k}{2k+1}x_{1,-4} + \frac{2^k}{2k+1}x_{1,-8} + \frac{16^k}{(2k+1)^2}\binom{2k}{k}^{-1}x_{2,1}$$
$$+ \frac{16^k}{(2k+1)^2}\binom{2k}{k}^{-1}\sum_{l=0}^{k}\binom{2l}{l}2^{-3l}x_{2,8}\Big)w_{-1}^{2k+1}.$$

Consequently, we obtain

$$\Lambda = \sum_{k=0}^{\infty}\frac{1}{(2k+1)^2}\Big(z_{2k}w_1^{2k+1} + z_{2k+1}v_{2k+1}\Big),$$

where

$$v_{2k+1} = v_{2k+1}(\xi) = \frac{1}{2}\big((-1)^k w_{-1}^{2k+1} - w_1^{2k+1}\big)$$

is a 2-adic integer and the numbers z_n ($n \geq 0$) are defined in section 4.6.

Denote by 2^λ the highest power of 2 dividing all the z_n ($n \geq 0$). Then the lemma follows from the definition of the sequence (z_n). If $n \geq 2c(L) - 1$, we have by definition

$$z_n \equiv 0 \,(\bmod\, 2^{c(L)})$$

and 2^λ is the greatest common divisor of

$$2^{c(L)} \text{ and } z_n, \quad 0 \leq n \leq 2c(L) - 2.$$

Now it suffices to use the congruences

$$z_{2l+1} \equiv 2^{l+1}\gamma \,(\bmod\, 2^{l+2}), \quad \tilde{z}_{2l+1} \equiv 2^{l+1}\tilde{\gamma} \,(\bmod\, 2^{l+2}),$$

$$z_{2l} \equiv 2^l(\gamma + \tilde{\gamma}) \,(\mathrm{mod}\, 2^{l+1}),$$

where $l \geq 1$ and

$$\gamma = x_{-1,-8} + x_{0,8} + x_{1,-8} + x_{2,8}$$

(recall the definitions of z_n and $\tilde{z}_n$ in the proof of Lemma 7). In view of these congruences we have

$$z_{2c(L)-2} \equiv 2^{c(L)-1}(\gamma + \tilde{\gamma}) \,(\mathrm{mod}\, 2^{c(L)}),$$

$$z_{2c(L)-4} \equiv 2^{c(L)-2}(\gamma + \tilde{\gamma}) \,(\mathrm{mod}\, 2^{c(L)-1}),$$

$$z_{2c(L)-3} \equiv 2^{c(L)-1}\gamma \,(\mathrm{mod}\, 2^{c(L)}),$$

$$z_{2c(L)-5} \equiv 2^{c(L)-2}\gamma \,(\mathrm{mod}\, 2^{c(L)-1}),$$

provided $c(L)>2$. Hence we may ignore $z_{2c(L)-2}$ and $z_{2c(L)-3}$ if $c(L)>2$. ∎

LEMMA 11 ([Wójcik, 1998, Lemma 6]) *Let $m \geq 1$ be an integer and let $K = \{-m+2, -m+3, \ldots, 1\}$. Let $x_{k,e}$ ($k \in K$, $e \in T_8$) be integers in $\mathbb{C}_2$ not all even. Then in the notation of section 4.7 we have*

(i)

$$\sum_{(k,e)\in K\times T_8} x_{k,e}\mathcal{L}_{k,e}(\xi) \equiv 0 \,(\mathrm{mod}\, 2^\lambda),$$

where 2^λ is the greatest common divisor of

$$2^{c(u_n)} \text{ and } u_n, \ 0 \leq n \leq 4m - 1,$$

(ii) *for an arbitrary integer s*

$$\Lambda := \sum_{(k,e)\in K\times T_8} x_{k,e}\mathcal{L}_{k+s,e}(\xi) \equiv 0 \,(\mathrm{mod}\, 2^\lambda).$$

PROOF. We prove congruence (ii). Congruence (i) is a special case of this congruence for $s = 0$. By definition of the numbers $W_{k,e}(r)$ and Lemma 6 (with $\alpha = (-1)^{k+s+1}\mathrm{sgn}(e)$) we have

$$\Lambda = \sum_{(k,e)\in K\times T_8} x_{k,e} \sum_{r=0}^{\infty} \frac{\alpha^r W_{k+s,e}(r)}{2r+1} w_\alpha^{2r+1}$$

$$= \sum_{r=0}^{\infty} \frac{1}{2r+1} \sum_{(k,e)\in K\times T_8} x_{k,e}$$

$$\times \sum_{l=0}^{r} \alpha^r (-1)^{l(k+s+1)}(2l+1)^{1-k-s}\gamma_{l,e}\binom{2r+1}{r-l} w_\alpha^{2r+1}.$$

Consequently, we obtain

$$\Lambda = \sum_{r=0}^{\infty} g_{2r} w_{(-1)^s}^{2r+1} + \sum_{r=0}^{\infty} g_{2r+1} v_{2r+1}, \tag{45}$$

where

$$g_{2r+\varrho} = \frac{1}{2r+1} \sum_{l=0}^{r} \binom{2r+1}{r-l} (-1)^{ls} (2l+1)^{-s} u_{2l+\varrho}$$

and

$$v_{2r+1} = \frac{(-1)^s}{2} \left((-1)^r w_{-1}^{2r+1} - w_1^{2r+1} \right),$$

for $r \geq 0$ and $\varrho \in \{0,1\}$.

Note that both the series on the right hand side of equation (45) converge 2-adically because

$$g_{2r+\varrho} = \frac{2^\varrho}{2r+1} \sum_{(k,e) \in K \times T_8} W_{k+s,e}(r) x_{k,e}$$

and by (28) $\mathrm{ord}_2\left(W_{k+s,e}(r)\right) \geq r$.

By definition the numbers $g_{2r+\varrho}$ and v_{2r+1} are 2-adic integers. Moreover, it is easily seen that

$$c(g_n) = c(u_n).$$

Indeed, if for 2-adic integers $x_{k,e}$ not all even we have

$$u_n(x) \equiv 0 \,(\bmod \, 2^\nu), \quad n \geq 0 \tag{46}$$

with $\nu = c(u_n)$ then by definition of the sequence (g_n)

$$g_n(x) \equiv 0 \,(\bmod \, 2^\nu), \quad n \geq 0, \tag{47}$$

and so $c(g_n) \geq c(u_n)$. In order to prove the inequality $c(g_n) \leq c(u_n)$, we note that by definition there exist 2-adic integers $\{x_{k,e}\}$ not all even such that congruences (47) with $\nu = c(g_n)$ hold. Then congruences (46) with $\nu = c(g_n)$ follow by a simple induction on n from the obvious identity

$$u_n = (-1)^{s(n-\varrho)/2}(n-\varrho+1)^{s+1} g_n$$
$$- \sum_{0 \leq l \leq (n-\varrho-2)/2} \binom{n-\varrho-1}{(n-\varrho-2l)/2} (-1)^{sl}(2l+1)^{-s} u_{2l+\varrho},$$

where $\varrho = 0$ if n is even, and $\varrho = 1$ if n is odd.

Thus in order to prove Lemma 11 it suffices to prove that congruences (47) with $\nu = \lambda$ hold for any $n \geq 4m$. Analysis similar to that in the proof of Lemma 9 shows that if congruences (46) hold for $n \leq 4m - 1$ then the congruences also hold for $n \geq 4m$. In the same manner as in the proof of Lemma 9 we first prove this for n odd, and then for n even.   ∎

COROLLARY *Let $m \geq 1$ be an integer and let K be an arbitrary set consisting of m consecutive integers. Then we have*

$$\sum_{(k,e) \in K \times \mathcal{T}_8} x_{k,e} \mathcal{L}_{k,e}(\xi) \equiv 0 \, (\bmod \, 2^\lambda),$$

where 2^λ is the greatest common divisor of

$$2^c \text{ and } u_n, \; 0 \leq n \leq 4m - 1,$$

and

$$c = 3m - 1 + \mathrm{ord}_2((m - 1)!).$$

PROOF. The corollary is an immediate consequence of Lemmas 11(i), (ii) and 9. Note that the u_n in the statement of the corollary differ from the u_n given in the statement of Lemma 11(i) by an odd factor.   ∎

5.   LINEAR CONGRUENCE RELATIONS

5.1   *Congruences of the Gras-Uehara Type*

In this section we give a further generalization of the Gras-Uehara type congruence for linear combinations of the values of 2-adic L-functions $L_2(k, \chi\omega^{1-k})$, where χ is a quadratic Dirichlet character. We restrict our attention to the following cases.

(a) When k is taken over an arbitrary nonempty subset L of the set $K = \{-1, 0, 1, 2\}$ (see [Urbanowicz and Wójcik, 1995/1996]).

(b) When k is taken over an arbitrary finite set of consecutive integers (see [Wójcik, 1998]).

It appears to be still an open problem to find the Gras-Uehara type congruence when k is taken over any finite subset of the rational integers.

Let d be an odd fundamental discriminant and let $m > 1$ be a natural number. Throughout this section let Ψ, $\Theta \colon \mathbb{N} \to \mathbb{C}_2$ be multiplicative functions such that $\Psi(s) \equiv \Theta(s) \equiv 1 \, (\bmod \, 2)$ if $s \mid m$. Let $\delta_{d,1}$ denote the Kronecker delta function, that is, $\delta_{d,1} = 1$ if $d = 1$ and is zero otherwise. For $k \in \mathbb{Z}$ and $e \in \mathcal{T}_8$ we set

(i) if $d = e = k = 1$

$$L_2^{[m,\Theta]}(k, \chi_{ed}\omega^{1-k}) = 0,$$

(ii) otherwise

$$L_2^{[m,\Theta]}(k, \chi_{ed}\omega^{1-k})$$

$$= \Big(\prod_{\substack{p \mid m \\ p \text{ prime}}} (1 - \chi_{ed}(p)\Theta(p)p^{1-k}) - \delta_{d,1} \prod_{\substack{p \mid m \\ p \text{ prime}}} (1 - \Theta(p))\Big) L_2(k, \chi_{ed}\omega^{1-k}).$$

In the case when $\Theta(s) = 1$ for $s \mid m$, write

$$L_2^{[m]}(k, \chi_{ed}\omega^{1-k}) = L_2^{[m,\Theta]}(k, \chi_{ed}\omega^{1-k}),$$

that is, $L_2^{[m]}(k, \chi_{ed}\omega^{1-k}) = 0$ if $e = d = k = 1$, and

$$L_2^{[m]}(k, \chi_{ed}\omega^{1-k}) = \prod_{\substack{p \mid m \\ p \text{ prime}}} (1 - \chi_{ed}(p)p^{1-k}) L_2(k, \chi_{ed}\omega^{1-k}),$$

otherwise.

The following theorem gives a simpler version (when $\Theta(s) = 1$ for $s \mid m$) of the Gras-Uehara type congruences for class numbers and the orders of K_2-groups of the integers of appropriate quadratic fields.

THEOREM 45 [Urbanowicz and Wójcik, 1995/1996, Main Theorem] *Let $m > 1$ be a squarefree odd natural number having $\nu := \nu(m)$ prime factors and let $\Psi : \mathbb{N} \to \mathbb{C}_2$ be a multiplicative function satisfying $\Psi(s) \equiv 1 \pmod 2$, if $s \mid m$. Set $K = \{-1, 0, 1, 2\}$. Let L be a nonempty subset of K and let $x = \{x_{k,e}\}_{(k,e) \in K \times \mathcal{T}_8}$ be a set of 2-adic integers not all even defined on L. Write*

$$\Lambda_1(m) = \begin{cases} -(\log_2 m)/2, & \text{if } m \text{ is a prime number,} \\ 0, & \text{otherwise.} \end{cases}$$

Then the number

$$\Lambda(x, m, \Psi) := \sum_{(k,e) \in K \times \mathcal{T}_8} x_{k,e} \sum_{d \in \mathcal{T}_m} \Psi(|d|) L_2^{[m]}(k, \chi_{ed}\omega^{1-k}) + x_{1,1}\Lambda_1(m)$$

is a 2-adic integer divisible by $2^{\nu+\lambda}$, where 2^λ is the greatest common divisor of

$$2^{c(L)} \text{ and } z_n, \ 0 \leq n \leq \max(2c(L) - 4, 2),$$

and

$$c(L) = 12, 9, 5, \text{ resp. } 2,$$

if $\mathrm{card}(L) = 4,\ 3,\ 2,$ *resp.* $1,$ *unless* $L = \{-1, 1\}$ *or* $\{0, 2\},$ *in which cases*

$$c(L) = 6.$$

PROOF. As usual we denote by ζ_n a primitive nth root of unity in $\mathbb{C}_2$. For any $d \in \mathcal{T}_m$ assume that

$$\zeta_{|d|} = \prod_{\substack{p \mid d \\ p\,\text{prime}}} \zeta_p .$$

By Lemma 2 and a Möbius inversion argument, we obtain

$$\Lambda(x, m, \Psi) = (-1)^\nu \sum_{(k,e) \in K \times \mathcal{T}_8} x_{k,e} \sum_{d \in \mathcal{T}_m} \Psi(|d|)\mu(|d|)$$

$$\times \tau(\chi_d, \zeta_{|d|})|d|^{-1} \sum_{a=1}^{m}{}' \chi_d(a)\mathcal{L}_{k,e}(\zeta_m^a)$$

$$= (-1)^\nu \sum_{a=1}^{m}{}' \Big(\sum_{\substack{k \in K \\ e \in \mathcal{T}_8}} x_{k,e}\mathcal{L}_{k,e}(\zeta_m^a) \Big)$$

$$\times \Big(\sum_{d \in \mathcal{T}_m} \Psi(|d|)\mu(|d|)\tau(\chi_d, \zeta_{|d|})|d|^{-1}\chi_d(a) \Big)$$

$$= (-1)^\nu \sum_{a=1}^{m}{}' \Big(\sum_{\substack{k \in K \\ e \in \mathcal{T}_8}} x_{k,e}\mathcal{L}_{k,e}(\zeta_m^a) \Big)$$

$$\times \Big(\prod_{\substack{p \mid m \\ p\,\text{prime}}} (1 - \Psi(p)\tau(\chi_{p^*}, \zeta_p)p^{-1}\chi_{p^*}(a)) \Big),$$

where $p^* = (-1)^{(p-1)/2}p$.

Therefore in view of Lemma 5 the numbers $\Lambda(x, m, \Psi)$ are 2-adic integers. Moreover by Lemma 10 and

$$\Psi(p)\tau(\chi_{p^*})p^{-1}\chi_{p^*}(a) - 1 \equiv 1 + \zeta_p + \ldots + \zeta_p^{p-1} \equiv 0 \,(\bmod\, 2),$$

the $\Lambda(x, m, \Psi)$ are divisible by $2^{\nu+\lambda}$. The rest of the theorem follows at once from Lemma 7. ∎

Wójcik [Wójcik, 1998] has extended the congruence of Theorem 45.

THEOREM 46 ([Wójcik, 1998, Theorem]) *Let* $m > 1$ *be a squarefree odd natural number having* ν *prime factors and let* $\Psi : \mathbb{N} \to \mathbb{C}_2$ *be a multiplicative function with odd values at divisors of* m. *Let* K *denote a finite set consisting of consecutive integers and write* $\delta = \mathrm{card}(K)$. *Let* $x = \{x_{k,e}\}_{(k,e)\in K\times\mathcal{T}_8}$ *be a set of 2-adic integers not all even. Write* $\Lambda_1(m) = -(\log_2 m)/2$ *if* m *is a prime number, and* $\Lambda_1(m) = 0$ *otherwise. Then the number*

$$\sum_{(k,e)\in K\times\mathcal{T}_8} x_{k,e} \sum_{d\in\mathcal{T}_m} \Psi(|d|) L_2^{[m]}(k, \chi_{ed}\omega^{-1}) + x_{1,1}\Lambda_1(m)$$

is a 2-adic integer divisible by $2^{\nu+\lambda}$, *where* 2^λ *is the greatest common divisor of*

$$2^{c(\delta)} \text{ and } u_n, \ 0 \le n \le 4\delta - 1$$

and

$$c(\delta) = 4\delta - 1 - s_2(\delta) - \mathrm{ord}_2(\delta),$$

where $s_2(t)$ *denotes the sum of the digits of the 2-adic expansion of* t.

PROOF. Set $c(\delta) = c(u_n)$. We apply the same arguments as in the proof of Theorem 45. Now the theorem is a consequence of Lemmas 2, 6, 9 and Corollary to Lemma 11. Here we have used the obvious identities

$$\mathrm{ord}_2((t-1)!) = t - 1 - s_2(t-1) = t - s_2(t) - \mathrm{ord}_2(t). \qquad \blacksquare$$

REMARK In [Urbanowicz and Wójcik, 1995/1996] (and [Wójcik, 1998]) there is a small error in the formulation of the main theorem. The factor $(-1)^{k+1}$ occuring in the $\Lambda_2(x, m)$ (denoted by $\Lambda(x, m, \Psi)$ in Theorem 45) should be deleted. See Remark after Lemma 1.

5.2 Further Extensions of Theorems 45 and 46

Theorems 45 and 46 are special cases of more general results. Let Φ, Θ : $\mathbb{N} \to \mathbb{C}_2$ be multiplicative functions. Let $s > 1$ be a squarefree natural number. Following [Uehara, 1990, (3.1)] we have

$$\sum_{t\,|\,s} \Theta(t) \prod_{\substack{p\,|\,(s/t)\\p\,\mathrm{prime}}} (1-\Theta(p)) \prod_{\substack{p\,|\,t\\p\,\mathrm{prime}}} (1-\Phi(p)) = \prod_{\substack{p\,|\,s\\p\,\mathrm{prime}}} (1-\Phi(p)\Theta(p)). \quad (48)$$

(To prove the above identity we proceed by a simple induction on the number of prime factors of s.)

THEOREM 47 ([Urbanowicz, 1999, Theorem 1]) *Let* $m > 1$, ν, K, $\Psi : \mathbb{N} \to \mathbb{C}_2$, $x = \{x_{k,e}\}_{(k,e)\in K\times\mathcal{T}_8}$ *have the same meaning as in Theorem 45 (resp. 46). Let* $\Theta : \mathbb{N} \to \mathbb{C}_2$ *be a multiplicative function such that* $\Theta(s) = 1 \ (\mathrm{mod}\ 2)$, *if* $s\,|\,m$. *Set*

$$\Lambda_1(m, \Theta) = -\frac{1}{2} \sum_{\substack{p\,|\,m\\p\,\mathrm{prime}}} \Theta(p) \log_2 p \prod_{\substack{q\,|\,(m/p)\\q\,\mathrm{prime}}} (1-\Theta(q)).$$

Then the number

$$\Lambda(x, m, \Psi, \Theta)$$

$$:= \sum_{(k,e) \in K \times \mathcal{T}_8} x_{k,e} \sum_{d \in \mathcal{T}_m} \Psi(|d|) L_2^{[m,\Theta]}(k, \chi_{ed}\omega^{1-k}) + x_{1,1}\Lambda_1(m, \Theta)$$

is a 2-adic integer divisible by $2^{\nu+\lambda}$, where λ has the same meaning as in Theorem 45 if $K = \{-1, 0, 1, 2\}$ and $\{x_{k,e}\}$ is defined on $\emptyset \neq L \subseteq K$ (resp. Theorem 46 if K is a set of consecutive integers).

PROOF. Write

$$\Lambda_2(x, m, \Theta) = \prod_{\substack{p \mid m \\ p \text{ prime}}} (1 - \Theta(p)) \sum_{\substack{(k,e) \in K \times \mathcal{T}_8 \\ (k,e) \neq (1,1)}} x_{k,e} L_2(k, \chi_e \omega^{1-k}).$$

and

$$L_2'(k, \chi_{ed}\omega^{1-k}) = \begin{cases} 0, & \text{if } e = d = k = 1, \\ L_2(k, \chi_{ed}\omega^{1-k}), & \text{otherwise}. \end{cases}$$

We proceed in the same manner as in the proof of Theorem 45 (resp. Theorem 46). Making use of (48), for any multiplicative function $\Phi : \mathbb{N} \to \mathbb{C}_2$ and fixed u, s with $u \mid s$ we obtain

$$\Theta^{-1}(u) \sum_{u \mid t \mid s} \Theta(t) \prod_{\substack{p \mid (s/t) \\ p \text{ prime}}} (1 - \Theta(p)) \prod_{\substack{p \mid (t/u) \\ p \text{ prime}}} (1 - \Phi(p)) = \prod_{\substack{p \mid (s/u) \\ p \text{ prime}}} (1 - \Phi(p)\Theta(p)).$$

This follows from (48) by a simple induction on the number of prime factors of s/u. We observe that for any functions f and g

$$\sum_{d \mid m} f(d) \sum_{c \mid d} g(c) h(d, c) = \sum_{d \mid m} g(d) \sum_{d \mid c \mid m} f(c) h(c, d).$$

Therefore we have

$$\Lambda(x, m, \Psi, \Theta) - x_{1,1}\Lambda_1(m, \Theta) + \Lambda_2(x, m, \Theta)$$

$$= \sum_{(k,e) \in K \times \mathcal{T}_8} x_{k,e} \sum_{d \in \mathcal{T}_m} \Psi(|d|) \prod_{\substack{p \mid (m/d) \\ p \text{ prime}}} (1 - \Theta(p)\chi_{ed}(p)p^{1-k}) L_2'(k, \chi_{ed}\omega^{1-k})$$

$$= \sum_{(k,e) \in K \times \mathcal{T}_8} x_{k,e} \sum_{d \in \mathcal{T}_m} \Psi(|d|)\Theta^{-1}(|d|) L_2'(k, \chi_{ed}\omega^{1-k})$$

$$\times \sum_{\substack{c \in \mathcal{T}_m \\ d \in \mathcal{T}_c}} \Theta(|c|) \prod_{\substack{p \mid (m/c) \\ p \text{ prime}}} (1 - \Theta(p)) \prod_{\substack{p \mid (c/d) \\ p \text{ prime}}} (1 - \chi_{ed}(p)p^{1-k})$$

$$= \sum_{(k,e) \in K \times \mathcal{T}_8} x_{k,e} \sum_{d \in \mathcal{T}_m} \Theta(|d|) \prod_{\substack{p \,|\, (m/d) \\ p \,\text{prime}}} (1 - \Theta(p))$$

$$\times \sum_{c \in \mathcal{T}_d} \Psi(|c|) \Theta^{-1}(|c|) \prod_{\substack{p \,|\, (d/c) \\ p \,\text{prime}}} (1 - \chi_{ec}(p) p^{1-k}) L_2'(k, \chi_{ec} \omega^{1-k}).$$

Consequently appealing to Lemma 2 we obtain

$$\Lambda(x, m, \Psi, \Theta) = \sum_{1 \neq d \in \mathcal{T}_m} \Theta(|d|) \mu(|d|) \prod_{\substack{p \,|\, (m/d) \\ p \,\text{prime}}} (1 - \Theta(p))$$

$$\times \sum_{a=1}^{|d|}{}' \Big(\sum_{\substack{k \in K \\ e \in \mathcal{T}_8}} x_{k,e} \mathcal{L}_{k,e}(\zeta_{|d|}^a) \Big)$$

$$\times \Big(\sum_{c \in \mathcal{T}_d} \mu(|c|) \Psi(|c|) \Theta^{-1}(|c|) \tau(\chi_c, \zeta_{|c|}) |c|^{-1} \chi_c(a) \Big)$$

$$= \sum_{1 \neq d \in \mathcal{T}_m} \Theta(|d|) \mu(|d|) \prod_{\substack{p \,|\, (m/d) \\ p \,\text{prime}}} (1 - \Theta(p)) \sum_{a=1}^{|d|}{}' \Big(\sum_{\substack{k \in K \\ e \in \mathcal{T}_8}} x_{k,e} \mathcal{L}_{k,e}(\zeta_{|d|}^a) \Big)$$

$$\times \Big(\prod_{\substack{p \,|\, d \\ p \,\text{prime}}} \big(1 - \tau(\chi_{p^*}, \zeta_p) p^{-1} \Psi(p) \Theta^{-1}(p) \chi_{p^*}(a) \big) \Big),$$

where $p^* = (-1)^{(p-1)/2} p$ and $\zeta_{|d|} = \prod_{\substack{p \,|\, d \\ p \,\text{prime}}} \zeta_p$.

Now Theorem 47 follows from Lemma 10 when $K = \{-1, 0, 1, 2\}$ or from Lemma 11 when K is a set of consecutive integers. ∎

5.3 Applications

In this section we shall look more closely at some special cases of Theorems 45 and 47. When $L = \{0, 1\}$ these theorems give the congruences of Gras and Uehara for class numbers of quadratic fields which are modulo $2^{\nu+\lambda}$, where $\lambda \leq 5$. When $L = \{-1, 0\}$ (resp. $L = \{0\}$) we obtain congruences for the same objects as those in [Urbanowicz, 1990b] (resp. [Hardy and Williams, 1986]). The obtained congruences are modulo $2^{\nu+\lambda}$, where $\lambda \leq 5$ (resp. $\lambda \leq 2$). When $2 \in L$ the congruences implied by Theorems 45 and 47 are quite new and especially interesting. They produce, via a 2-adic version of the Lichtenbaum conjecture, some new congruences for the conjectured orders of K_2-groups of the integers of imaginary quadratic fields. We present these congruences in a general form in Theorem 48.

For the discriminant $\mathcal{D}$ of a quadratic field, we write

$$
H(\mathcal{D}) = \begin{cases} 2w^{-1}(1 - \chi_{\mathcal{D}}(2))h(\mathcal{D}), & \text{if } \mathcal{D} < 0, \\ (2 - \chi_{\mathcal{D}}(2))\mathcal{D}^{-1/2}h(\mathcal{D})\log_2 \varepsilon_{\mathcal{D}}, & \text{if } \mathcal{D} > 1, \end{cases}
$$

and

$$
K_2(\mathcal{D}) = \begin{cases} -24w_2^{-1}(1 - \chi_{\mathcal{D}}(2)2)k_2(\mathcal{D}), & \text{if } \mathcal{D} > 1, \\ (4 - \chi_{\mathcal{D}}(2))|\mathcal{D}|^{-3/2}R_{2,2}(\mathcal{D})k_2(\mathcal{D}), & \text{if } \mathcal{D} < 0, \end{cases}
$$

where $w = w(\mathcal{D})$, $w_2 = w_2(\mathcal{D})$ are defined in section 8, Chapter I, and $\varepsilon_{\mathcal{D}}$ (resp. $R_{2,2}(\mathcal{D})$) denotes the fundamental unit (resp. the second 2-adic regulator) of a corresponding quadratic field with discriminant $\mathcal{D}$. We have

$$
L_2(k, \chi_{\mathcal{D}}\omega^{1-k}) = H(\mathcal{D}) \text{ (resp. } L_2(k, \chi_{\mathcal{D}}\omega^{1-k}) = \frac{1}{2}K_2(\mathcal{D}) \text{)},
$$

if $k = 0$, $\mathcal{D} < 0$ or $k = 1$, $\mathcal{D} > 1$ (resp. $k = -1$, $\mathcal{D} > 1$ or $k = 2$, $\mathcal{D} < 0$). Making use of (6), (7), (8) and (9), we rewrite Theorem 47 with $K = \{-1, 0, 1, 2\}$ in the form:

THEOREM 48 [Urbanowicz, 1999, Theorem 3], cf. [Urbanowicz and Wójcik, 1995/1996] *Let $m > 1$ be a squarefree odd natural number having ν prime factors and let $\Theta, \Psi : \mathbb{N} \to \mathbb{C}_2$ be multiplicative functions such that $\Theta(s) \equiv \Psi(s) \equiv 1 \,(\bmod\, 2)$ if $s \,|\, m$. Set $K = \{-1, 0, 1, 2\}$ and let L be a nonempty subset of K. Given a set $x = \{x_{k,e}\}_{(k,e)\in K \times T_8}$ of 2-adic integers not all even defined on L, set*

$$
\Lambda = \Lambda_{-1} + \Lambda_0 + \Lambda_1 + \Lambda_2 + \Lambda'_{-1} + \Lambda'_1,
$$

where

$$
\Lambda_{-1} = \frac{1}{2} \sum_{e \in T_8} x_{-1,e} \sum_{\substack{d \in T_m \\ ed > 1}} \Psi(|d|)
$$

$$
\times \Big(\prod_{\substack{p \,|\, m \\ p \text{ prime}}} (1 - \chi_{ed}(p)\Theta(p)p^2) - \delta_{d,1} \prod_{\substack{p \,|\, m \\ p \text{ prime}}} (1 - \Theta(p)) \Big) K_2(ed),
$$

$$
\Lambda_0 = \sum_{e \in T_8} x_{0,e} \sum_{\substack{d \in T_m \\ ed < 0}} \Psi(|d|)
$$

$$\times \Bigg(\prod_{\substack{p \mid m \\ p\,\text{prime}}} (1 - \chi_{ed}(p)\Theta(p)p) - \delta_{d,1} \prod_{\substack{p \mid m \\ p\,\text{prime}}} (1 - \Theta(p)) \Bigg) H(ed),$$

$$\Lambda_1 = \sum_{e \in \mathcal{T}_8} x_{1,e} \sum_{\substack{d \in \mathcal{T}_m \\ ed > 1}} \Psi(|d|)$$

$$\times \Bigg(\prod_{\substack{p \mid m \\ p\,\text{prime}}} (1 - \chi_{ed}(p)\Theta(p)) - \delta_{d,1} \prod_{\substack{p \mid m \\ p\,\text{prime}}} (1 - \Theta(p)) \Bigg) H(ed),$$

$$\Lambda_2 = \frac{1}{2} \sum_{e \in \mathcal{T}_8} x_{2,e} \sum_{\substack{d \in \mathcal{T}_m \\ ed < 0}} \Psi(|d|)$$

$$\times \Bigg(\prod_{\substack{p \mid m \\ p\,\text{prime}}} (1 - \chi_{ed}(p)\Theta(p)p^{-1}) - \delta_{d,1} \prod_{\substack{p \mid m \\ p\,\text{prime}}} (1 - \Theta(p)) \Bigg) K_2(ed),$$

$$\Lambda'_{-1} = \frac{1}{12} x_{-1,1} \Bigg(\prod_{\substack{p \mid m \\ p\,\text{prime}}} (1 - \Theta(p)p^2) - \prod_{\substack{p \mid m \\ p\,\text{prime}}} (1 - \Theta(p)) \Bigg),$$

$$\Lambda'_1 = -\frac{1}{2} x_{1,1} \sum_{\substack{p \mid m \\ p\,\text{prime}}} \Theta(p) \log_2 p \prod_{\substack{q \mid (m/p) \\ q\,\text{prime}}} (1 - \Theta(q)).$$

Assume in the case when $2 \in L$ *that the 2-adic Lichtenbaum conjecture for imaginary quadratic fields holds. Then the number* Λ *is a 2-adic integer divisible by* $2^{\nu+\lambda}$, *where* λ *has the same meaning as in Theorem 45 and 47.*

5.4   Linear Congruence Relations between Class Numbers of Quadratic Fields. The Case $L = \{0, 1\}$

In [Hardy and Williams, 1986] a new type of linear congruence relating class numbers of imaginary quadratic fields was discovered. This congruence extends those proved by Pizer [Pizer, 1976] and Kenku [Kenku, 1977]. A general linear congruence relating class numbers and units both of real and imaginary quadratic fields was discovered by Gras [Gras, 1989]. Gras derived his congruence using 2-adic measure theory. See also [Gras, 1987, 1991/1992], [Desnoux, 1987, 1988] and [Pioui, 1990, 1992]. Uehara [Uehara, 1990] reproved Gras' congruence using elementary 2-adic arguments. Both Gras and Uehara used the 2-adic analogue of Dirichlet's class number formulae. In [Urbanowicz and Wójcik, 1995/1996] and [Wójcik, 1998] the authors indicated how Uehara's techniques may be used to obtain more general congruences between the values of 2-adic L-functions. Gras and Uehara's congruences are special cases of Theorems 45, 46, 47 and 48. These congruences cover those of [Kudo, 1975], [Lang and Schertz, 1976], [Kaplan, 1981], [Williams, 1981a, 1982], [Kaplan and Williams, 1982a,b], [Lang, H., 1985], and [Hikita, 1986a].

Following [Uehara, 1990], we summarize without proofs the relevant material on the case when $L = \{0, 1\}$. We leave it to the reader to show that the obtained congruences are consequences of Theorem 48. In this case we have $c(L) = 5$ and the congruences are modulo $2^{\nu+\lambda}$, where $\lambda \leq 5$.

THEOREM 49 ([Uehara, 1990, Theorem 1]) *Let $m > 1$ be an odd squarefree integer having ν prime factors, and let $\Theta, \Psi : \mathbb{N} \to \mathbb{C}_2$ be multiplicative functions such that $\Psi(s) \equiv \Theta(s) \equiv 1 \,(\bmod\, 2)$ for any divisor $s \,|\, m$. In the notation of Theorem 48, for any 2-adic integers $x_{0,e}$, $x_{1,e}$ ($e \in \mathcal{T}_8$) not all even we have*

$$\sum_{e \in \mathcal{T}_8} x_{0,e} \sum_{\substack{d \in \mathcal{T}_m \\ ed < 0}} \Psi(|d|)$$

$$\times \left(\prod_{\substack{p \,|\, m \\ p \text{ prime}}} (1 - \chi_{ed}(p)\Theta(p)p) - \delta_{d,1} \prod_{\substack{p \,|\, m \\ p \text{ prime}}} (1 - \Theta(p)) \right) H(ed)$$

$$+ \sum_{e \in \mathcal{T}_8} x_{1,e} \sum_{\substack{d \in \mathcal{T}_m \\ ed > 1}} \Psi(|d|)$$

$$\times \left(\prod_{\substack{p \,|\, m \\ p \text{ prime}}} (1 - \chi_{ed}(p)\Theta(p)) - \delta_{d,1} \prod_{\substack{p \,|\, m \\ p \text{ prime}}} (1 - \Theta(p)) \right) H(ed)$$

$$- \frac{1}{2} x_{1,1} \sum_{\substack{p \,|\, m \\ p \text{ prime}}} \Theta(p) \log_2 p \prod_{\substack{q \,|\, (m/p) \\ q \text{ prime}}} (1 - \Theta(q)) \equiv 0 \,(\bmod\, 2^{\nu+\lambda}),$$

where 2^λ is the greatest common divisor of the eight integers s_i ($0 \leq i \leq 7$) defined by

$$s_0 = x_{0,-8} + x_{0,-4} + x_{0,1} + x_{0,8} + x_{1,-8} + x_{1,-4} + x_{1,1} + x_{1,8} \,,$$

$$s_1 = 2(x_{0,1} + x_{0,8} + x_{1,-8} + x_{1,-4}) \,,$$

$$s_2 = 2(3x_{0,-8} + 3x_{0,8} + x_{1,-8} + 2x_{1,-4} + 2x_{1,1} + x_{1,8}) \,,$$

$$s_3 = 4(3x_{0,8} + x_{1,-8} + 2x_{1,-4}) \,,$$

$$s_4 = 4(5x_{0,-8} + 5x_{0,8} + x_{1,-8} + 4x_{1,-4} + 4x_{1,1} + x_{1,8}) \,,$$

$$s_5 = 8(x_{0,8} + x_{1,-8}) \,,$$

$$s_6 = 8(x_{0,-8} + x_{0,8} - x_{1,-8} - x_{1,8}),$$

$$s_7 = 32.$$

Theorem 49 is the main result of [Uehara, 1990]. This theorem and its supplement stated in [Uehara, 1990, Theorem 2] include the congruences proved by [Gras, 1989, Théorèmes (1.3), (1.4)] and [Hardy and Williams, 1986]. For details see [Uehara, 1990, section 3]. Making use of the congruence of Theorem 49, Uehara [Uehara, 1990, section 4] has obtained a number of congruences between the class numbers and units of the fields $\mathbb{Q}(\sqrt{fm})$ and $\mathbb{Q}(\sqrt{-fm})$ with $f = 1, 2$. In the case when $f = 1$ define

$$x_{0,1} = x_{0,-4} = 1 \text{ and } x_{1,1} = x_{1,-4} = \alpha = \pm 1,$$

and

$$x_{0,e} = x_{1,e} = 0 \quad \text{if } e \neq 1, -4.$$

Then Theorem 49 shows that

$$\sum_{\substack{d \in T_m \\ d < 0}} \Psi(|d|) \prod_{\substack{p \mid m \\ p \text{ prime}}} (1 - \chi_d(p)\Theta(p)p)H(d)$$

$$+ \sum_{\substack{d \in T_m \\ d > 0}} \Psi(|d|)\Big(\prod_{\substack{p \mid m \\ p \text{ prime}}} (1 - \chi_{-4d}(p)\Theta(p)p) - \delta_{d,1} \prod_{\substack{p \mid m \\ p \text{ prime}}} (1 - \Theta(p))\Big)H(-4d)$$

$$+ \alpha \sum_{\substack{d \in T_m \\ d > 1}} \Psi(|d|) \prod_{\substack{p \mid m \\ p \text{ prime}}} (1 - \chi_d(p)\Theta(p))H(d)$$

$$+ \alpha \sum_{\substack{d \in T_m \\ d < 0}} \Psi(|d|) \prod_{\substack{p \mid m \\ p \text{ prime}}} (1 - \chi_{-4d}(p)\Theta(p))H(-4d)$$

$$- \frac{\alpha}{2} \sum_{\substack{p \mid m \\ p \text{ prime}}} \Theta(p) \log_2 p \prod_{\substack{q \mid (m/p) \\ q \text{ prime}}} (1 - \Theta(q)) \equiv 0 \, (\bmod \, 2^{\nu + \lambda}),$$

where $\lambda = 3$ if $\alpha = -1$, and $\lambda = 2$ if $\alpha = 1$.

In the case when $f = 2$ define

$$x_{0,8} = x_{0,-8} = 1 \text{ and } x_{1,8} = x_{1,-8} = \alpha = \pm 1,$$

and

$$x_{0,e} = x_{1,e} = 0 \text{ if } e \neq \pm 8.$$

Then Theorem 49 shows that

$$\sum_{\substack{d \in \mathcal{T}_m \\ d<0}} \Psi(|d|) \prod_{\substack{p \mid m \\ p\,\text{prime}}} (1 - \chi_{8d}(p)\Theta(p)p)H(8d)$$

$$+ \sum_{\substack{d \in \mathcal{T}_m \\ d>0}} \Psi(|d|) \left(\prod_{\substack{p \mid m \\ p\,\text{prime}}} (1 - \chi_{-8d}(p)\Theta(p)p) - \delta_{d,1} \prod_{\substack{p \mid m \\ p\,\text{prime}}} (1 - \Theta(p)) \right) H(-8d)$$

$$+ \alpha \sum_{\substack{d \in \mathcal{T}_m \\ d>0}} \Psi(|d|) \left(\prod_{\substack{p \mid m \\ p\,\text{prime}}} (1 - \chi_{8d}(p)\Theta(p)) - \delta_{d,1} \prod_{\substack{p \mid m \\ p\,\text{prime}}} (1 - \Theta(p)) \right) H(8d)$$

$$+ \alpha \sum_{\substack{d \in \mathcal{T}_m \\ d<0}} \Psi(|d|) \prod_{\substack{p \mid m \\ p\,\text{prime}}} (1 - \chi_{-8d}(p)\Theta(p))H(-8d) \equiv 0 \,(\bmod\, 2^{\nu+\lambda}),$$

where $\lambda = 3$ if $\alpha = -1$ and $\lambda = 2$ if $\alpha = 1$.

We now turn to the case when $\Theta(p) = \Psi(p) = 1$ for $p \mid m$. By virtue of the above congruences, if d is an odd discriminant of a quadratic field having ν prime factors we obtain by induction on ν

$$H(d) \equiv H(-4d) \equiv 0 \,(\bmod\, 2^{\nu+\lambda}),$$

$$H(8d) \equiv H(-8d) \equiv 0 \,(\bmod\, 2^{\nu+\lambda}),$$

where $\lambda = 1$ if any prime p dividing d satisfies $p \equiv \pm 1 \,(\bmod\, 8)$ or $\lambda = 0$ otherwise. In [Uehara, 1990, Theorem 3] the following congruences involving $H(\pm p)$, $H(\pm 4p)$ and $H(\pm 8p)$ are given. We leave it to the reader to show that they are simple consequences of the above congruences with $\nu = 1$. Note that

$$\log_2 p \equiv -\frac{1}{2} \left(p - \left(\frac{-1}{p} \right) \right) \left(p - 3 \left(\frac{-1}{p} \right) \right) \,(\bmod\, 32).$$

THEOREM 50 ([Uehara, 1990, Theorem 3]) *In the above notation we have*

(i) *if* $p \equiv 3 \,(\bmod\, 4)$

$$H(-p) - H(4p) \equiv \frac{(p+1)(p+5)}{4} \,(\bmod\, 16),$$

$$H(-p) + H(4p) \equiv \left(\frac{p+1}{4} \right)^2 \,(\bmod\, 8),$$

(ii) *if* $p \equiv 1 \,(\bmod\, 4)$

$$H(-4p) - H(p) \equiv \frac{(p-1)(p-5)}{4} \,(\bmod\, 16),$$

$$H(-4p) + H(p) \equiv \left(\frac{p-1}{4} \right)^2 \,(\bmod\, 8),$$

(iii) $H(-8p) - H(8p) \equiv \left(\dfrac{2}{p}\right)\left(\left(\dfrac{-1}{p}\right)p - 3\right) + 2 \,(\mathrm{mod}\,16)$,

(iv) $H(-8p) + H(8p) \equiv \left(\dfrac{2}{p}\right)\left(\left(\dfrac{-1}{p}\right)p + 3\right) + 4 \,(\mathrm{mod}\,8)$.

Making use of results of Theorem 49 Uehara has deduced many other interesting congruences in the case when $\Theta(d) = \Psi(d) = 1$ $(d \mid m)$. As an example we give the following theorem. The proofs, by induction on ν, are left to the reader as an exercise.

THEOREM 51 ([Uehara, 1990, Theorem 4]) *Let d be an odd discriminant of a quadratic field having ν prime factors.*

(i) *We have*

$$H(d) \equiv H(-4d)\,(\mathrm{mod}\,2^{\nu+1+\lambda}),$$

$$H(8d) \equiv H(-8d)\,(\mathrm{mod}\,2^{\nu+1+\lambda}),$$

where $\lambda = 1$ if any prime p dividing d satisfies $p \equiv \pm 1\,(\mathrm{mod}\,8)$ or $\lambda = 0$ otherwise.

(ii) *If there exists a prime divisor q of d such that $q \equiv 5\,(\mathrm{mod}\,8)$ and $p \equiv 1$ $(\mathrm{mod}\,8)$ for any prime p dividing d/q, we have*

$$H(-4d) \equiv H(d)\,(\mathrm{mod}\,2^{\nu+2}),$$

(iii) *If $p \equiv 1\,(\mathrm{mod}\,4)$ for any prime p dividing d, we have*

$$H(-8d) \equiv -\chi_d(2)H(8d)\,(\mathrm{mod}\,2^{\nu+2}),$$

(iv) *If $p \equiv \pm 1\,(\mathrm{mod}\,16)$ for any prime p dividing d, we have*

$$H(d) \equiv H(-4d)\,(\mathrm{mod}\,2^{\nu+3}),$$

$$H(-8d) \equiv H(8d)\,(\mathrm{mod}\,2^{\nu+3}).$$

Let $d > 1$ be the discriminant of a quadratic field. Denote as usual by ε_d the fundamental unit of this field. For $8 \nmid d$ write

$$\varepsilon_d = \frac{t_0 + u_0\sqrt{\mathcal{D}}}{2}, \quad t_0, u_0 \in \mathbb{Z}, \qquad \eta_d = \varepsilon_d^\mu = T + U\sqrt{\mathcal{D}}, \; T, U \in \mathbb{Z},$$

where $\mathcal{D} = d$ if $d \equiv 1\,(\mathrm{mod}\,4)$, $\mathcal{D} = d/4$ if $4 \| d$, $\mu = 3$ if $\mathcal{D} \equiv 5\,(\mathrm{mod}\,8)$ and $t_0 \equiv u_0 \equiv 1\,(\mathrm{mod}\,2)$, and $\mu = 1$ otherwise. For $8 \mid d$ and $\mathcal{D} = d/4$ write

$$\varepsilon_d = t + u\sqrt{\mathcal{D}}, \; t, u \in \mathbb{Z}.$$

Making use of the congruences of Theorems 50 and 51, Uehara has shown the following congruences. We leave it to the reader to prove them as an exercise.

THEOREM 52 ([Uehara, 1990, Corollary 1]) *Given a prime number* $p \geq 3$, *let* T, U *and* μ *be defined as above for the fundamental unit* ε_p *(resp.* ε_{4p}*) if* $p \equiv 1 \,(\mathrm{mod}\, 4)$ *(resp.* $p \equiv 3 \,(\mathrm{mod}\, 4)$*), and let* t, u *be defined for the fundamental unit* ε_{8p}.

(i) *If* $p \equiv 1 \,(\mathrm{mod}\, 8)$ *we have*

$$h(-4p) \equiv TUh(p) + (p-1) \,(\mathrm{mod}\, 16),$$

$$h(-8p) \equiv N(\varepsilon_{8p})tuh(8p) + (p-1) \,(\mathrm{mod}\, 16).$$

(ii) *If* $p \equiv 3 \,(\mathrm{mod}\, 8)$ *we have*

$$2h(-p) \equiv (-1)^U TUh(4p) + (p+5) \,(\mathrm{mod}\, 16),$$

$$h(-8p) \equiv tuh(8p) + (p+5) \,(\mathrm{mod}\, 16).$$

(iii) *If* $p \equiv 5 \,(\mathrm{mod}\, 8)$ *we have*

$$h(-4p) \equiv 3\mu TUh(p) + (p-5) \,(\mathrm{mod}\, 16),$$

$$h(-8p) \equiv -(tu+4)h(8p) + (p-5) \,(\mathrm{mod}\, 16).$$

(iv) *If* $p \equiv 7 \,(\mathrm{mod}\, 8)$ *we have*

$$TUh(4p) \equiv p+1 \,(\mathrm{mod}\, 16),$$

$$h(-8p) \equiv tuh(8p) + (p+1) \,(\mathrm{mod}\, 16).$$

The first congruence in Theorem 52(i) is equivalent to that given in [Williams, 1981a]. The second congruence is equivalent to that given in [Kaplan and Williams, 1982a]. The congruences given in Theorem 52(iii) are equivalent to those of [Kaplan and Williams, 1982b].

THEOREM 53 ([Uehara, 1990, Corollary 2]) *Let* d *be an odd discriminant of a quadratic field having* ν *prime factors and let let* T, U *and* μ *be defined as above for the fundamental unit* ε_d *(resp.* ε_{-4d}*) if* $d > 0$ *(resp.* $d < 0$*), and let* t, u *be defined for the fundamental unit* ε_{8d} *(resp.* ε_{-8d}*) if* $d > 0$ *(resp.* $d < 0$*).*

(i) *We have*

$$h(-4d) \equiv \pm TUh(d) \,(\mathrm{mod}\, 2^{\nu+1+\lambda}), \quad if \ d > 0,$$

$$(1 - \chi_d(2))h(d) \equiv \pm TUh(-4d) \,(\mathrm{mod}\, 2^{\nu+1+\lambda}), \quad if \ d < 0,$$

$$h(-8d) \equiv \pm tuh(8d) \,(\mathrm{mod}\, 2^{\nu+1+\lambda}), \quad if \ d > 0,$$

$$h(8d) \equiv \pm tuh(-8d) \,(\mathrm{mod}\, 2^{\nu+1+\lambda}), \quad if \ d < 0,$$

where $\lambda = 1$ *if any prime p dividing d satisfies* $p \equiv \pm 1$ (mod 8) *or* $\lambda = 0$ *otherwise.*

(ii) *If* $d > 0$ *and there exists a prime divisor q of d such that* $q \equiv 5$ (mod 8) *and* $p \equiv 1$ (mod 8) *for any prime p dividing* d/q, *we have*

$$h(-4d) \equiv -\mu TUh(d) \,(\, \mathrm{mod}\, 2^{\nu+2}\,).$$

(iii) *If* $d > 0$ *and* $p \equiv 1$ (mod 4) *for any prime p dividing d, we have*

$$h(-8d) \equiv -\chi_d(2)N(\varepsilon_{8d})tuh(8d) \,(\, \mathrm{mod}\, 2^{\nu+2}\,),$$

(iv) *If* $p \equiv \pm 1$ (mod 16) *for any prime p dividing d, we have*

$$
\begin{aligned}
TUh(d) &\equiv h(-4d) \,(\, \mathrm{mod}\, 2^{\nu+3}\,), & \text{if } d > 0, \\
TUh(-4d) &\equiv 0 \,(\, \mathrm{mod}\, 2^{\nu+3}\,)), & \text{if } d < 0, \\
h(-8d) &\equiv N(\varepsilon_{8d})tuh(8d) \,(\, \mathrm{mod}\, 2^{\nu+3}\,), & \text{if } d > 0, \\
h(8d) &\equiv N(\varepsilon_{-8d})tuh(-8d) \,(\, \mathrm{mod}\, 2^{\nu+3}\,), & \text{if } d < 0.
\end{aligned}
$$

Appealing to Theorem 52 Uehara derived in an easy way the congruences proved in [Williams, 1981a], [Kaplan, 1981], [Kaplan and Williams, 1982a,b] and [Lang and Schertz, 1976]. As an example we derive the congruence proved in [Williams, 1981a]. Let $p \equiv 1$ (mod 8) be a prime number. In virtue of $T \equiv 0$ (mod 4) and $U \equiv 1$ (mod 4), Theorem 52(i) implies that

$$h(-4p) \equiv TUh(p) + (p-1) \equiv T + (p-1) + T(h(p) - 1) \,(\, \mathrm{mod}\, 16\,),$$

which gives easily

$$
h(-4p) \equiv
\begin{cases}
T + (p-1) \,(\, \mathrm{mod}\, 16\,), & \text{if } h(-4p) \equiv 0 \,(\, \mathrm{mod}\, 8\,), \\
T + (p-1) + 4(h(p)-1) \,(\, \mathrm{mod}\, 16\,), & \text{if } h(-4p) \equiv 4 \,(\, \mathrm{mod}\, 8\,).
\end{cases}
$$

Making use of Uehara's congruences, one can derive in an elementary way almost all the congruences of Chapter II. In fact Uehara has provided a general method of producing such congruences.

5.5 Optimal Linear Congruences

The congruences in the hypothesis of Theorems 45, 48 and 49 (resp. 46)

$$\sum_{(k,e)\in K\times T_8} x_{k,e} \sum_{d\in T_m} \Psi(|d|)L_2^{[m,\Theta]}(k,\chi_{ed}\omega^{1-k}) + x_{1,1}\Lambda_1(m,\Theta)$$

$$\equiv 0 \,(\, \mathrm{mod}\, 2^{\nu+\lambda}\,)$$

are said to be optimal if $\lambda = c(L)$ (resp. $\lambda = c(u_n)$). The 2-adic integers $x_{k,e}$ ($k \in K$, $e \in \mathcal{T}_8$) determining an optimal linear congruence are called optimal for K. We introduce this definition following [Urbanowicz, 1999]. The congruences proved in [Hardy and Williams, 1986], [Urbanowicz, 1990b] or resp. [Szmidt, Urbanowicz and Zagier, 1995] are optimal for $K = \{0\}$, $K = \{-1, 0\}$ or resp. $K = \{-m, \ldots, -1, 0\}$ ($m \geq 0$).

5.6   *The Case $L = \{-1, 0\}$*

We summarize without proofs the relevant material on the case when $L = \{-1, 0\}$. In this case the obtained congruences extend those of [Urbanowicz, 1990b] for the orders of K_2-groups of the integers of real quadratic fields and class numbers of imaginary quadratic fields. We leave it to the reader to show that Theorem 54 follows from Theorem 48 and implies Theorem 37. In the case when $L = \{-1, 0\}$ we have $c(L) = 5$ and the congruences are modulo $2^{\nu + \lambda + 1}$, where $\lambda \leq 5$.

THEOREM 54 ([Urbanowicz, 1999, Theorem 5]) *Let $m > 1$ be an odd squarefree integer having ν prime factors, and let Θ, $\Psi : \mathbb{N} \to \mathbb{C}_2$ be multiplicative functions such that $\Psi(s) \equiv \Theta(s) \equiv 1 \, (\bmod \, 2)$ for any divisor $s \mid m$. In the notation of Theorem 48, for any 2-adic integers $x_{-1,e}$, $x_{0,e}$ ($e \in \mathcal{T}_8$) not all even we have*

$$\sum_{e \in \mathcal{T}_8} x_{-1,e} \sum_{\substack{d \in \mathcal{T}_m \\ ed > 1}} \Psi(|d|)$$

$$\times \left(\prod_{\substack{p \mid m \\ p \, \text{prime}}} (1 - \chi_{ed}(p)\Theta(p)p^2) - \delta_{d,1} \prod_{\substack{p \mid m \\ p \, \text{prime}}} (1 - \Theta(p)) \right) K_2(ed)$$

$$+ 2 \sum_{e \in \mathcal{T}_8} x_{0,e} \sum_{\substack{d \in \mathcal{T}_m \\ ed < 0}} \Psi(|d|)$$

$$\times \left(\prod_{\substack{p \mid m \\ p \, \text{prime}}} (1 - \chi_{ed}(p)\Theta(p)p) - \delta_{d,1} \prod_{\substack{p \mid m \\ p \, \text{prime}}} (1 - \Theta(p)) \right) H(ed)$$

$$+ \frac{1}{6} x_{-1,1} \left(\prod_{\substack{p \mid m \\ p \, \text{prime}}} (1 - \Theta(p)p^2) - \prod_{\substack{p \mid m \\ p \, \text{prime}}} (1 - \Theta(p)) \right) \equiv 0 \, (\bmod \, 2^{\nu + \lambda + 1}),$$

where 2^λ is the greatest common divisor of the eight integers s_i ($0 \leq i \leq 7$) defined by

$$s_0 = x_{-1,-8} + x_{-1,-4} + x_{-1,1} + x_{-1,8} + x_{0,-8} + x_{0,-4} + x_{0,1} + x_{0,8},$$

$$s_1 = 2(x_{-1,-8} + x_{-1,-4} + x_{0,1} + x_{0,8}),$$

$$s_2 = 2(-x_{-1,-8} + 2x_{-1,-4} + 2x_{-1,1} - x_{-1,8} + x_{0,-8} + x_{0,8}),$$

$$s_3 = 4(-x_{-1,-8} + 2x_{-1,-4} + x_{0,8}),$$

$$s_4 = 4(-3x_{-1,-8} + 4x_{-1,-4} + 4x_{-1,1} - 3x_{-1,8} + x_{0,-8} + x_{0,8}),$$

$$s_5 = 8(x_{-1,-8} + x_{0,8}),$$

$$s_6 = 8(-x_{-1,-8} - x_{-1,8} + x_{0,-8} + x_{0,8}),$$

$$s_7 = 32.$$

5.7 The Case $L = \{-1, 2\}$

In the case when $L = \{-1, 2\}$ we derive linear congruences between the conjectured orders of K_2-groups of the integers of quadratic fields. We leave it to the reader to show that Theorem 55 follows from Theorem 48. In this case the obtained congruence provides an analogue of the Gras and Uehara congruence in K_2-theory. Here $c(L) = 5$ and the congruences are modulo $2^{\nu+\lambda+1}$, where $\lambda \leq 5$.

THEOREM 55 ([Urbanowicz, 1999, Theorem 6]) *Let $m > 1$ be an odd squarefree integer having ν prime factors, and let $\Theta, \Psi : \mathbb{N} \to \mathbb{C}_2$ be multiplicative functions such that $\Psi(s) \equiv \Theta(s) \equiv 1 \pmod 2$ for any divisor $s \mid m$. Assume that the 2-adic Lichtenbaum conjecture for imaginary quadratic fields holds. In the notation of Theorem 48, for any 2-adic integers $x_{-1,e}$, $x_{2,e}$ $(e \in \mathcal{T}_8)$ not all even we have*

$$\sum_{e \in \mathcal{T}_8} x_{-1,e} \sum_{\substack{d \in \mathcal{T}_m \\ ed > 1}} \Psi(|d|)$$

$$\times \left(\prod_{\substack{p \mid m \\ p \text{ prime}}} (1 - \chi_{ed}(p)\Theta(p)p^2) - \delta_{d,1} \prod_{\substack{p \mid m \\ p \text{ prime}}} (1 - \Theta(p)) \right) K_2(ed)$$

$$+ \sum_{e \in \mathcal{T}_8} x_{2,e} \sum_{\substack{d \in \mathcal{T}_m \\ ed < 0}} \Psi(|d|)$$

$$\times \left(\prod_{\substack{p \mid m \\ p \text{ prime}}} (1 - \chi_{ed}(p)\Theta(p)p^{-1}) - \delta_{d,1} \prod_{\substack{p \mid m \\ p \text{ prime}}} (1 - \Theta(p)) \right) K_2(ed),$$

$$+ \frac{1}{6} x_{-1,1} \left(\prod_{\substack{p \mid m \\ p \text{ prime}}} (1 - \Theta(p)p^2) - \prod_{\substack{p \mid m \\ p \text{ prime}}} (1 - \Theta(p)) \right) \equiv 0 \pmod{2^{\nu+\lambda+1}},$$

where 2^λ is the greatest common divisor of the eight integers s_i $(0 \le i \le 7)$ defined by

$$s_0 = x_{-1,-8} + x_{-1,-4} + x_{-1,1} + x_{-1,8} + x_{2,-8} + x_{2,-4} + x_{2,1} + x_{2,8},$$

$$s_1 = 2(x_{-1,-8} + x_{-1,-4} + x_{2,1} + x_{2,8}),$$

$$s_2 = 2(-9x_{-1,-8} + 2x_{-1,-4} + 2x_{-1,1} - 9x_{-1,8} \\ + 5x_{2,-8} + 4x_{2,-4} + 4x_{2,1} + 5x_{2,8}),$$

$$s_3 = 4(-x_{-1,-8} + 2x_{-1,-4} + 4x_{2,1} + 5x_{2,8}),$$

$$s_4 = 4(-3x_{-1,-8} + 4x_{-1,-4} + 4x_{-1,1} - 3x_{-1,8} + x_{2,-8} + x_{2,8}),$$

$$s_5 = 8(x_{-1,-8} + x_{2,8}),$$

$$s_6 = 8(-x_{-1,-8} - x_{-1,8} + x_{2,-8} + x_{2,8}),$$

$$s_7 = 32.$$

5.8   The Case $L = \{1, 2\}$

In the case when $L = \{1, 2\}$ we obtain linear congruences for class numbers of real quadratic fields and the orders of K_2-groups of the integers of imaginary quadratic fields. We leave it to the reader to show that Theorems 56 from Theorem 48. In this case $c(L) = 5$ and the obtained congruences are modulo $2^{\nu+\lambda+1}$, where $\lambda \le 5$.

THEOREM 56 ([Urbanowicz, 1999, Theorem 7]) *Let $m > 1$ be an odd squarefree integer having ν prime factors, and let Θ, $\Psi : \mathbb{N} \to \mathbb{C}_2$ be multiplicative functions such that $\Psi(s) \equiv \Theta(s) \equiv 1 \,(\mathrm{mod}\, 2)$ for any divisor $s \,|\, m$. Assume that the 2-adic Lichtenbaum conjecture for imaginary quadratic fields holds. In the notation of Theorem 48, for any 2-adic integers $x_{1,e}$, $x_{2,e}$ $(e \in \mathcal{T}_8)$ not all even we have*

$$2 \sum_{e \in \mathcal{T}_8} x_{1,e} \sum_{\substack{d \in \mathcal{T}_m \\ ed > 1}} \Psi(|d|)$$

$$\times \Big(\prod_{\substack{p \,|\, m \\ p\,\mathrm{prime}}} (1 - \chi_{ed}(p)\Theta(p)) - \delta_{d,1} \prod_{\substack{p \,|\, m \\ p\,\mathrm{prime}}} (1 - \Theta(p)) \Big) H(ed)$$

$$+ \sum_{e \in \mathcal{T}_8} x_{2,e} \sum_{\substack{d \in \mathcal{T}_m \\ ed < 0}} \Psi(|d|)$$

$$\times \left(\prod_{\substack{p \mid m \\ p \text{ prime}}} (1 - \chi_{ed}(p)\Theta(p)p^{-1}) - \delta_{d,1} \prod_{\substack{p \mid m \\ p \text{ prime}}} (1 - \Theta(p)) \right) K_2(ed)$$

$$- x_{1,1} \sum_{\substack{p \mid m \\ p \text{ prime}}} \Theta(p) \log_2 p \prod_{\substack{q \mid (m/p) \\ q \text{ prime}}} (1 - \Theta(q)) \equiv 0 \,(\bmod\, 2^{\nu + \lambda + 1}),$$

where 2^λ is the greatest common divisor of the eight integers s_i $(0 \le i \le 7)$ defined by

$$s_0 = x_{1,-8} + x_{1,-4} + x_{1,1} + x_{1,8} + x_{2,-8} + x_{2,-4} + x_{2,1} + x_{2,8}\,,$$

$$s_1 = 2(x_{1,-8} + x_{1,-4} + x_{2,1} + x_{2,8})\,,$$

$$s_2 = 2(3x_{1,-8} + 6x_{1,-4} + 6x_{1,1} + 3x_{1,8}$$
$$+ 5x_{2,-8} + 4x_{2,-4} + 4x_{2,1} + 5x_{2,8})\,,$$

$$s_3 = 4(3x_{1,-8} + 6x_{1,-4} + 4x_{2,1} + 5x_{2,8})\,,$$

$$s_4 = 4(5x_{1,-8} + 4x_{1,-4} + 4x_{1,1} + 5x_{1,8} + x_{2,-8} + x_{2,8})\,,$$

$$s_5 = 8(x_{1,-8} + x_{2,8})\,,$$

$$s_6 = 8(-x_{1,-8} - x_{1,8} + x_{2,-8} + x_{2,8})\,,$$

$$s_7 = 32\,.$$

5.9 The Cases $L = \{-1, 1\}$ and $L = \{0, 2\}$

In the case when $L = \{-1, 1\}$ (resp. $L = \{0, 2\}$) we obtain linear congruences between class numbers and the orders of K_2-groups of the integers of real (resp. imaginary) quadratic fields. We leave it to the reader to show that Theorems 57 and 58 follow from Theorem 48. In both the cases $c(L) = 6$ and the obtained congruences are modulo $2^{\nu + \lambda + 1}$, where $\lambda \le 6$.

THEOREM 57 ([Urbanowicz, 1999, Theorem 8]) *Let $m > 1$ be an odd squarefree integer having ν prime factors, and let Θ, $\Psi : \mathbb{N} \to \mathbb{C}_2$ be multiplicative functions such that $\Psi(s) \equiv \Theta(s) \equiv 1 \,(\bmod\, 2)$ for any divisor $s \mid m$. In the notation of Theorem 48, for any 2-adic integers $x_{1,e}$, $x_{1,e}$ $(e \in T_8)$ not all even we have*

$$\sum_{e \in T_8} x_{-1,e} \sum_{\substack{d \in T_m \\ ed > 1}} \Psi(|d|)$$

$$\times \left(\prod_{\substack{p\,|\,m \\ p\,\text{prime}}} (1 - \chi_{ed}(p)\Theta(p)p^2) - \delta_{d,1} \prod_{\substack{p\,|\,m \\ p\,\text{prime}}} (1 - \Theta(p)) \right) K_2(ed)$$

$$+ 2 \sum_{e \in \mathcal{T}_8} x_{1,e} \sum_{\substack{d \in \mathcal{T}_m \\ ed > 1}} \Psi(|d|)$$

$$\times \left(\prod_{\substack{p\,|\,m \\ p\,\text{prime}}} (1 - \chi_{ed}(p)\Theta(p)) - \delta_{d,1} \prod_{\substack{p\,|\,m \\ p\,\text{prime}}} (1 - \Theta(p)) \right) H(ed)$$

$$+ \frac{1}{6} x_{-1,1} \left(\prod_{\substack{p\,|\,m \\ p\,\text{prime}}} (1 - \Theta(p)p^2) - \prod_{\substack{p\,|\,m \\ p\,\text{prime}}} (1 - \Theta(p)) \right)$$

$$- x_{1,1} \sum_{\substack{p\,|\,m \\ p\,\text{prime}}} \Theta(p) \log_2 p \prod_{\substack{q\,|\,(m/p) \\ q\,\text{prime}}} (1 - \Theta(q)) \equiv 0 \,(\,\text{mod}\, 2^{\nu + \lambda + 1}\,),$$

where 2^λ is the greatest common divisor of the eight integers s_i $(0 \le i \le 7)$ defined by

$$s_0 = x_{-1,-8} + x_{-1,-4} + x_{-1,1} + x_{-1,8} + x_{1,-8} + x_{1,-4} + x_{1,1} + x_{1,8},$$

$$s_1 = 2(x_{-1,-8} + x_{-1,-4} + x_{1,-8} + x_{1,-4}),$$

$$s_2 = 2(-3x_{-1,-8} + 6x_{-1,-4} + 6x_{-1,1} - 3x_{-1,8}$$
$$+ x_{1,-8} + 2x_{1,-4} + 2x_{1,1} + x_{1,8}),$$

$$s_3 = 4(-9x_{-1,-8} + 2x_{-1,-4} + 3x_{1,-8} + 6x_{1,-4}),$$

$$s_4 = 4(x_{-1,-8} + 4x_{-1,-4} + 4x_{-1,1} + x_{-1,8}$$
$$+ x_{1,-8} + 4x_{1,-4} + 4x_{1,1} + x_{1,8}),$$

$$s_5 = 8(x_{-1,-8} + 4x_{-1,-4} + x_{1,-8} + 4x_{1,-4}),$$

$$s_6 = 8(3x_{-1,-8} + 3x_{-1,8} - x_{1,-8} - x_{1,8}),$$

$$s_7 = 64.$$

REMARK Note that in the case when $L = \{-1, 1\}$ we have

$$z_8 \equiv -2z_6 \,(\,\text{mod}\, 64\,), \quad z_7 \equiv -2z_5 \,(\,\text{mod}\, 64\,),$$

and in consequence we may ignore z_8 and z_7 (the z_n with $n = 2c(L) - 4$, $2c(L) - 5$).

THEOREM 58 ([Urbanowicz, 1999, Theorem 9]) *Let* $m > 1$ *be an odd squarefree integer having* ν *prime factors, and let* $\Theta, \Psi : \mathbb{N} \to \mathbb{C}_2$ *be multiplicative functions such that* $\Psi(s) \equiv \Theta(s) \equiv 1 \,(\mathrm{mod}\,2)$ *for any divisor* $s \,|\, m$. *Assume that the 2-adic Lichtenbaum conjecture for imaginary quadratic fields holds. In the notation of Theorem 48, for any 2-adic integers* $x_{0,e}, x_{2,e}$ $(e \in \mathcal{T}_8)$ *not all even we have*

$$
2 \sum_{e \in \mathcal{T}_8} x_{0,e} \sum_{\substack{d \in \mathcal{T}_m \\ ed < 0}} \Psi(|d|)
$$

$$
\times \Big(\prod_{\substack{p \,|\, m \\ p \,\text{prime}}} (1 - \chi_{ed}(p)\Theta(p)p) - \delta_{d,1} \prod_{\substack{p \,|\, m \\ p \,\text{prime}}} (1 - \Theta(p)) \Big) H(ed)
$$

$$
+ \sum_{e \in \mathcal{T}_8} x_{2,e} \sum_{\substack{d \in \mathcal{T}_m \\ ed < 0}} \Psi(|d|)
$$

$$
\times \Big(\prod_{\substack{p \,|\, m \\ p \,\text{prime}}} (1 - \chi_{ed}(p)\Theta(p)p^{-1}) - \delta_{d,1} \prod_{\substack{p \,|\, m \\ p \,\text{prime}}} (1 - \Theta(p)) \Big) K_2(ed),
$$

$$
\equiv 0 \,(\mathrm{mod}\, 2^{\nu+\lambda+1}),
$$

where 2^λ *is the greatest common divisor of the eight integers* s_i $(0 \le i \le 7)$ *defined by*

$$
s_0 = x_{0,-8} + x_{0,-4} + x_{0,1} + x_{0,8} + x_{2,-8} + x_{2,-4} + x_{2,1} + x_{2,8},
$$

$$
s_1 = 2(x_{0,1} + x_{0,8} + x_{2,1} + x_{2,8}),
$$

$$
s_2 = 2(9x_{0,-8} + 9x_{0,8} + 5x_{2,-8} + 4x_{2,-4} + 4x_{2,1} + 5x_{2,8}),
$$

$$
s_3 = 4(9x_{0,8} + 4x_{2,1} + 5x_{2,8}),
$$

$$
s_4 = 4(x_{0,-8} + x_{0,8} + x_{2,-8} + x_{2,8}),
$$

$$
s_5 = 8(x_{0,8} + x_{2,8}),
$$

$$
s_6 = 8(5x_{0,-8} + 5x_{0,8} + x_{2,-8} + x_{2,8})
$$

$$
s_7 = 64.
$$

REMARK Note that in the case when $L = \{0,2\}$ we have

$$
z_8 \equiv 2z_6 \,(\mathrm{mod}\,64), \quad z_7 \equiv 2z_5 \,(\mathrm{mod}\,64),
$$

and in consequence we may ignore the z_8 and z_7 (the z_n with $n = 2c(L) - 4$, $2c(L) - 5$).

5.10   *Concluding Remarks*

Uehara's approach used in [Urbanowicz and Wójcik, 1995/1996] and [Wójcik, 1998] gives a method of producing linear congruences. It would be interesting to use this method to find for given λ explicit formulae for the $x_{k,e}$ such that the linear congruences are valid modulo $2^{\nu+\lambda}$. This approach should yield many new congruences between class numbers and the orders of K_2-groups of the rings of integers of quadratic fields. In the case of the orders of K_2-groups for imaginary quadratic fields such congruences would be completely new. For more details, see [Urbanowicz, 1999].

Another direction for further investigation would be to extend Wójcik's congruence [Wójcik, 1998] by giving a congruence for a linear combination of the values $L_2(k, \chi\omega^{1-k})$, where the numbers k are taken from any finite subset of the integers. Wójcik's congruence involved the case when this subset consisted of consecutive integers. In [Urbanowicz and Wójcik, 1995/1996] the authors found such a congruence for any subset of the set $\{-1, 0, 1, 2\}$.

Chapter V

APPLICATIONS OF ZAGIER'S FORMULA (I)

In this chapter we present results from the paper [Schinzel, Urbanowicz and van Wamelen, 1999], which were proved using Zagier's identity. A. Granville has communicated to the authors that he also deduced the main theorem of the paper from Zagier's formula in a somewhat different way. Another application of Zagier's identity is given in Chapter VI.

1. INTRODUCTION

1.1 Notation

We shall consider sums of the form

$$S(\mathcal{D}, q_1, q_2) = \sum_{q_1|\mathcal{D}|<n<q_2|\mathcal{D}|} \left(\frac{\mathcal{D}}{n}\right),$$

where $\mathcal{D}$ belongs to the set $\mathcal{F}$ of fundamental discriminants different from 1,

$$q_1, q_2 \in \mathbb{Q}; \quad 0 \leq q_1 < q_2 \leq 1. \tag{1}$$

Let r be the least common denominator of q_1 and q_2, and

$$\chi = \chi_{\mathcal{D}} = \left(\frac{\mathcal{D}}{\cdot}\right).$$

It follows from Theorem 2, section 9.2 of Chapter I that if $\mathcal{D} \in \mathcal{F}$ and $(\mathcal{D}, r) = 1$ we have

$$S(\mathcal{D}, q_1, q_2) = \sum_{\psi} c_\psi B_{1,\chi\psi}, \tag{2}$$

where the sum is over all primitive characters ψ with conductor f_ψ such that

$$f_\psi \,|\, r, \text{ and } \psi(-1) = -\chi(-1), \tag{3}$$

$B_{1,\chi\psi}$ is the generalized first Bernoulli number attached to the character $\chi\psi$ and $c_\psi = c_\psi(\mathcal{D}, q_1, q_2)$ are given explicitly. In particular, for all ψ satisfying (3)

$$c_\psi(\mathcal{D}, 1 - q_2, 1 - q_1) = \chi(-1) c_\psi(\mathcal{D}, q_1, q_2), \tag{4}$$

$$c_\psi(\mathcal{D}, 0, 1 - q_2) = -\chi(-1) c_\psi(\mathcal{D}, 0, q_2), \tag{5}$$

$$c_\psi(\mathcal{D}, q_1, 1 - q_2) = c_\psi(\mathcal{D}, q_1, q_2), \text{ if } q_1 + q_2 \le 1, \mathcal{D} < 0, \tag{6}$$

$$c_\psi(\mathcal{D}, q_1, q_2) = c_\psi(\mathcal{D}_0, q_1, q_2) \tag{7}$$

provided

$$\mathcal{D}, \mathcal{D}_0 \in \mathcal{F}, \text{ sgn } \mathcal{D} = \text{sgn } \mathcal{D}_0, \text{ and } \mathcal{D} \equiv \mathcal{D}_0 \left(\bmod\, r \frac{\gcd(r^3, 8)}{\gcd(r, 8)} \right), \tag{8}$$

and

$$c_{\overline{\psi}}(\mathcal{D}, q_1, q_2) = \overline{c}_\psi(\mathcal{D}, q_1, q_2), \tag{9}$$

where the bar denotes complex conjugation.

Denote by $C(\mathcal{D}, q_1, q_2)$ the set of all primitive characters ψ satisfying (3) such that $c_\psi \ne 0$.

For $\varepsilon = 0$ or 1, let $\mathcal{P}_\varepsilon$ be the set of all pairs $\langle q_1, q_2 \rangle$ satisfying both (1) and

$$q_1 + q_2 \le 1 \quad \text{and} \quad q_2 \le \frac{1}{2} \text{ if } q_1 = 0 \text{ or } \mathcal{D} < 0 \tag{10}$$

such that card $C(\mathcal{D}, q_1, q_2) = \varepsilon$ for at least one $\mathcal{D} \in \mathcal{F}$ prime to r.

The main result of [Schinzel, Urbanowicz and van Wamelen, 1999] is the following theorem.

THEOREM 59 ([Schinzel, Urbanowicz and van Wamelen, 1999, Theorem])
The sets $\mathcal{P}_0$ and $\mathcal{P}_1$ are both finite. More precisely, $\mathcal{P}_0$ has 55 elements and $\mathcal{P}_1$ has 116 elements, as listed in section 5.

The sets $\mathcal{P}_\varepsilon$ are of interest for the following reason.

For each $\varepsilon = 0, 1$ and $\langle q_1, q_2 \rangle \in \mathcal{P}_\varepsilon$ the set of all $\mathcal{D} \in \mathcal{F}$ prime to r satisfying card $C(\mathcal{D}, q_1, q_2) = \varepsilon$ is, by virtue of (7), the intersection of $\mathcal{F}$ with the union of some arithmetic progressions with the first term $\mathcal{D}_0$ and the difference $r \dfrac{\gcd(r^3, 8)}{\gcd(r, 8)} \text{ sgn } \mathcal{D}_0$. Hence the condition card $C(\mathcal{D}, q_1, q_2) = \varepsilon$ if fulfilled by one $\mathcal{D} \in \mathcal{F}$ prime to r is fulfilled by infinitely many such $\mathcal{D}$. The condition (10) is justified by the formulae (4), (5) and (6), it permits us to retain in Theorem 59 only essentially different cases.

Moreover, card $C(\mathcal{D}_0, q_1, q_2) = 0$, $\gcd(\mathcal{D}_0, r) = 1$ and (8) imply, by virtue of (2), that

$$S(\mathcal{D}, q_1, q_2) = 0. \tag{11}$$

Further,

$$\text{card } C(\mathcal{D}_0, q_1, q_2) = 1, \quad \gcd(\mathcal{D}_0, r) = 1 \tag{12}$$

and (8) imply, by virtue of (3) and (9) that

$$C(\mathcal{D}, q_1, q_2) = C(\mathcal{D}_0, q_1, q_2) = \{\chi_E\},$$

where E is a fundamental discriminant satisfying

$$E \mid r, \quad \chi_E(-1) = -\chi_{\mathcal{D}}(-1).$$

We now recall (see formula (16) of Chapter I) that for every fundamental discriminant $\mathcal{D} < -4$ we have

$$B_{1, \chi_{\mathcal{D}}} = -h(\mathcal{D}),$$

where $h(\mathcal{D})$ is the class number of the field $\mathbb{Q}(\sqrt{\mathcal{D}})$ and infer that (8) and (12) imply

$$h(E\mathcal{D}) = cS(\mathcal{D}, q_1, q_2) \tag{13}$$

$(c = -c_{\chi_E}^{-1})$. Now, relations of the form (11) and (13) have been studied by many authors, see e.g. [Berndt, 1976], [Johnson and Mitchell, 1977], [Hudson and Williams, 1982], mostly, but not exclusively, in the case $q_2 - q_1 = \dfrac{1}{r}$, $|\mathcal{D}|$ a prime. All pairs $\langle q_1, q_2 \rangle$ known to satisfy (1), (10) and (11) for some $\mathcal{D}$ prime to r happen to belong to $\mathcal{P}_0$ and $\mathcal{P}_0$ does not contain any new pair with $q_2 - q_1 = \dfrac{1}{r}$. We do not know, however, whether the conditions (1), (10) and (11) for some $\mathcal{D}$ prime to r imply $\langle q_1, q_2 \rangle \in \mathcal{P}_0$. Also, we cannot say anything about relation (11) with $\gcd(\mathcal{D}_0, r) \neq 1$. On the other hand, as we shall show in section 4, if $\langle q_1, q_2 \rangle \in \mathcal{P}_0$, card $C(\mathcal{D}_0, q_1, q_2) = 0$ and $\gcd(\mathcal{D}_0, r) = 1$ then (8) implies (11) without the condition $\mathcal{D} \in \mathcal{F}$, for all nonsquare discriminants $\mathcal{D}$.

Further, all known pairs $\langle q_1, q_2 \rangle$ satisfying (1), (10) and (13) for fixed E, c and some $\mathcal{D}_0 \in \mathcal{F}$ prime to r happen to belong to $\mathcal{P}_1$, and again we do not know whether these conditions imply $\langle q_1, q_2 \rangle \in \mathcal{P}_1$. In this case, however, there is a new relation with $q_2 - q_1 = \dfrac{1}{r}$, namely

$$h(12\mathcal{D}) = 4 \sum_{\frac{1}{12}|\mathcal{D}| < n < \frac{1}{10}|\mathcal{D}|} \left(\frac{\mathcal{D}}{n}\right)$$

for all $\mathcal{D} \in \mathcal{F}$, $|\mathcal{D}| \equiv 11$ or $59 \ (\bmod \ 60)$.

1.2   *Organization of Chapter V*

The chapter is organized as follows. In section 2 we prove three theorems on Dirichlet characters, in section 3 we express $S(\mathcal{D}, q_1, q_2)$ in the form (2) obtaining the formulae for c_ψ, and in section 4 we apply the results of sections 2 and 3 to prove the finiteness of the sets $\mathcal{P}_0$ and $\mathcal{P}_1$. Section 5 consists of two tables, which for every $\langle q_1, q_2 \rangle \in \mathcal{P}_\varepsilon$ ($\varepsilon = 0, 1$) satisfying (10) give arithmetic progressions relevant to (11) or (13). In the former case r and in the latter case $e = E \operatorname{sgn} \mathcal{D}$ and c are also given.

2.   DIRICHLET CHARACTERS WITH CERTAIN PROPERTIES

2.1   *Three Useful Lemmas*

In section 2 we prove three theorems giving the existence of Dirichlet characters with certain properties. For definitions and basic facts on Dirichlet characters we refer the reader to [Hasse, 1964] or [Davenport, 1980].

For a prime power P let g_P be a generator of the group of characters mod P, even if $P = 2^\alpha$, $\alpha \geq 3$. The following lemmas are implicit in [Hasse, 1964, §13, no. 6].

LEMMA 1 ([Schinzel, Urbanowicz and van Wamelen, 1999, Lemma 1]) *For every positive integer* $m \not\equiv 2 \,(\operatorname{mod} 4)$ *there exists a primitive character of conductor* m.

LEMMA 2 ([Schinzel, Urbanowicz and van Wamelen, 1999, Lemma 2]) *If* P *equals* p^α *(p an odd prime), then* g_P^k *has conductor* P *unless*

$$\frac{\phi(p^\alpha)}{\phi(p^{\alpha-1})} \Big| k.$$

LEMMA 3 ([Schinzel, Urbanowicz and van Wamelen, 1999, Lemma 3]) *For every positive integer* $f \not\equiv 2 \,(\operatorname{mod} 4)$, $f \nmid 12$ *and each* $\varepsilon = \pm 1$ *there exists a primitive character* ξ *of conductor* f *such that* $\xi(-1) = \varepsilon$.

PROOF.   By the assumption there exists a prime power P such that $P \mid f$, $\gcd(P, f/P) = 1$ and $P > 4$. By Lemma 1 there exists a primitive character ξ_0 of conductor f/P. We now take

$$\xi = \xi_0 g_P^{(3+\varepsilon\xi_0(-1))/2}, \quad \text{if } P \text{ is odd},$$

$$\xi = \xi_0 g_4^{(3+\varepsilon\xi_0(-1))/2} g_P, \quad \text{if } P \text{ is even}.$$

By Lemma 2, if $P > 4$ is odd, g_P^2 has conductor P, hence ξ has conductor f. Moreover $\xi(-1) = \varepsilon\xi_0(-1)^2 = \varepsilon$.   ∎

2.2   *The Main Theorems*

Throughout this chapter ζ_m denotes a fixed primitive mth root of unity and ϕ denotes the Euler phi function.

THEOREM 60 [Schinzel, Urbanowicz and van Wamelen, 199, Proposition 1] *Let f be a positive integer with $f \not\equiv 2 \ (\bmod\, 4)$, and either $f \nmid 120$ or $40 \mid f$. Assume $\varepsilon = \pm 1$. Then there exists a nonreal primitive character ψ of conductor f such that*

$$\psi(-1) = \varepsilon \quad and \quad \psi(2)^2 \neq 1.$$

THEOREM 61 [Schinzel, Urbanowicz and van Wamelen, 1999, Proposition 2] *Let $k, f \in \mathbb{N}$, $\gcd(k, f) = 1$, $k \not\equiv \pm 1 \ (\bmod\, f)$ and f odd, $f \nmid 3 \cdot 5 \cdot 17$. For each $\varepsilon = \pm 1$ and $\eta = \pm 1$ there exists a nonreal primitive character ψ of conductor d such that $d \mid f$ and*

$$\psi(-1) = \varepsilon, \quad \psi(k) \neq \eta, \quad and \quad \psi(p)^2 \neq 1 \quad for\ all\ primes \quad p$$

such that

$$p \mid 2f \quad and \quad p \nmid d. \tag{14}$$

If the condition (14) is restricted to $p \mid f$ and $p \nmid d$ then the condition $f \nmid 3 \cdot 5 \cdot 17$ can be relaxed to $f \nmid 3 \cdot 5$.

THEOREM 62 [Schinzel, Urbanowicz and van Wamelen, 1999, Proposition 3] *Let $k, f \in \mathbb{N}$, $\gcd(k, f) = 1$, $k \not\equiv \pm 1 \ (\bmod\, f)$ and either $f \nmid 16 \cdot 3 \cdot 5$ or $16 \cdot 5 \mid f$. For each $\varepsilon = \pm 1$ there exists a nonreal primitive character ψ of conductor d, where $d \mid f$ such that*

$$\psi(-1) = \varepsilon, \quad \psi(k) \neq 1, \quad and \quad \psi(p)^2 \neq 1 \quad for\ all\ primes \quad p$$

satisfying

$$p \mid f \quad and \quad p \nmid d. \tag{15}$$

2.3   *Proof of Theorem 60*

Assume first that $f \nmid 120$. Then there exists a prime power P such that

$$P \mid f, \quad P \nmid 120, \quad \gcd\left(P, \frac{f}{P}\right) = 1. \tag{16}$$

We distinguish two cases

(i) $\dfrac{f}{P} \mid 12$,

(ii) $\dfrac{f}{P} \nmid 12$.

In the case (i) there exists a primitive real character ξ of conductor f/P. We take

$$\psi = \begin{cases} \xi g_P^{(3+\varepsilon\xi(-1))/2}, & \text{if } P \text{ is odd,} \\ \xi g_4^{(3+\varepsilon\xi(-1))/2} g_P, & \text{if } P \text{ is even} \end{cases}$$

and obtain $\psi(-1) = \varepsilon$. Also ψ is not real since g_P^2 (P odd) and g_P (P even) is not real. The same characters, by Lemma 2, have conductor P, hence ψ is primitive of conductor f. If we had $\psi(2)^2 = 1$ it would follow that P is odd, $g_P(2)^4 = 1$, and hence $2^4 \equiv 1 \pmod P$, contrary to (16).

In the case (ii) by Lemma 3 there exists a primitive character ξ of conductor f/P such that $\xi(-1) = \varepsilon g_P(-1)$. We put

$$\psi_\pm = \xi g_P^{\pm 1}.$$

$\psi_\pm$ are primitive characters of conductor f, nonreal since g_P is not real. Also $\psi_\pm(-1) = \varepsilon$. If we had $\psi_\pm(2)^2 = 1$ for both signs, it would follow that

$$1 = \psi_+(2)^2 \psi_-(2)^{-2} = g_P(2)^4,$$

hence $2^4 \equiv 1 \pmod P$, contrary to (16).

It remains to consider the case $40 \mid f \mid 120$. Here we take

$$\psi = \begin{cases} g_5 g_4^{(3-\varepsilon)/2} g_8, & \text{if } f = 40, \\ g_3 g_5 g_4^{(3+\varepsilon)/2} g_8, & \text{if } f = 120. \end{cases} \qquad \blacksquare$$

2.4   *Proof of Theorem 61*

Let

$$f = \prod_{i=1}^h p_i^{\alpha_i},$$

where the p_i are distinct primes. We put $p_i^{\alpha_i} = P_i$ and

$$\mathcal{P} = \{P_i : 1 \leq i \leq h\},$$
$$\mathcal{T} = \{P_i : 1 \leq i \leq h, \ k^2 \not\equiv 1 \pmod{P_i}\}$$

and we shall consider successively three cases

(i) $\mathcal{T} \not\subseteq \{5, 17\}$.

(ii) $\emptyset \neq \mathcal{T} \subseteq \{5, 17\}$.

(iii) $\mathcal{T} = \emptyset$.

Case (i). Here there exists a $P \in \mathcal{P}$ such that

$$k^2 \not\equiv 1 \pmod P \quad \text{and} \quad 2^8 \not\equiv 1 \pmod P. \tag{17}$$

We put

$$\psi = \psi_1 \psi_2,$$

where

$$\psi_1(x) = \prod_{\substack{1 \leq i \leq h \\ p_i \nmid P}} \left(\frac{x}{p_i}\right),$$

and

$$\psi_2 = g_P, \text{ if } \psi_1(-1) = -\varepsilon, \tag{18}$$

$$\psi_2 = g_P^2, \text{ if } \psi_1(-1) = \varepsilon \text{ and either } \psi_1(k) = \eta \text{ or } k^2 \not\equiv -1 \,(\bmod P), \tag{19}$$

$$\psi_2 = g_P^4, \text{ if } \psi_1(-1) = \varepsilon \text{ and } \psi_1(k) = -\eta \text{ and } k^2 \equiv -1 \,(\bmod P). \tag{20}$$

The characters ψ_1 and ψ_2 are primitive mod $\prod_{p_i \nmid P} p_i$ and P, respectively, and ψ_2 is not real since $\zeta_{\phi(P)}^4 \notin \mathbb{R}$. Hence ψ is not real and is primitive mod $d = \prod_{p_i \nmid P} p_i P$. Thus the only prime satisfying (14) is 2 and the equality $\psi(2)^2 = 1$ would give

$$g_P(2)^8 = 1,$$

hence $2^8 \equiv 1 \,(\bmod P)$, contrary to (17). Moreover

$$\psi(-1) = \psi_1(-1)\psi_2(-1) = \varepsilon$$

and $\psi(k) = \eta$ would imply

$$\psi_2(k) = \eta \psi_1(k),$$

which gives in the case (18) $g_P(k) = \pm 1$, in the case (19) either $g_P(k)^2 = 1$ or $g_P(k)^2 = -1$ and $k^2 \not\equiv -1 \,(\bmod P)$, and in the case (20) $g_P(k)^4 = -1$ and $k^2 \equiv -1 \,(\bmod P)$. In cases (18) and (19) this contradicts $k^2 \not\equiv 1 \,(\bmod P)$. In case (20) we have $k^4 \equiv 1 \,(\bmod P)$ and so $g_P(k)^4 = 1$, a contradiction.

Before we proceed to the cases (ii) and (iii) we make the following observation. Since $f \nmid 2^8 - 1$, there exists a least $P \in \mathcal{P}$ such that

$$P \nmid 2^8 - 1. \tag{21}$$

Then if ψ_1 is a character mod f/P and $c \mid 2$, at least one of the characters $\psi_\pm = \psi_1 g_P^{\pm c}$ satisfies $\psi(2)^2 \neq 1$, otherwise we would have

$$1 = \psi_+(2)^2 \psi_-(2)^{-2} = g_P^{4c}(2),$$

hence $2^8 \equiv 1 \,(\bmod P)$, contrary to (21).

Therefore, whenever in the sequel the exponent of g_P divides 2, we obtain $\psi(2)^2 \neq 1$ by replacing g_P by g_P^{-1}, if necessary.

Case (ii). Here we put

$$
\psi = \begin{cases}
g_5^2 g_{17} g_P^{(3-\varepsilon)/2} \displaystyle\prod_{q\in\mathcal{P}\setminus\{5,17,P\}} g_q^2, & \text{if } \mathcal{T} = \{5,17\} \text{ and } 3 \notin \mathcal{P}, \\[2ex]
g_3 g_5^2 g_{17} g_P^{(3+\varepsilon)/2} \displaystyle\prod_{q\in\mathcal{P}\setminus\{3,5,17,P\}} g_q^2, & \text{if } \mathcal{T} = \{5,17\} \text{ and } 3 \in \mathcal{P}, \\[2ex]
g_p g_P^{(3-\varepsilon)/2} \displaystyle\prod_{q\in\mathcal{P}\setminus\{p,P\}} g_q^2, & \text{if } \mathcal{T} \cap \{5,17\} = \{p\} \text{ and } 3 \notin \mathcal{P}, \\[2ex]
g_3 g_p g_P^{(3+\varepsilon)/2} \displaystyle\prod_{q\in\mathcal{P}\setminus\{3,p,P\}} g_q^2, & \text{if } \mathcal{T} \cap \{5,17\} = \{p\} \text{ and } 3 \in \mathcal{P}.
\end{cases}
$$

By Lemma 2 ψ is a primitive character of conductor f. The only prime satisfying (14) is 2. Since $((3\pm\varepsilon)/2)\,|\,2$ a proper choice of g_P gives $\psi(2)^2 \neq 1$ and ψ is not real. Moreover

$$
\psi(-1) = -(-1)^{(3-\varepsilon)/2} = (-1)^{(3+\varepsilon)/2} = \varepsilon
$$

and $\psi(k) = \eta$ would imply

$$
g_{17}(k)^2 = \psi(k)^2 = 1, \quad k^2 \equiv 1\,(\bmod\,17), \quad \text{if } \mathcal{T} = \{5,17\},
$$

or

$$
g_p(k)^2 = \psi(k)^2 = 1, \quad k^2 \equiv 1\,(\bmod\,p), \quad \text{if } \mathcal{T} \cap \{5,17\} = \{p\},
$$

contrary to $17 \in \mathcal{T}$, or $p \in \mathcal{T}$, respectively.

Case (iii). Here we assume without loss of generality that $k \equiv 1\,(\bmod\,P_i)$ for $i \leq j$, $k \equiv -1\,(\bmod\,P_i)$ for $i > j$ and if $3 \in \{P_1, \ldots, P_h\}$ then $3 = P_1$ or $3 = P_h$. Since $k \not\equiv \pm 1\,(\bmod\,f)$ we have $1 \leq j < h$.

If either $3 \notin \{P_1, \ldots, P_h\}$ or $3 = P_1, \eta = \varepsilon$ or $3 = P_h, \eta = 1$ we put

$$
\psi = g_{P_1}^{(3-\varepsilon\eta)/2} g_{P_h}^{(3-\eta)/2} \prod_{i=2}^{h-1} g_{P_i}. \tag{22}
$$

By Lemma 2, ψ is a primitive character of conductor f thus the only prime satisfying (14) is 2. Since all the exponents on the right hand side of (22) divide 2, a proper choice of g_P gives $\psi(2)^2 \neq 1$ and ψ is not real. Moreover

$$
\psi(-1) = (-1)^{(3-\varepsilon\eta)/2}(-1)^{(3-\eta)/2} = \varepsilon,
$$
$$
\psi(k) = (-1)^{(3-\eta)/2} = -\eta.
$$

If $3 = P_1, \eta = -\varepsilon$ and $j > 1$ we put

$$\psi = g_3 g_{P_2} g_{P_h}^{(3-\eta)/2} \prod_{i=3}^{h-1} g_{P_i}^2 . \qquad (23)$$

By Lemma 2, ψ is a primitive character of conductor f, thus the only prime satisfying (14) is 2. Since all the exponents on the right hand side of (23) divide 2, a proper choice of g_P gives $\psi(2)^2 \neq 1$ and ψ is not real. Similarly to the above we obtain $\psi(-1) = \varepsilon$ and $\psi(k) = -\eta$.

Likewise if $P_h = 3, \eta = -1$ and $j < h - 1$ we put

$$\psi = g_3 g_{P_1}^{(3+\varepsilon)/2} g_{P_{h-1}} \prod_{i=2}^{h-2} g_{P_i}^2 . \qquad (24)$$

By Lemma 2, ψ is a primitive character of conductor f, thus the only prime satisfying (14) is 2. Since all the exponents on the right hand side of (24) divide 2, a proper choice of g_P gives $\psi(2)^2 \neq 1$ and ψ is not real. Moreover,

$$\psi(-1) = (-1)^{(3+\varepsilon)/2} = \varepsilon, \quad \psi(k) = 1 .$$

If either

$$3 = P_1, \quad \eta = -\varepsilon \quad \text{and} \quad j = 1 \qquad (25)$$

or

$$3 = P_h, \quad \eta = -1 \quad \text{and} \quad j = h - 1, \qquad (26)$$

we put

$$\psi = \psi_1 \psi_2 ,$$

where

$$\psi_1(x) = \prod_{\substack{1 \leq i \leq h \\ p_i \nmid 3P}} \left(\frac{x}{p_i} \right), \quad \psi_2 = g_P^{(3+\varepsilon\psi_1(-1))/2} .$$

ψ_1, ψ_2 are primitive characters mod $\prod_{p_i \nmid 3P} p_i$ and P, respectively. Thus ψ is a primitive character mod $\prod_{p_i \nmid 3P} p_i P$, nonreal since $\zeta_{\phi(P)}^2 \notin \mathbb{R}$. We have

$$\psi(-1) = \psi_1(-1)\psi_2(-1) = \psi_1(-1)(-1)^{(3+\varepsilon\psi_1(-1))/2} = \varepsilon .$$

Moreover, (25) implies $k \equiv -1 \,(\bmod\,(f/3))$, $\psi(k) = \psi_1(-1)\psi_2(-1) = \varepsilon \neq \eta$; (26) implies $k \equiv 1 \,(\bmod\,(f/3))$, $\psi(k) = 1 \neq \eta$. The only primes satisfying (14) are 2 and 3 and the equalities $\psi(2)^2 = 1, \psi(3)^2 = 1$ would give $2^4 \equiv 1 \,(\bmod\,P)$ or $3^4 \equiv 1 \,(\bmod\,P)$, $P = 5$, contrary to (21).

This completes the proof of the proposition except for the last statement. That follows by inspection of the argument, where the prime 17 is avoided only because of the condition $\psi(2)^2 \neq 1$. ∎

2.5   *Proof of Theorem 62*

We shall distinguish four cases

(i) $f \not\equiv 0 \,(\mathrm{mod}\,4)$,

(ii) $f \equiv 4 \,(\mathrm{mod}\,8)$,

(iii) $f = 2^\alpha f_1$, where $\alpha \geq 3$, f_1 odd, $f_1 \nmid 15$,

(iv) $f = 2^\alpha f_1$, where $f_1 \mid 15$ and either $\alpha \geq 5$ or $\alpha = 4, 5 \mid f_1$.

For the sake of brevity in each case we only define a character ψ with the required properties, leaving to the reader the actual verification.

Case (i). If f is odd it suffices to take in Theorem 61 $\eta = 1$. If $f \equiv 2 \,(\mathrm{mod}\,4)$ it suffices in view of Theorem 61 to consider the case $17 \mid f \mid 510$. If $k \not\equiv \pm 1$ $(\mathrm{mod}\,17)$ or $k \equiv -1 \,(\mathrm{mod}\,17)$, $\varepsilon = -1$ we put

$$\psi = g_{17}^{(3+\varepsilon)/2}.$$

If $k \equiv -1 \,(\mathrm{mod}\,17)$, $\varepsilon = 1$ there is a prime $p \mid f$ such that $k \not\equiv -1$ $(\mathrm{mod}\,p)$. We put

$$\psi = g_p g_{17}.$$

If $k \equiv 1 \,(\mathrm{mod}\,17)$, there is a prime $p \mid f$ such that $k \not\equiv 1 \,(\mathrm{mod}\,p)$, $p = 3$ or 5. We put

$$\psi = \begin{cases} g_3 g_{17}^{(3-\varepsilon)/2}, & \text{if } p = 3, \\ g_5^2 g_{17}^{(3+\varepsilon)/2}, & \text{if } p = 5,\ k \equiv \pm 2 \,(\mathrm{mod}\,5), \\ g_5 g_{17}^{(7-\varepsilon)/2}, & \text{if } p = 5,\ k \equiv -1 \,(\mathrm{mod}\,5) \\ & \text{and either } f = 170 \text{ or } \varepsilon = 1, \\ g_3 g_5 g_{17}, & \text{otherwise.} \end{cases}$$

Case (ii). In Theorem 61 we replace f by $f/4$, ε by $-\varepsilon$, η by $g_4(k)$. If $k \not\equiv \pm 1$ $(\mathrm{mod}\, f/4)$ there exists by virtue of Theorem 61 a nonreal primitive character ψ' of conductor d such that $d \mid f/4$ and

$$\psi'(-1) = -\varepsilon, \quad \psi'(k) \neq g_4(k) \tag{27}$$

and

$$\psi'(p)^2 \neq 1 \quad \text{for all primes } p \mid f/4,\ p \nmid d. \tag{28}$$

If $k \equiv -1 \,(\mathrm{mod}\,4)$, $k \equiv 1 \,(\mathrm{mod}\, f/4)$ or if $k \equiv 1 \,(\mathrm{mod}\,4)$, $k \equiv -1$ $(\mathrm{mod}\, f/4)$, $\varepsilon = 1$ the properties (27) and (28) belong to every primitive character $\psi' \bmod f/4$ with $\psi'(-1) = -\varepsilon$. In each case the character

$$\psi = g_4 \psi'$$

satisfies the condition of the theorem.

In the remaining case $k \equiv 1 \,(\bmod\, 4)$, $k \equiv -1 \,(\bmod\, f/4)$, $\varepsilon = -1$ there exists a prime power P such that

$$P \Big| \frac{f}{4}, \quad \gcd\Big(P, \frac{f}{4P}\Big) = 1, \quad P \neq 3, 5.$$

We put

$$\psi = \psi_1 \psi_2,$$

where

$$\psi_1(x) = \prod_{\substack{p \,|\, f/4P \\ p \text{ prime}}} \Big(\frac{x}{p}\Big), \quad \psi_2 = g_P^{(3-\psi_1(-1))/2}.$$

Case (iii). If $k \not\equiv \pm 1 \,(\bmod\, f_1)$ we argue as in the case $\alpha = 2$ above. If $k \equiv 1 \,(\bmod\, f_1)$ we have $k \not\equiv 1 \,(\bmod\, 2^\alpha)$. We choose a character $\psi_2 \bmod 2^\alpha$ such that $\psi_2(k) \neq 1$ and then a nonreal primitive character $\psi_1 \bmod f_1$ such that

$$\psi_1(-1) = \varepsilon \psi_2(-1).$$

Then $\psi = \psi_1 \psi_2$ has the required properties.

If $k \equiv -1 \,(\bmod\, f_1)$ we have $k \not\equiv -1 \,(\bmod\, 2^\alpha)$. We choose a nontrivial character $\psi_2 \bmod 2^\alpha$ such that $\psi_2(-k) \neq \varepsilon$ and then a nonreal primitive character $\psi_1 \bmod f_1$ such that

$$\psi_1(-1) = \varepsilon \psi_2(-1),$$

The character $\psi = \psi_1 \psi_2$ has the required properties.

Case (iv). If $k \not\equiv \pm 1 \,(\bmod\, 2^{\alpha-1})$ we put

$$\psi = g_4^{(3+\varepsilon)/2} g_{2^\alpha}.$$

The same formula is good if $k \equiv 1 + 2^{\alpha-1} \,(\bmod\, 2^\alpha)$, or $k \equiv -1 + 2^{\alpha-1}$ $(\bmod\, 2^\alpha)$, $\varepsilon = 1$, or $k \equiv -1 \,(\bmod\, 2^\alpha)$, $\varepsilon = -1$.

If $k \equiv -1 + 2^{\alpha-1} \,(\bmod\, 2^\alpha)$, $\varepsilon = -1$ we put

$$\psi = \begin{cases} g_4 g_{2^\alpha}^2, & \text{if } \alpha \geq 5, \\ g_5 g_{2^\alpha}^2, & \text{if } \alpha = 4, k \not\equiv 1 \,(\bmod\, 5), \\ g_5 g_{2^\alpha}, & \text{if } \alpha = 4, k \equiv 1 \,(\bmod\, 5), f = 80, \\ g_3 g_5^2, & \text{if } \alpha = 4, k \equiv 1 \,(\bmod\, 5), f = 240, k \equiv 1 \,(\bmod\, 3), \\ g_3 g_4 g_5 g_{2^\alpha}, & \text{if } \alpha = 4, k \equiv 1 \,(\bmod\, 5), f = 240, k \equiv -1 \,(\bmod\, 3). \end{cases}$$

If $k \equiv \eta \,(\bmod\, 2^\alpha)$, $\eta = \pm 1$, $\varepsilon = 1$, then f_1 has a prime factor p such that $k \not\equiv \eta \,(\bmod\, p)$.

If $\alpha \geq 5$, or if $p = 3$, or if $p = 5 = f_1$ we put

$$\psi = g_4 g_p g_{2^\alpha}.$$

If $f = 240$, $k \equiv \eta \,(\bmod\, 48)$, $k \not\equiv \eta \,(\bmod\, 5)$, $\varepsilon = 1$ we put

$$\psi = g_3 g_5 g_{2^\alpha}.$$

In the remaining case $k \equiv 1 \,(\bmod\, 2^\alpha)$, $\varepsilon = -1$, f_1 has a prime factor p such that $k \not\equiv 1 \,(\bmod\, p)$. We put

$$\psi = \begin{cases} g_p g_{2^\alpha}, & \text{if } \alpha \geq 5, \text{ or if } p = 3, \text{ or if } p = 5 = f_1, \\ g_5 g_{2^\alpha}^2, & \text{otherwise.} \end{cases}$$ ∎

3.   CHARACTER SUMS IN TERMS OF BERNOULLI NUMBERS

3.1   *Consequences of the Zagier Formula*

Let χ be a Dirichlet character mod M and let N be a multiple of M. For any integer $r > 1$ prime to N and any natural number m, we recall the following formula given in section 9.2 of Chapter I, Theorem 2

$$mr^{m-1} \sum_{0 < n < (N/r)} \chi(n) n^{m-1} = -B_{m,\chi} r^{m-1} + \frac{\bar{\chi}(r)}{\phi(r)} \sum_{\psi} \bar{\psi}(-N) B_{m,\chi\psi}(N), \quad (29)$$

where the last sum is over all Dirichlet characters $\psi \bmod r$. Here for a Dirichlet character θ, $B_{m,\theta}$ denotes the generalized Bernoulli number attached to θ and $B_{m,\theta}(X)$ denotes the generalized Bernoulli polynomial defined in sections 3.1 and 3.2 of Chapter I respectively. We recall that if the character θ modulo T is induced from a character θ_1 modulo some divisor of T then we have

$$B_{m,\theta} = B_{m,\theta_1} \prod_{\substack{p \mid T \\ p \text{ prime}}} \left(1 - \theta_1(p) p^{m-1}\right), \quad (30)$$

where the product is over all primes p dividing T, see section 1.8, Chapter II.

If $m = 1$ and χ is not the trivial character, then formula (29) becomes

$$\sum_{0 < n < (N/r)} \chi(n) = -B_{1,\chi} + \frac{\bar{\chi}(r)}{\phi(r)} \sum_{\psi} \bar{\psi}(-N) B_{1,\chi\psi}$$

because for nontrivial characters θ we have $B_{0,\theta} = 0$ and so $B_{1,\theta}(X) = B_{1,\theta}$.

Let χ be a nontrivial character of conductor M. Then the above identity and (30) give

$$\sum_{0 < n < (N/r)} \chi(n) = -B_{1,\chi} + \frac{\bar{\chi}(r)}{\phi(r)} \sum_{\psi} \bar{\psi}(-N) \prod_{\substack{q \mid r \\ q \text{ prime}}} (1 - \chi\psi(q)) B_{1,\chi\psi}, \quad (31)$$

where the last sum is over all primitive characters ψ with conductor f_χ such that $f_\chi \mid r$ and where ψ has parity opposite to that of χ (recall that $B_{1,\theta} = 0$ for even nontrivial θ).

3.2   *Short Sums of Kronecker Symbols*

Let $\chi = \left(\dfrac{\mathcal{D}}{\cdot}\right)$, where $\mathcal{D} \neq 1$ is a fundamental discriminant. We shall consider sums of the form

$$S(\mathcal{D}, q_1, q_2) = \sum_{q_1|\mathcal{D}|<n<q_2|\mathcal{D}|} \left(\frac{\mathcal{D}}{n}\right),$$

where

$$q_1, q_2 \in \mathbb{Q}; \ 0 \leq q_1 < q_2 \leq 1,$$

$$q_i = \frac{k_i}{r_i}, \ r_i \in \mathbb{N}, \ k_i \in \mathbb{Z}, \ \gcd(k_i, r_i) = 1, \ \gcd(r_i, \mathcal{D}) = 1 \ (i = 1, 2).$$

We put $r = \mathrm{lcm}(r_1, r_2)$ and $\rho = \gcd(r_1, r_2)$, so that

$$r = \frac{r_1}{\rho} \cdot \rho \cdot \frac{r_2}{\rho} \ \text{ and } \ \gcd\left(\frac{r_1}{\rho}, \frac{r_2}{\rho}\right) = 1.$$

The formula (31) implies

THEOREM 63   [Schinzel, Urbanowicz and van Wamelen, 1999, Proposition 4]
For $r > 1$

$$S(\mathcal{D}, q_1, q_2) = \sum_\psi c_\psi B_{1,\chi\psi},$$

where the sum is over all primitive characters ψ with conductor f_ψ such that $f_\psi \mid r$ and $\psi(-1) = -\chi(-1)$ and where the c_ψ are given by the formulae

1. *if $q_1 = 0$ $(r_1 = 1, k_2 = k, r_2 = r)$*

   (a) *if ψ is trivial*

   $$c_\psi = -1 + \frac{\chi(r)}{\phi(r)} \prod_{\substack{q \mid r \\ q \text{ prime}}} (1 - \chi(q)),$$

   (b) *otherwise*

   $$c_\psi = \frac{\chi(r)\bar\psi(-k|\mathcal{D}|)}{\phi(r)} \prod_{\substack{q \mid r \\ q \text{ prime}}} (1 - \chi(q)\psi(q)). \tag{32}$$

2. *if $q_1 > 0$*

(a) *if $f_\psi \mid \rho$*

$$c_\psi = c'_\psi \cdot c''_\psi,$$ (33)

where

$$c'_\psi = \chi(\rho)\bar{\psi}(-|\mathcal{D}|) \prod_{\substack{q \mid \rho \\ q \, \mathrm{prime}}} (1 - \chi(q)\psi(q))$$

and

$$c''_\psi = \frac{\chi(r_2/\rho)\bar{\psi}(k_2)}{\phi(r_2)} \prod_{\substack{q \mid r_2, \, q \nmid \rho \\ q \, \mathrm{prime}}} (1 - \chi(q)\psi(q))$$

$$- \frac{\chi(r_1/\rho)\bar{\psi}(k_1)}{\phi(r_1)} \prod_{\substack{q \mid r_1, \, q \nmid \rho \\ q \, \mathrm{prime}}} (1 - \chi(q)\psi(q)),$$

(b) *if $f_\psi \mid r_2$ but $f_\psi \nmid \rho$*

$$c_\psi = \frac{\chi(r_2)\bar{\psi}(-k_2|\mathcal{D}|)}{\phi(r_2)} \prod_{\substack{q \mid r_2 \\ q \, \mathrm{prime}}} (1 - \chi(q)\psi(q)),$$ (34)

(c) *if $f_\psi \mid r_1$ but $f_\psi \nmid \rho$*

$$c_\psi = -\frac{\chi(r_1)\bar{\psi}(-k_1|\mathcal{D}|)}{\phi(r_1)} \prod_{\substack{q \mid r_1 \\ q \, \mathrm{prime}}} (1 - \chi(q)\psi(q)),$$ (35)

(d) *if $f_\psi \nmid r_1$, $f_\psi \nmid r_2$*

$$c_\psi = 0.$$

COROLLARY 1 *For all characters ψ satisfying (3) we have (4), (5) and (6).*

PROOF. It follows by calculation from the theorem that

$$c_\psi(\mathcal{D}, 1 - q_2, 1 - q_1) = -\bar{\psi}(-1)c_\psi(\mathcal{D}, q_1, q_2),$$
$$c_\psi(\mathcal{D}, 0, 1 - q_2) = \bar{\psi}(-1)c_\psi(\mathcal{D}, 0, q_2),$$
$$c_\psi(\mathcal{D}, q_1, 1 - q_2) = c_\psi(\mathcal{D}, q_1, q_2) \text{ if } q_1 + q_2 \leq 1 \text{ and } \psi(-1) = 1,$$

and then we use the condition $\psi(-1) = -\chi(-1)$. ∎

COROLLARY 2 *For all characters ψ satisfying* (3), (8) *implies* (7).

PROOF. It follows from the theorem that

$$c_\psi(\mathcal{D}, q_1, q_2) = c_\psi(\mathcal{D}_0, q_1, q_2),$$

provided $\chi_{\mathcal{D}}(-1) = \chi_{\mathcal{D}_0}(-1)$ and $\chi_{\mathcal{D}}(q) = \chi_{\mathcal{D}_0}(q)$ for all primes $q \mid r$. The first condition gives $\operatorname{sgn}\mathcal{D} = \operatorname{sgn}\mathcal{D}_0$, the second $\mathcal{D} \equiv \mathcal{D}_0 \,(\operatorname{mod} 8)$ if $q = 2$ and $\mathcal{D} \equiv \mathcal{D}_0 \,(\operatorname{mod} q)$ if $q > 2$.   ∎

COROLLARY 3 *For all characters ψ satisfying* (3) *we have* (9).

PROOF. This follows from the theorem since χ is real.   ∎

COROLLARY 4 *If $c_\psi(\mathcal{D}, q_1, q_2) = 0$ for all characters ψ with $f_\psi \mid r$ and $\psi(-1) = -\chi_{\mathcal{D}}(-1)$, then*

$$S(\chi, q_1, q_2) = \sum_{q_1 m < n < q_2 m} \chi(n) = 0$$

for every character $\chi \bmod m$ induced from $\chi_{\mathcal{D}}$ provided $\gcd(m, r) = 1$.

PROOF. It follows from (30) that the coefficient of $B_{1,\chi\psi}$ in the representation of $S(\chi, q_1, q_2)$ as the linear combination of generalized Bernoulli numbers derived from (31) is divisible by $c_\psi(\mathcal{D}, q_1, q_2)$, hence 0.   ∎

The last corollary justifies the remark made in the introduction, about nonfundamental discriminants $\mathcal{D}$. Indeed if $\mathcal{D} = \mathcal{D}_1 s^2$, where $\mathcal{D}_1 \in \mathcal{F}$ then $\left(\dfrac{\mathcal{D}}{\cdot}\right)$ is induced from $\left(\dfrac{\mathcal{D}_1}{\cdot}\right)$ and, assuming $\gcd(s, r) = 1$, the residue class of $\mathcal{D}_1 s^2 \bmod r \dfrac{\gcd(r^3, 8)}{\gcd(r, 8)}$ is admissible in the sense of (8).

4.   APPLICATIONS

4.1   Main Result

In this section we shall determine when card $C(D, q_1, q_2) = 0$ or 1. Recall that $r = \operatorname{lcm}(r_1, r_2)$ and $\rho = \gcd(r_1, r_2)$.

THEOREM 64 [Schinzel, Urbanowicz and van Wamelen, 1999, Proposition 5] *Let $C = C(D, q_1, q_2)$, where $D \in \mathcal{F}$, $\gcd(D, r) = 1$ and $\langle q_1, q_2 \rangle$ satisfy* (1). *Assume that $r \nmid 8 \cdot 3 \cdot 5$ and besides*

(a) $r \neq 14, 18$ *if* $\max\{r_1, r_2\} = 2\min\{r_1, r_2\} \equiv 2\,(\operatorname{mod} 4)$,

(b) $r \neq 16, 48$ *if* $r_1 = r_2 = r$, $q_1 + q_2 \neq 1$,

(c) $D > 0$ if $q_1 + q_2 = 1$.

Then C contains a nonreal character and

$$\text{card } C \geq 2.$$

PROOF. It is enough to show that under the assumption of the theorem there exists both an even and an odd primitive nonreal character ψ, such that $c_\psi \neq 0$. In view of (4) we may assume without loss of generality that $r_1 \leq r_2$. If $k_1 = 0$ a required ψ exists by virtue of Theorem 60 and formula (32). If $k_1 > 0$ we consider four cases

(i) $r_1 \nmid r_2$,

(ii) $r_1 \mid r_2$, $r_1 \neq r_2$,

(iii) $r_1 = r_2$, $q_1 + q_2 \neq 1$,

(iv) $r_1 = r_2$, $q_1 + q_2 = 1$.

Case (i). In this case we have $r_1 < r_2$. If $r_2 \nmid 120$ and r_2 is odd or divisible by 4 then by virtue of Theorem 60 we can find both an even and an odd primitive nonreal character ψ of conductor $f_\psi = r_2$. Moreover $f_\psi \nmid \rho$ (recall $r_1 < r_2$) and it follows by (34) that for this ψ, $c_\psi \neq 0$. A similar argument shows that we get a nonzero c_ψ if $r_1 \nmid 120$ and r_1 is odd or divisible by 4. Then by virtue of Theorem 60 we can find a primitive nonreal character of prescribed parity of conductor $f_\psi = r_1$ and by assumption (i) we have $f_\psi \nmid \rho$. Here we can use formula (35) and it remains to consider the cases when $r_2 \nmid 120$, $2 \parallel r_2$ or $r_1 \nmid 120$, $2 \parallel r_1$.

If $r_2 \nmid 120$ and $2 \parallel r_2$ by Theorem 60 we can find both an even and an odd primitive nonreal character ψ of conductor $f_\psi = (r_2/2)$ such that $\psi(2) \neq \pm 1$. Moreover the divisibility $(r_2/2) \mid \rho$ would imply $r_1 = r_2$ or $r_2 = 2r_1$, which is not the case. Thus $f_\psi \nmid \rho$ and by virtue of (34) $c_\psi \neq 0$.

If $r_1 \nmid 120$ and $2 \parallel r_1$ by Theorem 60 we can find a primitive nonreal character of prescribed parity of conductor $f_\psi = (r_1/2)$ such that $\psi(2) \neq \pm 1$. If $f_\psi \nmid \rho$ we can use formula (35) and $c_\psi \neq 0$. If $(r_1/2) \mid \rho$ we have $(r_1/2) \mid r_2$ and in consequence $r_2 \nmid 120$. This case was considered above.

Case (ii). Here we may assume that $r_2 \nmid 120$ and by Theorem 60 we can find both an odd and an even primitive nonreal character of conductor $f_\psi = (r_2/2)$ such that $\psi(2) \neq \pm 1$. If $f_\psi \nmid \rho$ we can use formula (34) and $c_\psi \neq 0$. The divisibility $(r_2/2) \mid \rho$ implies $(r_2/2) \mid r_1$ and by assumption we obtain $r_2 = 2r_1$.

Let $r_2 = 2r_1$ and $r_2 \nmid 120$. Then r_2 is even. If r_1 is divisible by 4 in virtue of Theorem 60 we can find a nonreal primitive character of prescribed parity of conductor $f_\psi = r_2$. By assumption we have $f_\psi \nmid \rho$. Consequently we can use formula (34) and $c_\psi \neq 0$.

It remains to consider the case $2 \parallel r_2$ (r_1 odd). By virtue of Theorem 61 (for $k = 8$) since by (a) $r_1 \nmid 7$ and $r_1 \nmid 9$ we can find a primitive nonreal character ψ of prescribed parity of conductor $d \mid r_1$ such that $\psi(8) \neq -\chi(2)$ and $\psi(q)^2 \neq 1$ for all primes q such that $q \mid r_1$ and $q \nmid d$. By virtue of formula (33) it follows easily that $c'_\psi \neq 0$ for this ψ. We shall prove the same for c''_ψ by contradiction. The equality $c''_\psi = 0$ would imply

$$\chi(2)\bar\psi(k_2)\left(1 - \chi(2)\psi(2)\right) - \bar\psi(k_1) = 0,$$

and in consequence

$$\psi(k) + \psi(2) = \chi(2), \qquad (36)$$

where $k_2 \equiv kk_1 \pmod d$. Therefore we obtain

$$|\psi(t) + 1| = 1,$$

where $k \equiv 2t \pmod d$ and hence

$$\operatorname{Re}\psi(t) = -\frac{1}{2}.$$

Therefore $\psi(t) = \zeta_3^{\pm 1}$ and $\psi(k) = \zeta_3^{\pm 1}\psi(2)$. Substituting it into formula (36) gives

$$\psi(2) = -\chi(2)\zeta_3^{\pm 1},$$

which implies

$$\psi(8) = -\chi(2),$$

a contradiction.

Case (iii). Here in view of (33) we have

$$c'_\psi = \chi(r)\bar\psi(-|\mathcal{D}|) \prod_{\substack{q \mid r \\ q \text{ prime}}} (1 - \chi(q)\psi(q))$$

and

$$\phi(r)c''_\psi = \bar\psi(k_2) - \bar\psi(k_1).$$

Let k denote a natural number such that $1 < k < r - 1$ and $k_2 \equiv kk_1 \pmod r$. Since $q_1 < q_2$, $q_1 + q_2 \neq 1$ we have $k_2 \not\equiv \pm k_1 \pmod r$, $k \not\equiv \pm 1 \pmod r$. By (b) $r \nmid 240$ or $80 \mid r$, hence by virtue of Theorem 62 we can find a nonreal primitive character ψ of prescribed parity of conductor $d \mid r$ such that $\psi(k) \neq 1$ and $\psi(q)^2 \neq 1$ for all primes $q \mid r$ and $q \nmid d$. Hence we have $c'_\psi \neq 0$ and $c''_\psi \neq 0$, and in consequence $c_\psi \neq 0$.

Case (iv). Here by (c) we have $\chi(-1) = 1$, hence all characters ψ in question are odd and $\phi(r)c''_\psi = 2\bar\psi(k_2) \neq 0$. Since $r \nmid 8 \cdot 3 \cdot 5$, by Theorem 60 there exists an odd primitive nonreal character ψ of conductor $r\dfrac{\gcd(r,4)}{\gcd(r^2,4)}$ with $\psi(2)^2 \neq 1$ if $2 \| r$. For this character we have $c'_\psi \neq 0$ and in consequence $c_\psi \neq 0$.   ∎

4.2   *Proof of Theorem 59*

Theorem 64 implies that there is only a small finite set of $\langle q_1, q_2\rangle$ satisfying (1) for which $\operatorname{card} C(\mathcal{D}, q_1, q_2) < 2$ for at least one $\mathcal{D} \in \mathcal{F}$ prime to r (with $\mathcal{D} > 0$ if $q_1 + q_2 = 1$). It is now an easy matter to write a computer program to find, given q_1 and q_2, the congruence $|\mathcal{D}|$ must satisfy in order to make at most one of the $c_\psi \neq 0$. The results of such a search are presented in the next section. It follows from them that if we only require $\operatorname{card} C \geq 2$, the conditions on r in Theorem 64 can be relaxed to $r \nmid 4 \cdot 3 \cdot 5$, $r \neq 8$, 24 and both (a), (c).   ∎

4.3   *Remarks*

Using Theorem 64 one can find all pairs $\langle q_1, q_2\rangle$ such that $C(\mathcal{D}, q_1, q_2)$ consists only of real characters for at least one $\mathcal{D}$ in $\mathcal{F}$ prime to r. For such pairs and for all $\mathcal{D}$ in $\mathcal{F}$ from a certain arithmetic progression

$$S(\mathcal{D}, q_1, q_2) = \sum_{i=1}^{k} a_i h(-e_i|\mathcal{D}|),$$

where the coefficients a_i, e_i do not depend on $\mathcal{D}$.

5.   TABLES

5.1   *Class Numbers as Single Sums of Kronecker Symbols*

In the following table we list values for e, q_1, q_2, c, s_0 and m such that $0 \leq q_1 < q_2$, $q_1 + q_2 \leq 1$, $q_2 \leq \dfrac{1}{2}$ if $q_1 = 0$

$$h(-e|\mathcal{D}|) = c \sum_{q_1|\mathcal{D}| < n < q_2|\mathcal{D}|} \left(\frac{\mathcal{D}}{n}\right),$$

for all fundamental discriminants $\mathcal{D} \neq 1$ relatively prime to r such that $|\mathcal{D}| \equiv s_0$ ($\operatorname{mod} m$), $e|\mathcal{D}| > 4$ and if m is odd, $\mathcal{D}$ has the prescribed sign. For $q_2 > \dfrac{1}{2}$ we exclude $\mathcal{D} < 0$. We list 222 such formulae involving 116 different pairs $\langle q_1, q_2\rangle$. The section number sign § (resp. dagger †) means that $\mathcal{D} < 0$ (resp. $\mathcal{D} > 0$).

TABLE 3

Dirichlet type class number formulae (revisited)

e	q_1, q_2	c	$s_0;m$	e	q_1, q_2	c	$s_0;m$	e	q_1, q_2	c	$s_0;m$
1	$0, \frac{1}{2}$	$\frac{1}{3}$	$3;8$	1	$\frac{5}{14}, \frac{3}{7}$	1	$11, 43;56$	3	$0, \frac{5}{12}$	-2	$17;24$
1	$0, \frac{1}{2}$	1	$7;8$				$51;56$	3	$\frac{1}{12}, \frac{1}{6}$	2	$17;24$
1	$0, \frac{1}{3}$	$\frac{1}{2}$	$1;3^{\S}$	1	$\frac{1}{18}, \frac{2}{9}$	-1	$19;24$	3	$\frac{1}{12}, \frac{1}{4}$	-2	$1;24$
1	$0, \frac{1}{3}$	1	$2;3^{\S}$	1	$\frac{1}{9}, \frac{5}{18}$	1	$19;24$	3	$\frac{1}{12}, \frac{1}{3}$	2	$5;24$
1	$0, \frac{1}{4}$	1	$7;8$	1	$\frac{7}{18}, \frac{4}{9}$	-1	$19;24$	3	$\frac{1}{12}, \frac{5}{12}$	-1	$1;8$
1	$\frac{1}{4}, \frac{1}{2}$	$\frac{1}{3}$	$3;8$	1	$\frac{1}{10}, \frac{1}{4}$	-1	$11, 19;40$	3	$\frac{1}{12}, \frac{1}{2}$	-2	$17;24$
1	$0, \frac{1}{6}$	-1	$19;24$	1	$\frac{1}{4}, \frac{3}{10}$	1	$11, 19;40$	3	$\frac{1}{12}, \frac{2}{3}$	-2	$5;24$
1	$0, \frac{1}{6}$	1	$7, 11;24$	1	$\frac{1}{10}, \frac{1}{6}$	-1	$43, 67;120$	3	$\frac{1}{12}, \frac{2}{3}$	-1	$17;24$
			$23;24$	1	$\frac{1}{10}, \frac{1}{6}$	$-\frac{1}{2}$	$19, 91;120$	3	$\frac{1}{12}, \frac{5}{6}$	$-\frac{2}{3}$	$17;24$
1	$\frac{1}{6}, \frac{1}{3}$	$\frac{1}{3}$	$19;24$	1	$\frac{1}{10}, \frac{1}{6}$	1	$83, 107;120$	3	$\frac{1}{12}, \frac{11}{12}$	-1	$17;24$
1	$\frac{1}{6}, \frac{1}{3}$	1	$7;24$	1	$\frac{1}{10}, \frac{1}{3}$	$\frac{1}{2}$	$43, 67;120$	3	$\frac{1}{6}, \frac{5}{12}$	$-\frac{2}{3}$	$17;24$
1	$\frac{1}{6}, \frac{1}{2}$	$\frac{1}{4}$	$19;24$	1	$\frac{1}{10}, \frac{1}{3}$	1	$19, 83;120$	3	$\frac{1}{6}, \frac{7}{12}$	-2	$17;24$
1	$\frac{1}{6}, \frac{1}{2}$	$\frac{1}{2}$	$11;24$				$91, 107;120$	3	$\frac{1}{4}, \frac{5}{12}$	-2	$1;24$
1	$\frac{1}{3}, \frac{1}{2}$	-1	$7;24$	1	$\frac{1}{6}, \frac{3}{10}$	-1	$83, 107;120$	3	$\frac{1}{3}, \frac{5}{12}$	-2	$5;24$
1	$\frac{1}{3}, \frac{1}{2}$	$\frac{1}{2}$	$11;24$	1	$\frac{1}{6}, \frac{3}{10}$	$\frac{1}{2}$	$19, 91;120$	3	$\frac{1}{3}, \frac{5}{12}$	-1	$17;24$
1	$\frac{1}{3}, \frac{1}{2}$	1	$19;24$	1	$\frac{1}{6}, \frac{3}{10}$	1	$43, 67;120$	3	$\frac{1}{3}, \frac{7}{12}$	-2	$5;24$
1	$0, \frac{1}{10}$	1	$11, 19;40$	1	$\frac{3}{10}, \frac{1}{3}$	$\frac{1}{2}$	$43, 67;120$	3	$\frac{5}{12}, \frac{1}{2}$	2	$17;24$
1	$0, \frac{3}{10}$	1	$11, 19;40$	1	$\frac{3}{10}, \frac{1}{3}$	1	$19, 83;120$	3	$\frac{5}{12}, \frac{7}{12}$	1	$17;24$
1	$\frac{1}{10}, \frac{1}{2}$	$\frac{1}{3}$	$3, 27;40$				$91, 107;120$	3	$\frac{1}{18}, \frac{2}{9}$	-2	$1;24$
1	$\frac{1}{10}, \frac{1}{2}$	$\frac{1}{2}$	$11, 19;40$	3	$0, \frac{1}{3}$	2	$0;1^{\dagger}$	3	$\frac{1}{18}, \frac{2}{9}$	2	$17;24$
1	$\frac{3}{10}, \frac{1}{2}$	$\frac{1}{3}$	$3, 27;40$	3	$\frac{1}{3}, \frac{2}{3}$	-1	$0;1^{\dagger}$	3	$\frac{1}{9}, \frac{5}{18}$	-2	$1;24$
1	$\frac{3}{10}, \frac{1}{2}$	$\frac{1}{2}$	$11, 19;40$	3	$0, \frac{1}{6}$	1	$1;8$	3	$\frac{1}{9}, \frac{5}{18}$	2	$17;24$
1	$\frac{1}{6}, \frac{1}{4}$	-1	$11;24$	3	$\frac{1}{6}, \frac{1}{3}$	-2	$1;8$	3	$\frac{7}{18}, \frac{5}{9}$	-2	$1;24$
1	$\frac{1}{6}, \frac{1}{4}$	1	$19;24$	3	$\frac{1}{6}, \frac{1}{3}$	2	$5;8$	3	$\frac{7}{18}, \frac{5}{9}$	2	$17;24$
1	$\frac{1}{4}, \frac{1}{3}$	$\frac{1}{2}$	$19;24$	3	$\frac{1}{6}, \frac{1}{2}$	-1	$1;8$	3	$\frac{1}{24}, \frac{5}{24}$	-2	$1;24$
1	$\frac{1}{4}, \frac{1}{3}$	1	$7, 11;24$	3	$\frac{1}{6}, \frac{2}{3}$	-2	$5;8$	3	$\frac{1}{24}, \frac{11}{24}$	-2	$17;24$
1	$\frac{1}{14}, \frac{2}{7}$	1	$11, 43;56$	3	$\frac{1}{6}, \frac{2}{3}$	$-\frac{2}{3}$	$1;8$	3	$\frac{5}{24}, \frac{7}{24}$	2	$17;24$
			$51;56$	3	$\frac{1}{6}, \frac{5}{6}$	$-\frac{1}{2}$	$1;8$	3	$\frac{7}{24}, \frac{11}{24}$	-2	$1;24$
1	$\frac{1}{7}, \frac{3}{14}$	-1	$11, 43;56$	3	$\frac{1}{3}, \frac{1}{2}$	-2	$1;4$	3	$\frac{1}{30}, \frac{1}{5}$	-2	$1, 49;120$
			$51;56$	3	$0, \frac{1}{12}$	2	$17;24$				

e	q_1, q_2	c	$s_0;m$	e	q_1, q_2	c	$s_0;m$	e	q_1, q_2	c	$s_0;m$
3	$\frac{1}{30}, \frac{11}{30}$	-1	$1,41;120$	4	$\frac{1}{6}, \frac{7}{12}$	-2	$13;24$	5	$\frac{1}{3}, \frac{2}{5}$	4	$11,14;15^{\S}$
			$49,89;120$	4	$\frac{1}{6}, \frac{3}{4}$	-2	$5;8$	5	$\frac{1}{10}, \frac{1}{4}$	2	$7;8$
3	$\frac{1}{30}, \frac{3}{5}$	-2	$41,89;120$	4	$\frac{1}{4}, \frac{5}{12}$	-2	$5;24$	5	$\frac{1}{5}, \frac{1}{4}$	4	$31,39;40$
3	$\frac{1}{5}, \frac{7}{30}$	2	$41,89;120$	4	$\frac{1}{4}, \frac{7}{12}$	-2	$5;24$	5	$\frac{1}{4}, \frac{3}{10}$	2	$7;8$
3	$\frac{1}{5}, \frac{11}{30}$	-2	$1,49;120$	4	$\frac{1}{4}, \frac{7}{12}$	-1	$13;24$	5	$\frac{1}{4}, \frac{2}{5}$	4	$31,39;40$
3	$\frac{1}{5}, \frac{17}{30}$	-2	$41,89;120$	4	$\frac{1}{3}, \frac{7}{12}$	-2	$1;24$	5	$\frac{1}{10}, \frac{1}{6}$	2	$7;8$
3	$\frac{7}{30}, \frac{2}{5}$	-2	$1,49;120$	4	$\frac{5}{12}, \frac{1}{2}$	-2	$13;24$	5	$\frac{1}{10}, \frac{1}{3}$	2	$23;24$
3	$\frac{7}{30}, \frac{17}{30}$	-1	$1,41;120$	4	$\frac{5}{12}, \frac{7}{12}$	-1	$13;24$	5	$\frac{1}{6}, \frac{1}{5}$	-4	$11,31;120$
			$49,89;120$	4	$\frac{1}{24}, \frac{7}{24}$	-2	$1;24$				$59,71;120$
3	$\frac{11}{30}, \frac{3}{5}$	2	$41,89;120$	4	$\frac{1}{24}, \frac{7}{24}$	2	$13;24$				$79,119;120$
3	$\frac{2}{5}, \frac{17}{30}$	-2	$1,49;120$	4	$\frac{1}{24}, \frac{11}{24}$	2	$13;24$	5	$\frac{1}{6}, \frac{3}{10}$	2	$7;8$
3	$\frac{1}{60}, \frac{11}{60}$	-2	$1,49;120$	4	$\frac{5}{24}, \frac{7}{24}$	2	$13;24$	5	$\frac{1}{6}, \frac{2}{5}$	4	$11,31;120$
3	$\frac{7}{60}, \frac{17}{60}$	-2	$1,49;120$	4	$\frac{5}{24}, \frac{11}{24}$	-2	$1;24$				$59,71;120$
3	$\frac{13}{60}, \frac{23}{60}$	-2	$1,49;120$	4	$\frac{5}{24}, \frac{11}{24}$	2	$13;24$				$79,119;120$
3	$\frac{19}{60}, \frac{29}{60}$	-2	$1,49;120$	4	$\frac{1}{60}, \frac{11}{60}$	2	$73,97;120$	5	$\frac{3}{10}, \frac{1}{3}$	-2	$23;24$
4	$0, \frac{1}{4}$	2	$1;4$	4	$\frac{7}{60}, \frac{17}{60}$	-2	$73,97;120$	7	$\frac{1}{14}, \frac{2}{7}$	2	$5;8$
4	$\frac{1}{4}, \frac{1}{2}$	-2	$1;4$	4	$\frac{13}{60}, \frac{23}{60}$	2	$73,97;120$	7	$\frac{1}{7}, \frac{11}{14}$	-2	$5;8$
4	$\frac{1}{4}, \frac{3}{4}$	-1	$1;4$	4	$\frac{19}{60}, \frac{29}{60}$	-2	$73,97;120$	7	$\frac{5}{14}, \frac{3}{7}$	-2	$5;8$
4	$\frac{1}{8}, \frac{3}{8}$	-2	$1;8$	5	$\frac{1}{5}, \frac{2}{5}$	2	$0;1^{\S}$	8	$\frac{1}{8}, \frac{1}{4}$	4	$7;8$
4	$\frac{1}{8}, \frac{3}{8}$	2	$5;8$	5	$\frac{1}{10}, \frac{1}{5}$	-4	$11,19;40$	8	$\frac{1}{8}, \frac{3}{8}$	2	$3;4$
4	$0, \frac{1}{12}$	2	$13;24$	5	$\frac{1}{10}, \frac{1}{5}$	4	$31,39;40$	8	$\frac{1}{8}, \frac{1}{2}$	4	$7;8$
4	$0, \frac{5}{12}$	2	$13;24$	5	$\frac{1}{10}, \frac{3}{10}$	1	$7;8$	8	$\frac{1}{8}, \frac{5}{8}$	-2	$1;4$
4	$\frac{1}{12}, \frac{1}{6}$	-2	$13;24$	5	$\frac{1}{10}, \frac{2}{5}$	$\frac{4}{3}$	$31,39;40$	8	$\frac{1}{4}, \frac{3}{8}$	4	$7;8$
4	$\frac{1}{12}, \frac{1}{4}$	2	$5;24$	5	$\frac{1}{10}, \frac{2}{5}$	4	$11,19;40$	8	$\frac{3}{8}, \frac{1}{2}$	-4	$7;8$
4	$\frac{1}{12}, \frac{1}{3}$	-2	$1;24$	5	$\frac{1}{10}, \frac{1}{2}$	2	$7;8$	8	$\frac{1}{24}, \frac{5}{24}$	2	$5;24$
4	$\frac{1}{12}, \frac{1}{2}$	-2	$13;24$	5	$\frac{1}{5}, \frac{3}{10}$	$\frac{4}{3}$	$31,39;40$	8	$\frac{1}{24}, \frac{7}{24}$	2	$5;12$
4	$\frac{1}{12}, \frac{7}{12}$	-1	$1;12$	5	$\frac{1}{5}, \frac{3}{10}$	4	$11,19;40$	8	$\frac{1}{8}, \frac{1}{6}$	4	$7;8$
4	$\frac{1}{12}, \frac{3}{4}$	-2	$5;24$	5	$\frac{1}{5}, \frac{1}{2}$	4	$31,39;40$	8	$\frac{1}{8}, \frac{1}{3}$	4	$23;24$
4	$\frac{1}{12}, \frac{3}{4}$	-1	$13;24$	5	$\frac{3}{10}, \frac{2}{5}$	-4	$31,39;40$	8	$\frac{1}{6}, \frac{3}{8}$	4	$7;8$
4	$\frac{1}{12}, \frac{5}{6}$	-2	$13;24$	5	$\frac{3}{10}, \frac{2}{5}$	4	$11,19;40$	8	$\frac{5}{24}, \frac{11}{24}$	-2	$5;12$
4	$\frac{1}{12}, \frac{11}{12}$	-1	$13;24$	5	$\frac{3}{10}, \frac{1}{2}$	-2	$7;8$	8	$\frac{7}{24}, \frac{11}{24}$	-2	$5;24$
4	$\frac{1}{6}, \frac{1}{4}$	2	$5;8$	5	$\frac{2}{5}, \frac{1}{2}$	-4	$31,39;40$	12	$\frac{1}{12}, \frac{1}{6}$	4	$7,11;24$
4	$\frac{1}{6}, \frac{5}{12}$	2	$13;24$	5	$\frac{1}{5}, \frac{1}{3}$	4	$11,14;15^{\S}$				$23;24$

e	q_1,q_2	c	$s_0;m$	e	q_1,q_2	c	$s_0;m$	e	q_1,q_2	c	$s_0;m$
12	$\frac{1}{12},\frac{1}{4}$	4	$7;8$	15	$\frac{2}{15},\frac{7}{15}$	-4	$\pm2;5^\dagger$	15	$\frac{2}{5},\frac{17}{30}$	-4	$29,53;120$
12	$\frac{1}{12},\frac{1}{3}$	4	$11,19;24$	15	$\frac{1}{30},\frac{1}{5}$	4	$29,53;120$				$77,101;120$
			$23;24$				$77,101;120$	15	$\frac{1}{60},\frac{11}{60}$	-4	$61,109;120$
12	$\frac{1}{12},\frac{5}{12}$	2	$7,11;24$	15	$\frac{1}{30},\frac{11}{30}$	2	$13,29;120$	15	$\frac{7}{60},\frac{17}{60}$	4	$61,109;120$
			$19,23;24$				$37,53;120$	15	$\frac{13}{60},\frac{23}{60}$	4	$61,109;120$
12	$\frac{1}{12},\frac{1}{2}$	4	$7;8$				$61,77;120$	15	$\frac{19}{60},\frac{29}{60}$	-4	$61,109;120$
12	$\frac{1}{6},\frac{5}{12}$	4	$7,11;24$				$101;120$	20	$\frac{1}{20},\frac{11}{20}$	-4	$1,9;20$
			$23;24$				$109;120$	20	$\frac{3}{20},\frac{13}{20}$	-4	$1,9;20$
12	$\frac{1}{4},\frac{5}{12}$	4	$7;8$	15	$\frac{1}{30},\frac{2}{5}$	4	$13;24$	20	$\frac{1}{60},\frac{11}{60}$	4	$17,113;120$
12	$\frac{1}{3},\frac{5}{12}$	4	$11,19;24$	15	$\frac{1}{5},\frac{11}{30}$	4	$29,53;120$	20	$\frac{7}{60},\frac{17}{60}$	4	$17,113;120$
			$23;24$				$77,101;120$	20	$\frac{13}{60},\frac{23}{60}$	-4	$17,113;120$
12	$\frac{5}{12},\frac{1}{2}$	-4	$7;8$	15	$\frac{1}{5},\frac{13}{30}$	4	$13;24$	20	$\frac{19}{60},\frac{29}{60}$	-4	$17,113;120$
12	$\frac{1}{24},\frac{5}{24}$	-4	$19;24$	15	$\frac{1}{5},\frac{23}{30}$	-4	$13;24$	24	$\frac{1}{24},\frac{11}{24}$	4	$7;12$
12	$\frac{1}{24},\frac{5}{24}$	4	$7;24$	15	$\frac{7}{30},\frac{2}{5}$	-4	$29,53;120$	24	$\frac{1}{24},\frac{13}{24}$	-4	$1;12$
12	$\frac{7}{24},\frac{11}{24}$	-4	$7;24$				$77,101;120$	24	$\frac{1}{24},\frac{17}{24}$	-4	$5;8$
12	$\frac{7}{24},\frac{11}{24}$	4	$19;24$	15	$\frac{7}{30},\frac{17}{30}$	-2	$13,29;120$	24	$\frac{1}{24},\frac{19}{24}$	-4	$5;12$
12	$\frac{1}{12},\frac{1}{10}$	4	$11,59;120$	15	$\frac{7}{30},\frac{17}{30}$	-2	$37,53;120$	24	$\frac{5}{24},\frac{7}{24}$	4	$7;12$
12	$\frac{1}{12},\frac{3}{10}$	4	$11,59;120$				$61,77;120$	24	$\frac{5}{24},\frac{13}{24}$	-4	$5;8$
12	$\frac{1}{10},\frac{5}{12}$	4	$11,59;120$				$101;120$	24	$\frac{5}{24},\frac{17}{24}$	-4	$1;12$
12	$\frac{3}{10},\frac{5}{12}$	4	$11,59;120$				$109;120$	24	$\frac{7}{24},\frac{13}{24}$	-4	$5;12$
15	$\frac{1}{15},\frac{11}{15}$	-4	$\pm2;5^\dagger$	15	$\frac{11}{30},\frac{2}{5}$	-4	$13;24$				

5.2   Vanishing of Short Sums of Kronecker Symbols

The following table lists values for r, q_1, q_2, s_0 and m such that $0 \leq q_1 < q_2$, $q_1 + q_2 \leq 1$ and $q_2 \leq \frac{1}{2}$ if $q_1 = 0$, r is the least common denominator of q_1 and q_2, and

$$\sum_{q_1|\mathcal{D}|<n<q_2|\mathcal{D}|} \left(\frac{\mathcal{D}}{n}\right) = 0,$$

for all fundamental discriminants $\mathcal{D} \neq 1$ (in fact for all nonsquare discriminants $\mathcal{D}$) such that $|\mathcal{D}| \equiv s_0 \pmod{m}$. For $q_2 > \frac{1}{2}$ we exclude $\mathcal{D} < 0$. There are 55 such formulae.

TABLE 4

Cases where short sums of Kronecker symbols vanish (revisited)

r	q_1, q_2	$s_0;m$	r	q_1, q_2	$s_0;m$	r	q_1, q_2	$s_0;m$
2	$0, \frac{1}{2}$	1;4	12	$\frac{1}{12}, \frac{5}{6}$	5;24	24	$\frac{1}{24}, \frac{5}{24}$	13;24
4	$0, \frac{1}{4}$	3;8	12	$\frac{1}{12}, \frac{11}{12}$	5;24	24	$\frac{1}{24}, \frac{11}{24}$	5;24
4	$\frac{1}{4}, \frac{1}{2}$	7;8	12	$\frac{1}{6}, \frac{1}{4}$	7;8	24	$\frac{5}{24}, \frac{7}{24}$	5;24
6	$0, \frac{1}{6}$	5;8	12	$\frac{1}{6}, \frac{5}{12}$	5;24	24	$\frac{7}{24}, \frac{11}{24}$	13;24
6	$\frac{1}{6}, \frac{1}{3}$	11;12	12	$\frac{1}{6}, \frac{7}{12}$	5;24	30	$\frac{1}{30}, \frac{1}{5}$	73, 97;120
6	$\frac{1}{6}, \frac{1}{2}$	5, 7;8	12	$\frac{1}{4}, \frac{1}{3}$	23;24	30	$\frac{1}{30}, \frac{11}{30}$	17, 33;40
6	$\frac{1}{6}, \frac{5}{6}$	5;8	12	$\frac{1}{4}, \frac{5}{12}$	13;24	30	$\frac{1}{30}, \frac{3}{5}$	17, 113;120
6	$\frac{1}{3}, \frac{1}{2}$	23;24	12	$\frac{1}{3}, \frac{7}{12}$	17;24	30	$\frac{1}{10}, \frac{1}{6}$	11, 59;120
10	$0, \frac{1}{10}$	3, 27;40	12	$\frac{5}{12}, \frac{1}{2}$	5;24	30	$\frac{1}{10}, \frac{1}{3}$	11, 59;120
10	$0, \frac{3}{10}$	3, 27;40	12	$\frac{5}{12}, \frac{7}{12}$	5;24	30	$\frac{1}{6}, \frac{3}{10}$	11, 59;120
10	$\frac{1}{10}, \frac{3}{10}$	3;8	14	$\frac{1}{14}, \frac{2}{7}$	3, 19, 27;56	30	$\frac{1}{5}, \frac{7}{30}$	17, 113;120
12	$0, \frac{1}{12}$	5;24	14	$\frac{1}{7}, \frac{3}{14}$	3, 19, 27;56	30	$\frac{1}{5}, \frac{11}{30}$	73, 97;120
12	$0, \frac{5}{12}$	5;24	14	$\frac{5}{14}, \frac{3}{7}$	3, 19, 27;56	30	$\frac{1}{5}, \frac{17}{30}$	17, 113;120
12	$\frac{1}{12}, \frac{1}{6}$	5;24	18	$\frac{1}{18}, \frac{2}{9}$	11;24	30	$\frac{7}{30}, \frac{2}{5}$	73, 97;120
12	$\frac{1}{12}, \frac{1}{4}$	13;24	18	$\frac{1}{9}, \frac{5}{18}$	11;24	30	$\frac{7}{30}, \frac{17}{30}$	17, 33;40
12	$\frac{1}{12}, \frac{1}{3}$	17;24	18	$\frac{7}{18}, \frac{4}{9}$	11;24	30	$\frac{3}{10}, \frac{1}{3}$	11, 59;120
12	$\frac{1}{12}, \frac{5}{12}$	5;8	20	$\frac{1}{10}, \frac{1}{4}$	3, 27;40	30	$\frac{11}{30}, \frac{3}{5}$	17, 113;120
12	$\frac{1}{12}, \frac{1}{2}$	5;24	20	$\frac{1}{4}, \frac{3}{10}$	3, 27;40	30	$\frac{2}{5}, \frac{17}{30}$	73, 97;120
12	$\frac{1}{12}, \frac{7}{12}$	5;12						

REMARK The case $r = 6$, $q_1 = 1/6$, $q_2 = 1/2$, $s_0 = 5$, 7, $m = 8$ was listed in [Johnson and Mitchell, 1977] only for $\mathcal{D} = p \equiv 5 \pmod 8$. The cases $r = 14$, $q_1 = 1/7$, $q_2 = 3/14$ (resp. $q_1 = 5/14$, $q_2 = 3/7$), $s_0 = 3$, 19, 27, $m = 56$ were listed in [Johnson and Mitchell, 1977] for $\mathcal{D} = -p$ with a misprint, 9 instead of 19.

Chapter VI

APPLICATIONS OF ZAGIER'S FORMULA (II)

In this chapter we present some recent results from [Fox, Urbanowicz and Williams, 1999]. Let d denote the discriminant of a quadratic field. Let n be the number of distinct prime factors of d. Recall that χ_d and $h(d)$ denote the character and class number of the field respectively. Recall also that B_{k,χ_d} denotes the generalized Bernoulli number attached to χ_d. It is shown in an elementary manner how Gauss' congruence for imaginary quadratic fields $h(d) \equiv 0 \pmod{2^{n-1}}$ (see Chapter II) can be deduced from Dirichlet's formula for $h(d)$ (see Chapter I). We also generalize the Gauss congruence to 2-integral rational numbers $(B_{k,\chi_d}/k)$. We prove that $(B_{k,\chi_d}/k) \equiv 0$ $\pmod{2^{n-1}}$ if $\chi_d(-1) = (-1)^k$. This is a further application of Zagier's identity.

1. PRELIMINARIES

1.1 *Some Elementary Observations*

As a consequence of his theory of genera for imaginary quadratic fields, Gauss obtained algebraically the congruence

$$h(d) \equiv 0 \pmod{2^{n-1}}, \tag{1}$$

where n is the number of distinct prime factors of d ($d < 0$), see section 8.5 in Chapter I. Dirichlet showed analytically that

$$h(d) = \frac{w(d)}{2(2 - \chi_d(2))} \sum_{r=1}^{\lfloor |d|/2 \rfloor} \chi_d(r) \tag{2}$$

(see formula (17) of Chapter I). We show in an elementary manner how Dirichlet's formula (2) can be made to yield Gauss' congruence (1). We accomplish this by putting (2) into a form (see Theorem 65) from which (1) can be deduced by induction on n. The proof of Theorem 65 is based on three elementary lemmas. The first gives a congruence modulo a power of 2 for $\phi(|d|)$, where ϕ is Euler's phi function. The second evaluates a sum which occurs in the proof of Theorem 65. The third puts (2) into a more general form for use in the proof of Theorem 65. Before the lemmas we give some elementary observations. The detailed proofs of the lemmas are left to the reader.

As d is the discriminant of a quadratic field, we have $d \equiv 1 \,(\mathrm{mod}\,4)$, $d \equiv 8 \,(\mathrm{mod}\,16)$ or $d \equiv 12 \,(\mathrm{mod}\,16)$. Moreover, we have

$$d = \prod_{\substack{p \mid d \\ p \text{ prime}}} p^*,$$

where the prime discriminant p^* corresponding to the prime $p \mid d$ is given by

$$p^* = (-1)^{(p-1)/2} p,$$

if p is odd, and

$$2^* = \begin{cases} 8, & \text{if } d \equiv 8 \,(\mathrm{mod}\,32), \\ -8, & \text{if } d \equiv 24 \,(\mathrm{mod}\,32), \\ -4, & \text{if } d \equiv 12 \,(\mathrm{mod}\,16). \end{cases}$$

Write $2^* = 1$ if $d \equiv 1 \,(\mathrm{mod}\,4)$. If $d < 0$, we have

$$|d/2^*| \equiv \begin{cases} 3 \,(\mathrm{mod}\,4), & \text{if } d \equiv 1 \,(\mathrm{mod}\,4) \text{ or } d \equiv 8 \,(\mathrm{mod}\,32), \\ 1 \,(\mathrm{mod}\,4), & \text{if } d \equiv 12 \,(\mathrm{mod}\,16) \text{ or } d \equiv 24 \,(\mathrm{mod}\,32). \end{cases}$$

Let u denote the number of distinct prime divisors of d which are congruent to 1 modulo 4 and v the number of distinct prime divisors of d which are congruent to 3 modulo 4, so that

$$u + v = \begin{cases} n, & \text{if } d \equiv 1 \,(\mathrm{mod}\,4), \\ n - 1, & \text{if } d \equiv 0 \,(\mathrm{mod}\,4), \end{cases}$$

and

$$v \equiv \begin{cases} 1 \,(\mathrm{mod}\,2), & \text{if } d \equiv 1 \,(\mathrm{mod}\,4) \text{ or } d \equiv 8 \,(\mathrm{mod}\,32), \\ 0 \,(\mathrm{mod}\,2), & \text{if } d \equiv 12 \,(\mathrm{mod}\,16) \text{ or } d \equiv 24 \,(\mathrm{mod}\,32). \end{cases}$$

1.2   *Three Lemmas*

We now prove the three elementary lemmas mentioned above.

LEMMA 1 ([Fox, Urbanowicz and Williams, 1999, Lemma 1]) *Let d be the discriminant of an imaginary quadratic field. Let n denote the number of* *distinct prime divisors of d and u the number of prime divisors of d which are congruent to* 1 *modulo* 4. *Then*

$$\phi(|d|) \equiv \begin{cases} 0 \, (\bmod \, 2^{n+u}), & \text{if } d \equiv 1 \, (\bmod \, 4) \text{ or } d \equiv 12 \, (\bmod \, 16), \\ 0 \, (\bmod \, 2^{n+1+u}), & \text{if } d \equiv 8 \, (\bmod \, 16). \end{cases}$$

Moreover if $u = 0$ *then*

$$\phi(|d|) \equiv \begin{cases} 2^n \, (\bmod \, 2^{n+1}), & \text{if } d \equiv 1 \, (\bmod \, 4) \text{ or } d \equiv 12 \, (\bmod \, 16), \\ 2^{n+1} \, (\bmod \, 2^{n+2}), & \text{if } d \equiv 8 \, (\bmod \, 16). \end{cases}$$

PROOF. The proof is straightforward by an inspection of cases. We apply the observations made before the formulation of the lemma.   ∎

LEMMA 2 ([Fox, Urbanowicz and Williams, 1999, Lemma 2]) *If N is a positive integer with* $N \geq 3$ *then*

$$\sum_{\substack{1 \leq k \leq N/2 \\ \gcd(k,N)=1}} 1 = \frac{\phi(N)}{2}.$$

PROOF. Since for $1 \leq k \leq N$ we have $\gcd(k, N) = 1$ if and only if $\gcd(N - k, N) = 1$, the lemma follows at once.   ∎

LEMMA 3 ([Fox, Urbanowicz and Williams, 1999, Lemma 3]) *Let d be the discriminant of an imaginary quadratic field and let e be the discriminant of a quadratic field such that* $e \mid d$. *Then*

$$\sum_{\substack{1 \leq k \leq |d|/2 \\ \gcd(k,d)=1}} \chi_e(k) = \begin{cases} (2 - \chi_e(2)) \dfrac{2}{w(e)} \Big(\displaystyle\prod_{\substack{p \mid (d/e) \\ p \text{ prime}}} (1 - \chi_e(p)) \Big) h(e), \\ \qquad \text{if } e < 0 \text{ and } d/e \text{ is odd}, \\ 0, \quad \text{otherwise}. \end{cases}$$

PROOF. Denote by S the sum on the left hand side of the above equation. We have

$$S = \sum_{1 \leq k \leq |d|/2} \chi_e(k) \sum_{f \mid \gcd(k,d/e)} \mu(f) = \sum_{f \mid (d/e)} \mu(f) \sum_{\substack{1 \leq k \leq |d|/2 \\ f \mid k}} \chi_e(k)$$

(here as usual μ denotes the Möbius function). Replacing k by fg in the inner sum, we obtain

$$S = \sum_{f \mid (d/e)} \mu(f) \chi_e(f) \sum_{1 \leq g \leq |(d/e)/f||e|/2} \chi_e(g).$$

Therefore, in view of

$$\sum_{1\leq g\leq [|(d/e)/f|/2]|e|} \chi_e(g) = 0$$

we obtain

$$S = \sum_{f\,|\,(d/e)} \mu(f)\chi_e(f) \sum_{[|(d/e)/f|/2]|e|<g\leq|(d/e)/f||e|/2} \chi_e(g).$$

If $(d/e)/f$ is even the inner sum vanishes. Thus

$$S = \sum_{\substack{f\,|\,(d/e) \\ 2\,\nmid\,(d/e)/f}} \mu(f)\chi_e(f) \sum_{1\leq h\leq|e|/2} \chi_e\left(\frac{(|(d/e)/f|-1)}{2}|e|+h\right).$$

But $\chi_e(\lambda|e|+h) = \chi_e(h)$ for any integer λ so

$$S = \left(\sum_{\substack{f\,|\,(d/e) \\ 2\,\nmid\,(d/e)/f}} \mu(f)\chi_e(f)\right)\left(\sum_{1\leq h\leq|e|/2} \chi_e(h)\right). \tag{3}$$

Next we determine the left hand sum in the above product.

If (d/e) is odd then $(d/e)/f$ is odd for all $f\,|\,(d/e)$ and

$$\sum_{\substack{f\,|\,(d/e) \\ 2\,|\,(d/e)/f}} \mu(f)\chi_e(f) = 0.$$

If (d/e) is even, as (d/e) is the discriminant of a quadratic field, we have $(d/e) = 2^\alpha t$, where $\alpha = 2$ or 3 and t is odd. Writing $f = 2^\beta g$, where g is odd and $0 \leq \beta \leq \alpha - 1$, we see that

$$\sum_{\substack{f\,|\,(d/e) \\ 2\,|\,(d/e)/f}} \mu(f)\chi_e(f) = \left(\sum_{\beta=0}^{\alpha-1} \mu(2^\beta)\chi_e(2)^\beta\right)\left(\sum_{g\,|\,t} \mu(g)\chi_e(g)\right).$$

Hence, by a product expansion property of multiplicative functions, we deduce that

$$\sum_{\substack{f\,|\,(d/e) \\ 2\,\nmid\,(d/e)/f}} \mu(f)\chi_e(f) = (1-\chi_e(2)) \prod_{\substack{p\,|\,t \\ p\,\text{prime}}} (1-\chi_e(p)) = \prod_{\substack{p\,|\,(d/e) \\ p\,\text{prime}}} (1-\chi_e(p))$$

and

$$\sum_{\substack{f\,|\,(d/e) \\ 2\,\nmid\,(d/e)/f}} \mu(f)\chi_e(f) = \sum_{f\,|\,(d/e)} \mu(f)\chi_e(f) - \sum_{\substack{f\,|\,(d/e) \\ 2\,|\,(d/e)/f}} \mu(f)\chi_e(f)$$

$$= \begin{cases} \displaystyle\prod_{\substack{p \mid (d/e) \\ p\,\text{prime}}} (1 - \chi_e(p)), & \text{if } d/e \text{ is odd,} \\[2mm] 0, & \text{if } d/e \text{ is even.} \end{cases}$$

Now to prove Lemma 3 it suffices to make use of the above equation, formula (3), Dirichlet's formula (2) and the trivial equality

$$\sum_{1 \le h \le |e|/2} \chi_e(h) = 0,$$

which holds for $e > 0$.   ∎

We remark that when $e = d$ the formula of Lemma 3 reduces to (2).

1.3 A Useful Elementary Identity

We are now ready to prove the key identity. Throughout the chapter, as usual we denote by $\nu(n)$ the number of distinct prime divisors of an integer n ($n \ne 0$).

THEOREM 65 ([Fox, Urbanowicz and Williams, 1999, Theorem 1]) *Let d be the discriminant of an imaginary quadratic field. Then*

$$\sum_{\substack{e \mid d,\, e<0 \\ (d/e)\equiv 1\,(\text{mod}\,4)}} (-1)^{\nu(e)-1}(2 - \chi_e(2))\frac{2}{w(e)}\Big(\prod_{\substack{p \mid d,\, p \nmid e \\ p\,\text{prime}}} (1 - \chi_e(p))\Big)h(e)$$

$$= \frac{\phi(|d|)}{2} - \sum_{\substack{1 \le k \le |d|/2 \\ \gcd(k,d)=1}} \Big(\prod_{\substack{p \mid d \\ p\,\text{prime}}} (1 - \chi_{p^*}(k))\Big),$$

where e runs through fundamental discriminants dividing d such that $e < 0$ and $(d/e) \equiv 1 \,(\text{mod}\,4)$.

PROOF. From a product expansion property of multiplicative functions and Lemmas 2 and 3 we obtain

$$\sum_{\substack{1 \le k \le |d|/2 \\ \gcd(k,d)=1}} \Big(\prod_{\substack{p \mid d \\ p\,\text{prime}}} (1 - \chi_{p^*}(k))\Big) = \sum_{\substack{1 \le k \le |d|/2 \\ \gcd(k,d)=1}} 1 + \sum_{\substack{e \mid d \\ e \ne 1}} (-1)^{\nu(e)} \sum_{k=1}^{[|d|/2]} \chi_e(k)$$

$$= \frac{\phi(|d|)}{2} + \sum_{\substack{e \mid d,\, e<0 \\ (d/e)\equiv 1\,(\text{mod}\,4)}} (-1)^{\nu(e)}(2 - \chi_e(2))\frac{2}{w(e)}\Big(\prod_{\substack{p \mid d,\, p \nmid e \\ p\,\text{prime}}} (1 - \chi_e(p))\Big)h(e)$$

and the theorem follows.   ∎

2.   GAUSS' CONGRUENCE FROM DIRICHLET'S CLASS NUMBER FORMULA

2.1   *Some Preliminary Remarks*

Before proving Gauss' congruence we make some preliminary observations. Regarding the product in the left hand side of the equation of Theorem 65, we note that

$$\prod_{\substack{p \mid d, \, p \nmid e \\ p \text{ prime}}} (1 - \chi_e(p)) \equiv 0 \, (\bmod \, 2^{\nu(d) - \nu(e)}). \tag{4}$$

Similarly, for the sum on the right hand side of the equation of Theorem 65, we have

$$\sum_{\substack{1 \le k \le |d|/2 \\ \gcd(k,d)=1}} \Big(\prod_{\substack{p \mid d \\ p \text{ prime}}} (1 - \chi_{p^*}(k)) \Big) \equiv 0 \, (\bmod \, 2^{\nu(d)}). \tag{5}$$

We now discuss briefly the case when d $(d < 0)$ has exactly one prime factor. Then we have the classical congruence

$$h(d) \equiv 1 \, (\bmod \, 2).$$

In this case $d = -4$, -8 or $-p$, where p is a prime $\equiv 3 \, (\bmod \, 4)$. It is well known that $h(-4) = h(-8) = 1$. In the remaining case the above congruence follows from formula (2) easily. Indeed, if $d = -p$ we have $w(d) = 2$ if $p > 3$, and $w(3) = 6$, $2 - \chi_d(2) = 3$, resp. 1, if $p \equiv 3$, resp. $7 \, (\bmod \, 8)$, and

$$\sum_{r=1}^{(p-1)/2} \chi_d(r) \equiv \sum_{r=1}^{(p-1)/2} 1 = (p-1)/2 \equiv 1 \, (\bmod \, 2).$$

2.2   *Gauss' Congruence from Dirichlet's Class Number Formula*

We are now ready to deduce Gauss' congruence (1) from Dirichlet's class number formula (2) (see [Fox, Urbanowicz and Williams, 1999, Theorem 2]).

THEOREM 66   *Let d be the discriminant of an imaginary quadratic field. Then*

$$h(d) \equiv 0 \, (\bmod \, 2^{n-1}),$$

where n is the number of distinct prime factors of d.

PROOF.   We make use of Theorem 65 which was proved using Dirichlet's formula. We proceed by induction on the number n of prime divisors of the fundamental discriminant $d < 0$.

If $n = 1$ the congruence of Theorem 66 is trivial. We thus make the inductive hypothesis that $h(e) \equiv 0 \,(\bmod \, 2^{r-1})$ whenever e is a negative fundamental discriminant having r distinct prime factors where $r = 1, 2, \ldots, n - 1$ and $n \geq 2$. Let d be a negative fundamental discriminant with n prime factors.

First we treat the case when $d \equiv 1 \,(\bmod \, 4)$. In Theorem 65 the terms with $e \neq d$ have $\nu(e) = r$, $1 \leq r \leq n - 1$, and are congruent to 0 modulo $2^{n-r} \times 2^{r-1} = 2^{n-1}$ by (4) and the inductive hypothesis. The term with $e = d$ is

$$(-1)^{n-1}\frac{2}{w(d)}(2 - \chi_d(2))h(d) = (\text{odd number}) \times h(d).$$

By Lemma 1 and (5) the right hand side of the equation in the Theorem 65 is divisible by 2^{n-1}. Hence $h(d) \equiv 0 \,(\bmod \, 2^{n-1})$.

Secondly we treat the case $d \equiv 8 \,(\bmod \, 16)$. In this case the terms with $e \neq d$ in the equation of Theorem 65 have $\nu(e) = r$, $1 \leq r \leq n - 1$, and are divisible by $2 \times 2^{n-r} \times 2^{r-1} = 2^n$ by (4) and the inductive hypothesis. The term with $e = d$ is $(-1)^{n-1}2h(d)$. By Lemma 1 and (5) the right hand side of the equation of Theorem 65 is divisible by 2^n. Hence $h(d) \equiv 0 \,(\bmod \, 2^{n-1})$.

Finally we treat the case $d \equiv 12 \,(\bmod \, 16)$. Clearly the fundamental discriminants e in the summation on the left hand side of the equation of Theorem 65 contain -4 as one of their prime discriminant factors. The term with $e = -4$ is

$$\prod_{\substack{p \,|\, d, \, p \neq 2 \\ p \,\text{prime}}} (1 - \chi_e(p)) = \begin{cases} 2^{n-1}, & \text{if } u = 0, \\ 0, & \text{if } u \geq 1. \end{cases}$$

The term with $e = d$ is

$$(-1)^{n-1}2h(d).$$

The remaining terms in the sum each contribute by (4) and the inductive hypothesis

$$\pm 2(\text{multiple of } 2^{n-\nu(e)})(\text{multiple of } 2^{\nu(e)-1}) = \text{multiple of } 2^n.$$

Hence the left hand side of the equation of Theorem 65 is

$$\begin{cases} (-1)^{n-1}2h(d) + 2^{n-1} + k2^n, & \text{if } u = 0, \\ (-1)^{n-1}2h(d) + k2^n, & \text{if } u \geq 1, \end{cases}$$

for some integer k. By Lemma 1 and (5) the right hand side of the equation of Theorem 65 is

$$\begin{cases} 2^{n-1} + l2^n, & \text{if } u = 0, \\ l2^n, & \text{if } u \geq 1, \end{cases}$$

for some integer l. Hence $h(d) \equiv 0 \,(\bmod \, 2^{n-1})$. Gauss' congruence (1) now follows by the principle of mathematical induction. ∎

3.   CHARACTER POWER SUMS IN TERMS OF BERNOULLI NUMBERS

3.1   *Gauss Type Congruence for $k_2(d)$*

It follows from Dirichlet's class number formula that

$$h(d) = -\frac{w(d)}{2} B_{1,\chi_d},$$

where $B_{k,\chi}$ denotes the generalized Bernoulli number attached to χ (see formula (16) of Chapter I).

Let F denote a real quadratic field with the discriminant d and let O_F be the ring of integers in F. Let K_2 be the Milnor functor and $k_2(d)$ denote the order of the finite group K_2O_F. The corresponding formula for $k_2(d)$ is

$$k_2(d) = \frac{w_2(d)}{24} B_{2,\chi_d} \quad (d > 8) \tag{6}$$

(see formula (19) of Chapter I).

It is well known that $k_2(d)$ for positive d is always divisible by 4. Using Gauss' theory of genera Browkin and Schinzel in [Browkin and Schinzel, 1982] obtained algebraically the congruence

$$(k_2(d)/2) \equiv 0 \,(\bmod\, 2^{n-1}), \tag{7}$$

where d is the discriminant of a real quadratic field having n distinct prime factors, see Chapter III for more details.

We recall that for $d \neq -4$ the numbers $(B_{m,\chi_d}/m)$ $(m \geq 1)$ are 2-integral. In fact, these numbers are always integers unless $d = -4$ or $d = \pm p$, where p is an odd prime number such that $2m/(p-1)$ is an odd integer, in which case they have denominator 2 or p (for details see [Carlitz, 1959] or [Leopoldt, 1958]). The left hand side of the congruences (1), resp. (7) is equal to $(B_{m,\chi_d}/m)$ for $m = 1$, resp. 2. By analogy, one could expect that these congruences are special cases of a more general corresponding congruence for the generalized Bernoulli numbers attached to quadratic characters $(B_{m,\chi_d}/m)$ $(m \geq 1, \chi_d(-1) = (-1)^m)$. The main result of the chapter is Theorem 70 giving a congruence of this type.

It follows from (6) that

$$k_2(d) = \frac{w_2(d)}{6(\chi_d(2) - 4)} \sum_{r=1}^{[d/2]} \chi_d(r) r, \tag{8}$$

(see formula (20) of Chapter I). Formula (8) for $k_2(d)$ corresponds to formula (2) for $h(d)$, which yields the identities of Lemma 3 and Theorem 65, and in consequence Gauss' congruence. Thus one could expect that there should exist some corresponding identities for $k_2(d)$ implying Browkin and

Schinzel's congruence (7), see Chapter III. In this chapter we give such identities. In fact, we find identities of a more general form, for the generalized
Bernoulli numbers (Theorems 67, 68 and 69). Theorems 67, 68 and 69 are
further consequences of Zagier's identity proved in [Szmidt, Urbanowicz and
Zagier, 1995], expressing short character power sums of any length in terms of
generalized Bernoulli numbers.

3.2   *Some Consequences of Zagier's Identity*

Let χ be a Dirichlet character modulo M and let N be a multiple of M. For
any integer $r > 1$ prime to N and any natural number m we recall once again
the following formula

$$mr^{m-1} \sum_{0<a<(N/r)} \chi(a)a^{m-1}$$

$$= -B_{m,\chi}r^{m-1} + \frac{\bar{\chi}(r)}{\phi(r)} \sum_{\psi} \bar{\psi}(-N)B_{m,\chi\psi}(N), \qquad (9)$$

where the last sum is over all Dirichlet characters ψ modulo r. Here for a
Dirichlet character χ, we have

$$B_{m,\chi}(X) = \sum_{k=0}^{m} \binom{m}{k} B_{k,\chi}X^{m-k}.$$

Thus we have

$$mr^{m-1} \sum_{0<a<(N/r)} \chi(a)a^{m-1} = -B_{m,\chi}r^{m-1}$$

$$+ \frac{\bar{\chi}(r)}{\phi(r)} \sum_{k=0}^{m} \binom{m}{k} N^{m-k} \sum_{\psi} \bar{\psi}(-N)B_{k,\chi\psi}.$$

Note that if the character χ modulo T is induced from a character χ' modulo
some divisor of T then we have

$$B_{m,\chi} = B_{m,\chi'} \prod_{\substack{p|T \\ p\ \text{prime}}} (1 - \chi'(p)p^{m-1}), \qquad (10)$$

where the product is over all primes p dividing T. For more details see section
1.8 of Chapter II.

If N is odd and $r = 2$, 4, 8 formula (9) and formula (10) give some identities
for the generalized Bernoulli numbers corresponding to those in Lemma 3. The
case $m = 1$ is much simpler than the other cases. For $m \geq 2$ we also need
shorter character power sums.

If χ is a Dirichlet character modulo M and N is a multiple of M, then for $m \geq 0$ we set

$$B_{m,\chi}^{[N]} = B_{m,\chi} \prod_{\substack{p|(N/M) \\ p\,\text{prime}}} (1 - \chi(p)p^{m-1}).$$

Moreover, we write

$$B_{m,\chi}^{[N]}(X) = \sum_{k=0}^{m} \binom{m}{k} B_{k,\chi}^{[N]} X^{m-k}.$$

Set $B_m^{[N]} = B_{m,\chi_1}^{[N]}$, $B_m^{[N]}(X) = B_{m,\chi_1}^{[N]}(X)$ for any integer $N > 1$.

LEMMA 4 ([Fox, Urbanowicz and Williams, 1999, Lemma 4]) *Let χ be a Dirichlet character modulo M and let N be an odd positive multiple of M. Then for any natural number m we have the following formulae:*

(a) $m2^{m-1} \displaystyle\sum_{\substack{0<a<(N/2) \\ \gcd(a,N)=1}} \chi(a)a^{m-1} = (\bar{\chi}(2) - 2^m)B_{m,\chi}^{[N]}$

$$+ \bar{\chi}(2)\Big(B_{m,\chi}^{[2N]}(N) - B_{m,\chi}^{[2N]}\Big),$$

(b) $m2^{2m-2} \displaystyle\sum_{\substack{0<a<(N/4) \\ \gcd(a,N)=1}} \chi(a)a^{m-1} = \left(\frac{\bar{\chi}(4) - \bar{\chi}(2)2^{m-1} - 2^{2m-1}}{2}\right)B_{m,\chi}^{[N]}$

$$+ \frac{\bar{\chi}(4)}{2}\Big((B_{m,\chi}^{[2N]}(N) - B_{m,\chi}^{[2N]}) - \chi_{-4}(N)B_{m,\chi_{-4}\chi}^{[4N]}(N)\Big),$$

(c) $m2^{3m-3} \displaystyle\sum_{\substack{0<a<(N/8) \\ \gcd(a,N)=1}} \chi(a)a^{m-1} = \left(\frac{\bar{\chi}(8) - \bar{\chi}(4)2^{m-1} - 2^{3m-1}}{4}\right)B_{m,\chi}^{[N]}$

$$+ \frac{\bar{\chi}(8)}{4}\Big((B_{m,\chi}^{[2N]}(N) - B_{m,\chi}^{[2N]}) - \chi_{-8}(N)B_{m,\chi_{-8}\chi}^{[8N]}(N)$$

$$- \chi_{-4}(N)B_{m,\chi_{-4}\chi}^{[4N]}(N) + \chi_8(N)B_{m,\chi_8\chi}^{[8N]}(N)\Big).$$

PROOF. N is odd and so we can make use of formula (9) for $r = 2$, 4 and 8, respectively. Then we have

$$m2^{m-1} \sum_{0<a<(N/2)} \chi(a)a^{m-1} = (\bar{\chi}(2) - 2^m)B_{m,\chi} + \bar{\chi}(2)(B_{m,\chi}^{[2M]}(N) - B_{m,\chi}^{[2M]}),$$

$$m2^{2m-2} \sum_{0<a<(N/4)} \chi(a)a^{m-1} = \left(\frac{\bar{\chi}(4) - \bar{\chi}(2)2^{m-1} - 2^{2m-1}}{2}\right)B_{m,\chi}$$

$$+ \frac{\bar{\chi}(4)}{2} \left((B^{[2M]}_{m,\chi}(N) - B^{[2M]}_{m,\chi}) - \chi_{-4}(N) B_{m,\chi_{-4}\chi}(N) \right),$$

and

$$m 2^{3m-3} \sum_{0<a<(N/8)} \chi(a) a^{m-1} = \left(\frac{\bar{\chi}(8) - \bar{\chi}(4) 2^{m-1} - 2^{3m-1}}{4} \right) B_{m,\chi}$$

$$+ \frac{\bar{\chi}(8)}{4} \left((B^{[2M]}_{m,\chi}(N) - B^{[2M]}_{m,\chi}) - \chi_{-8}(N) B_{m,\chi_{-8}\chi}(N) \right.$$

$$\left. - \chi_{-4}(N) B_{m,\chi_{-4}\chi}(N) + \chi_8(N) B_{m,\chi_8\chi}(N) \right).$$

Now it is sufficient to replace the character χ modulo M in the above formulae by a character modulo N induced by χ and Lemma 4 follows immediately. ∎

4.   THE MAIN RESULTS

4.1   *Some Identities for Generalized Bernoulli Numbers*

We are now ready to deduce the main results of Chapter VI. First we prove some identities for the generalized Bernoulli numbers corresponding to that in Theorem 65 (Theorems 67, 68, 69). Next we deduce the main congruence of the chapter (Theorem 70) for the numbers $B_{m,\chi_d}/m$ for d odd, divisible by 4, or divisible by 8 from Theorems 67, 68, or 69 respectively. We will proceed by induction on the pairs (m, n), where n is the number of distinct prime factors of d.

THEOREM 67 ([Fox, Urbanowicz and Williams, 1999, Theorem 3]) *Let m be a natural number and let d be an odd discriminant of a quadratic field. Then we have*

$$\sum_{\substack{e \mid d \\ e \neq 1}} (-1)^{\nu(e)} (\chi_e(2) - 2^m) \frac{B^{[d]}_{m,\chi_e}}{m}$$

$$+ \sum_{\substack{e \mid d \\ e \neq 1}} (-1)^{\nu(e)} \chi_e(2) \left(\frac{B^{[2d]}_{m,\chi_e}(|d|) - B^{[2d]}_{m,\chi_e}}{m} \right)$$

$$= 2^{m-1} \left(\sum_{\substack{0<a<(|d|/2) \\ \gcd(a,d)=1}} a^{m-1} \left(\prod_{\substack{p \mid d \\ p \text{ prime}}} (1 - \chi_{p^*}(a)) \right) - \sum_{\substack{0<a<(|d|/2) \\ \gcd(a,d)=1}} a^{m-1} \right),$$

where e ($e \neq 1$) runs through fundamental discriminants dividing d.

PROOF. Denote by S the left hand side of the equality of Theorem 67. By Lemma 4(a) we have

$$S = 2^{m-1}\left(\sum_{e\mid d}\mu(e)\sum_{\substack{0<a<(|d|/2)\\ \gcd(a,d)=1}}\chi_e(a)a^{m-1} - \sum_{\substack{0<a<(|d|/2)\\ \gcd(a,d)=1}}a^{m-1}\right)$$

$$= 2^{m-1}\left(\sum_{\substack{0<a<(|d|/2)\\ \gcd(a,d)=1}}a^{m-1}\sum_{e\mid d}\mu(e)\chi_e(a) - \sum_{\substack{0<a<(|d|/2)\\ \gcd(a,d)=1}}a^{m-1}\right).$$

Now the theorem follows from a product expansion property of multiplicative functions. ∎

THEOREM 68 ([Fox, Urbanowicz and Williams, 1999, Theorem 4])   *Let m be a natural number and let d be an odd discriminant of a quadratic field. Then we have*

$$\sum_{\substack{e\mid d\\ e\neq 1}}(-1)^{\nu(e)}(1 - \chi_e(2)2^{m-1} - 2^{2m-1})\frac{B_{m,\chi_e}^{[d]}}{m}$$

$$+ \sum_{\substack{e\mid d\\ e\neq 1}}(-1)^{\nu(e)}\left(\frac{B_{m,\chi_e}^{[2d]}(|d|) - B_{m,\chi_e}^{[2d]}}{m} - \frac{\chi_{-4}(|d|)B_{m,\chi_{-4}\chi_e}^{[4d]}(|d|)}{m}\right)$$

$$= 2^{2m-1}\left(\sum_{\substack{0<a<(|d|/4)\\ \gcd(a,d)=1}}a^{m-1}\left(\prod_{\substack{p\mid d\\ p\,\text{prime}}}(1 - \chi_{p^*}(a))\right) - \sum_{\substack{0<a<(|d|/4)\\ \gcd(a,d)=1}}a^{m-1}\right),$$

where e ($e\neq 1$) runs through fundamental discriminants dividing d.

PROOF. Denote by S the left hand side of the equality of Theorem 68. By Lemma 4(b) we have

$$S = 2^{2m-1}\left(\sum_{e\mid d}\mu(e)\sum_{\substack{0<a<(|d|/4)\\ \gcd(a,d)=1}}\chi_e(a)a^{m-1} - \sum_{\substack{0<a<(|d|/4)\\ \gcd(a,d)=1}}a^{m-1}\right)$$

$$= 2^{2m-1}\left(\sum_{\substack{0<a<(|d|/4)\\ \gcd(a,d)=1}}a^{m-1}\sum_{e\mid d}\mu(e)\chi_e(a) - \sum_{\substack{0<a<(|d|/4)\\ \gcd(a,d)=1}}a^{m-1}\right).$$

Now the theorem follows from a product expansion property of multiplicative functions. ∎

THEOREM 69 ([Fox, Urbanowicz and Williams, 1999, Theorem 5]) *Let m be a natural number and let d be an odd discriminant of a quadratic field. Then we have*

$$\sum_{\substack{e \mid d \\ e \neq 1}} (-1)^{\nu(e)} (\chi_e(2) - 2^{m-1} - 2^{3m-1}) \frac{B^{[d]}_{m,\chi_e}}{m}$$

$$+ \sum_{\substack{e \mid d \\ e \neq 1}} (-1)^{\nu(e)} \chi_e(2) \times \left(\frac{B^{[2d]}_{m,\chi_e}(|d|) - B^{[2d]}_{m,\chi_e}}{m} - \frac{\chi_{-4}(|d|) B^{[4d]}_{m,\chi_{-4}\chi_e}(|d|)}{m} \right.$$

$$\left. - \frac{\chi_{-8}(|d|) B^{[8d]}_{m,\chi_{-8}\chi_e}(|d|)}{m} + \frac{\chi_8(|d|) B^{[8d]}_{m,\chi_8\chi_e}(|d|)}{m} \right)$$

$$= 2^{3m-1} \left(\sum_{\substack{0<a<(|d|/8) \\ \gcd(a,d)=1}} a^{m-1} \left(\prod_{\substack{p \mid d \\ p \text{ prime}}} (1 - \chi_{p^*}(a)) \right) - \sum_{\substack{0<a<(|d|/8) \\ \gcd(a,d)=1}} a^{m-1} \right),$$

where e ($e \neq 1$) runs through fundamental discriminants dividing d.

PROOF. Denote by S the left hand side of the equality of Theorem 69. By Lemma 4(c) we have

$$S = 2^{3m-1} \left(\sum_{e \mid d} \mu(e) \sum_{\substack{0<a<(|d|/8) \\ \gcd(a,d)=1}} \chi_e(a) a^{m-1} - \sum_{\substack{0<a<(|d|/8) \\ \gcd(a,d)=1}} a^{m-1} \right)$$

$$= 2^{3m-1} \left(\sum_{\substack{0<a<(|d|/8) \\ \gcd(a,d)=1}} a^{m-1} \sum_{e \mid d} \mu(e) \chi_e(a) - \sum_{\substack{0<a<(|d|/8) \\ \gcd(a,d)=1}} a^{m-1} \right).$$

Now the theorem follows from a product expansion property of multiplicative functions. ∎

4.2   Five Useful Lemmas

The proof of the main theorem (Theorem 70) is based on Theorems 67, 68, 69 and five lemmas (Lemmas 5, 6, 7, 8 and 9). In Lemma 5 we prove an elementary congruence. In Lemma 6 we give an extension (probably already known) of the von Staudt-Clausen theorem for Bernoulli numbers. These two lemmas together with Lemmas 7 and 8 imply Lemma 9. Theorems 67, 68, 69 and Lemma 9 give Theorem 70.

LEMMA 5 ([Fox, Urbanowicz and Williams, 1999, Lemma 5]) *Let $N > 1$ be a squarefree odd natural number. Then for an even integer $m \geq 2$ we have*

$$N \prod_{\substack{p \mid N \\ p \text{ prime}}} (1 + p + \ldots + p^{m-2}) \equiv (-1)^{\nu(N)} \, (\bmod \, 2^{\mathrm{ord}_2 m}).$$

Moreover, if $p \equiv 3 \, (\bmod \, 4)$ for any $p \mid N$ then the above congruence holds modulo $2^{\mathrm{ord}_2 m+1}$.

PROOF. If we prove that for any $p \mid N$ the congruence

$$p(1 + p + \ldots + p^{m-2}) \equiv -1 \, (\bmod \, 2^{\mathrm{ord}_2 m+\mathrm{ord}_2(p+1)-1}) \qquad (11)$$

holds, the lemma follows at once.

The task is now to prove congruence (11). Indeed, for u odd we have

$$u^m \equiv 1 \, (\bmod \, 2^{\mathrm{ord}_2 m+\mathrm{ord}_2(u^2-1)-1}). \qquad (12)$$

This follows from the equation

$$u^m = 1 + \sum_{k=1}^{m/2} \frac{m/2}{k} \binom{(m/2)-1}{k-1} (u^2 - 1)^k$$

because

$$k \, \mathrm{ord}_2(u^2 - 1) \geq \mathrm{ord}_2 k + \mathrm{ord}_2(u^2 - 1),$$

if $k \geq 1$.

Thus we obtain

$$p(1 + p + \ldots + p^{m-2}) = \frac{p^m - 1}{p - 1} - 1 \equiv -1 \, (\bmod \, 2^{\mathrm{ord}_2 m+\mathrm{ord}_2(p+1)-1}).$$

Hence congruence (11) follows at once.   ∎

Lemma 6 extends the von Staudt-Clausen theorem in case $p = 2$. The theorem asserts that for a prime number p satisfying $(p - 1) \mid m$, the rational numbers pB_m are p-integral and

$$pB_m \equiv -1 \, (\bmod \, p). \qquad (13)$$

If $(p - 1) \nmid m$ then B_m is p-integral, but we will not use this. In the proof of Lemma 6 we apply the power summation formula

$$\sum_{0 < a < N} a^{m-1} = \frac{1}{m}(B_m(N) - B_m), \qquad (14)$$

where $m \geq 1$ (see section 2.1 of Chapter I). In the proofs of Lemmas 7 and 8 we make use of the following formula for the generalized Bernoulli polynomials

$$B_{m,\chi}(X + Y) = \sum_{k=0}^{m} \binom{m}{k} B_{k,\chi}(X) Y^{m-k}, \qquad (15)$$

which easily follows from the formula

$$B'_{m,\chi}(X) = mB_{m-1,\chi}(X).$$

LEMMA 6 ([Fox, Urbanowicz and Williams, 1999, Lemma 6]) *For an even integer $m \geq 4$ we have*

$$2B_m \equiv 1 \,(\bmod\, 2^{\mathrm{ord}_2 m+1}).$$

PROOF. Let R be a natural number such that $R \equiv 2 \,(\bmod\, 4)$. From formula (14) we obtain

$$(m+1) \sum_{0<a<R} a^m = B_{m+1}(R) - B_{m+1},$$

and hence

$$\frac{Rm(m+1)}{2} \sum_{k=3}^{m+1} \frac{1}{k(k-1)} \binom{m-1}{k-2}(2B_{m+1-k})R^{k-1} + \frac{(m+1)R}{2}(2B_m)$$

$$\equiv (m+1) \sum_{\substack{0<a<R \\ a\;\mathrm{odd}}} a^m \,(\bmod\, 2^{\mathrm{ord}_2 m+1}).$$

Therefore by (12) and

$$k - 1 \geq \mathrm{ord}_2(k(k-1)) + 1,$$

if $k \geq 3$, we deduce that

$$\frac{(m+1)R}{2}(2B_m) \equiv \frac{(m+1)R}{2} \,(\bmod\, 2^{\mathrm{ord}_2 m+1}).$$

This implies Lemma 6 easily. ∎

REMARK Note that if $m = 2$ the above congruence holds modulo $2^{\mathrm{ord}_2 m}$ only, and if $m \geq 6$ we have $\mathrm{ord}_2(2B_m - 1) = \mathrm{ord}_2 m + 1$.

LEMMA 7 ([Fox, Urbanowicz and Williams, 1999, Lemma 7]) *Let $N > 1$ be a squarefree odd natural number having n distinct prime factors. Then for a natural number m we have*

$$\frac{B_m^{[2N]}(1)}{m} \equiv 0 \,(\bmod\, 2^n),$$

unless $m = 1, 2$ and $p \equiv 3 \,(\bmod\, 4)$ for any $p \mid N$, in which cases

$$\frac{B_m^{[2N]}(1)}{m} \equiv 2^{n-1} \,(\bmod\, 2^n).$$

PROOF. If $m = 1$ we have

$$B_1^{[2N]}(1) = B_0^{[2N]} + B_1^{[2N]} = \frac{1}{2} \prod_{\substack{p \mid N \\ p \text{ prime}}} (1 - p^{-1}) = \frac{\phi(N)}{2N}$$

and the lemma follows. If $m = 2$ we have

$$\frac{B_2^{[2N]}(1)}{2} = \frac{1}{2}\left(B_0^{[2N]} + B_2^{[2N]}\right) = \frac{\phi(N)}{4N}\left(1 - \frac{1}{3}(-1)^{\nu(N)}N\right).$$

Denote by u the number of distinct prime divisors of N which are congruent to 1 modulo 4. From now on, the proof will be divided into two cases

(i) $u > 0$,

(ii) $u = 0$.

Case (i). In this case

$$\frac{\phi(N)}{4N} \equiv 0 \,(\bmod\, 2^{n-1})$$

and

$$1 - \frac{1}{3}(-1)^{\nu(N)}N \equiv 0 \,(\bmod\, 2),$$

and hence the lemma follows immediately.

Case (ii). In this case

$$\frac{\phi(N)}{4N} \equiv 2^{n-2} \,(\bmod\, 2^{n-1})$$

and

$$1 - \frac{1}{3}(-1)^{\nu(N)}N \equiv 1 + (-1)^{2\nu(N)} \equiv 2 \,(\bmod\, 4),$$

and hence the lemma follows easily.

If $m \geq 3$ we proceed by induction on m. If $m = 3$ we have

$$\frac{B_3^{[2N]}(1)}{3} = \frac{1}{3}(B_0^{[2N]} + 3B_2^{[2N]}) = \frac{\phi(N)}{6N}(1 - (-1)^{\nu(N)}N) \equiv 0 \,(\bmod\, 2^{n-1}),$$

which gives the lemma.

The subsequent proof is based on the following observation. Putting in (15) $X = 1$ and $Y = -1$ we obtain

$$\frac{B_m^{[2N]}}{m} = \frac{1}{m} \sum_{k=0}^{m} \binom{m}{k} B_k^{[2N]}(1)(-1)^{m-k} = (-1)^m \frac{\phi(N)}{2mN} + (-1)^{m-1}\frac{\phi(N)}{2N}$$

$$+ (-1)^{m-2}(m-1)\frac{B_2^{[2N]}(1)}{2} + \sum_{k=3}^{m}(-1)^{m-k}\binom{m-1}{k-1}\frac{B_k^{[2N]}(1)}{k}.$$

Lemma 7 can be deduced from the above formula inductively if we prove that

$$(-1)^m \frac{B_m^{[2N]}}{m} - \frac{\phi(N)}{2mN} + \frac{\phi(N)}{2N} - (m-1)\frac{B_2^{[2N]}(1)}{2} \equiv 0 \,(\bmod\, 2^n). \quad (16)$$

Denote by L the left hand side of the above congruence. For odd m we have

$$L = -\frac{\phi(N)}{2mN} + \frac{\phi(N)}{2N} - (m-1)\frac{B_2^{[2N]}(1)}{2} \equiv \frac{\phi(N)}{2N}\left(1 - \frac{1}{m}\right) \equiv 0 \,(\bmod\, 2^n).$$

For even m (≥ 4) we have

$$L = \left(\frac{B_m^{[2N]}}{m} - \frac{\phi(N)}{2mN}\right) + \left(\frac{\phi(N)}{2N} - (m-1)\frac{B_2^{[2N]}(1)}{2}\right).$$

Case (i). In this case

$$\phi(N) \equiv 0 \,(\bmod\, 2^{n+1})$$

and we have

$$L \equiv \frac{B_m^{[2N]}}{m} - \frac{\phi(N)}{2mN} \,(\bmod\, 2^n).$$

Thus in order to prove (16) it suffices to show that

$$(2B_m)(1 - 2^{m-1})N \prod_{\substack{p\,|\,N \\ p\,\text{prime}}} (1 - p^{m-1}) \equiv \phi(N) \,(\bmod\, 2^{n+\text{ord}_2 m+1}).$$

The left hand side of the above congruence equals

$$\phi(N)(2B_m)(1 - 2^{m-1})N(-1)^{\nu(N)} \prod_{\substack{p\,|\,N \\ p\,\text{prime}}} (1 + p + \ldots + p^{m-2}),$$

and so the congruence follows easily from Lemmas 5 and 6.

Case (ii). In this case we have

$$L \equiv \left(\frac{B_m^{[2N]}}{m} - \frac{\phi(N)}{2mN}\right) + (2^{n-1} - 2^{n-1}) \,(\bmod\, 2^n),$$

and so

$$L \equiv \frac{B_m^{[2N]}}{m} - \frac{\phi(N)}{2mN} \,(\bmod\, 2^n).$$

Analysis similar to that in case (i) shows that $L \equiv 0 \,(\bmod\, 2^n)$. It suffices to make use of Lemmas 5, 6 and the congruence

$$\phi(N) \equiv 2^n \,(\bmod\, 2^{n+1}).$$

The congruence of Lemma 6 in this case has the form

$$N \prod_{\substack{p \mid N \\ p \text{ prime}}} (1 + p + \ldots + p^{m-2}) \equiv (-1)^{\nu(N)} \ (\bmod \ 2^{\operatorname{ord}_2 m + 1}). \qquad \blacksquare$$

LEMMA 8 ([Fox, Urbanowicz and Williams, 1999, Lemma 8])   *Let* m *be a natural number. Let* $N > 1$ *be an odd squarefree natural number having* n *distinct prime factors and let* $d = -4, \pm 8$. *Write* $\delta = |d|$. *Then we have*

$$\frac{B_{m,\chi_d}^{[\delta N]}(N)}{m} \equiv 0 \ (\bmod \ 2^n),$$

unless $m = 1, 2, d = -4$ *and* $p \equiv 3 \ (\bmod \ 4)$ *for any* $p \mid N$, *in which cases*

$$\frac{B_{m,\chi_{-4}}^{[4N]}(N)}{m} \equiv 2^{n-1} \ (\bmod \ 2^n).$$

PROOF.   The proof is straightforward and based on the following observations. For an even integer k ($k \geq 0$) we have

$$B_{k,\chi_{-4}} = 0,$$

and for an odd integer k ($k \geq 1$)

$$\frac{1}{k} B_{k,\chi_{-4}} = \frac{1}{2} E_{k-1},$$

where E_k denotes the kth classical Euler number, which is an odd integer.

Moreover, for an integer k ($k \geq 0$) we have

$$B_{k,\chi_{-8}} = 0,$$

if k is even, and

$$B_{k,\chi_8} = 0,$$

if k is odd, and

$$B_{0,\chi_8} = 0.$$

For $k \geq 1$ the numbers

$$\frac{1}{k} B_{k,\chi_{\pm 8}}$$

are rational and 2-integral (see [Carlitz, 1959] or [Leopoldt, 1958] for more details).

Let δ be a natural number relatively prime to N. For a nonprincipal Dirichlet character χ modulo δ we have

$$
\frac{B_{m,\chi}^{[\delta N]}(N)}{m}
$$

$$
= \sum_{\substack{1 \le k \le m \\ \chi(-1)=(-1)^k}} \binom{m-1}{k-1} \left(\prod_{\substack{p \mid N \\ p \text{ prime}}} (1 - \chi(p)p^{k-1}) \right) \frac{B_{k,\chi}}{k} N^{m-k}. \quad (17)
$$

We first consider the case $d = -4$. If $m = 1$, by virtue of (17), we have

$$
B_{1,\chi_{-4}}^{[4N]}(N) = B_{1,\chi_{-4}}^{[4N]} = -\frac{1}{2} \prod_{\substack{p \mid N \\ p \text{ prime}}} (1 - \chi_{-4}(p))
$$

$$
\equiv \begin{cases} 0 \ (\bmod\, 2^n), & \text{if } u > 0, \\ 2^{n-1} \ (\bmod\, 2^n), & \text{if } u = 0 \end{cases}
$$

(recall that $E_0 = 1$) and the lemma in this case follows. If $m = 2$, by virtue of (17), we have

$$
\frac{B_{2,\chi_{-4}}^{[4N]}(N)}{2} = B_{1,\chi_{-4}}^{[4N]} N
$$

and similar arguments to those in case $m = 1$ apply.

Denote by u the number of distinct prime divisors of N which are congruent to 1 modulo 4. By virtue of (17) if $d = -4$ and $m \ge 3$ we have

$$
\frac{B_{m,\chi_{-4}}^{[4N]}(N)}{m} = -\frac{1}{2} \sum_{\substack{1 \le k \le m \\ k \text{ odd}}} \binom{m-1}{k-1} \left(\prod_{\substack{p \mid N \\ p \text{ prime}}} (1 - \chi_{-4}(p)p^{k-1}) \right) E_{k-1} N^{m-k}
$$

$$
\equiv \begin{cases} 2^{n-1} \sum_{\substack{1 \le k \le m \\ k \text{ odd}}} \binom{m-1}{k-1} \equiv 0 \ (\bmod\, 2^n), & \text{if } u = 0, \\ 0 \ (\bmod\, 2^n), & \text{if } u > 0 \end{cases}
$$

because

$$
1 - \chi_{-4}(p)p^{k-1} \equiv 1 - \chi_{-4}(p) \ (\bmod\, 8)
$$

for k odd, and

$$
\sum_{\substack{1 \le k \le m \\ k \text{ odd}}} \binom{m-1}{k-1} = 2^{m-2}.
$$

We now turn to the case $d = \pm 8$. Then, by virtue of (17), we have

$$\frac{B_{m,\chi_{\pm 8}}^{[8N]}(N)}{m} = \sum_{\substack{1 \le k \le m \\ k \text{ odd}}} \binom{m-1}{k-1} \Big(\prod_{\substack{p \mid N \\ p \text{ prime}}} (1 - \chi_{\pm 8}(p)p^{k-1}) \Big) \frac{B_{k,\chi_{\pm 8}}}{k} N^{m-k}$$

$$\equiv 0 \, (\bmod \, 2^n)$$

because

$$\Big(\prod_{\substack{p \mid N \\ p \text{ prime}}} (1 - \chi_{\pm 8}(p)p^{k-1}) \Big) \frac{B_{k,\chi_{\pm 8}}}{k} \equiv 0 \, (\bmod \, 2^n)$$

for any $1 \le k \le m$. This completes the proof. ∎

LEMMA 9 ([Fox, Urbanowicz and Williams, 1999, Lemma 9]) *Let m be a natural number. Let $N > 1$ be an odd squarefree natural number having n distinct prime factors. Then we have*

$$2^{m-1} \sum_{\substack{0 < a < (N/2) \\ \gcd(a,N)=1}} a^{m-1} \equiv 2^{2m-1} \sum_{\substack{0 < a < (N/4) \\ \gcd(a,N)=1}} a^{m-1}$$

$$\equiv 2^{3m-1} \sum_{\substack{0 < a < (N/8) \\ \gcd(a,N)=1}} a^{m-1} \equiv 0 \, (\bmod \, 2^n),$$

except for one case

$$2^{m-1} \sum_{\substack{0 < a < (N/2) \\ \gcd(a,N)=1}} a^{m-1} \equiv 2^{n-1} \, (\bmod \, 2^n)$$

if $m = 1$ and $p \equiv 3 \, (\bmod \, 4)$ for any $p \mid N$.

PROOF. Denote by u the number of distinct prime divisors of N which are congruent to 1 modulo 4.

In order to prove that

$$2^{m-1} \sum_{\substack{0 < a < (N/2) \\ \gcd(a,N)=1}} a^{m-1} \equiv 0 \, (\bmod \, 2^n), \tag{18}$$

we make use of Lemma 4(a) for the principal character χ modulo N. Then, by (13) if $p = 2$, we have

$$2^{m-1} \frac{B_m^{[N]}}{m} = \frac{2^{m-1}}{2m} \prod_{\substack{p \mid N \\ p \text{ prime}}} (1 - p^{m-1})(2B_m)$$

$$\equiv \begin{cases} 2^{n-1} \ (\text{mod} \ 2^n), & \text{if } m = 2 \text{ and } u = 0, \\ 0 \ (\text{mod} \ 2^n), & \text{otherwise.} \end{cases}$$

Therefore Lemma 4(a) implies the congruence

$$2^{m-1} \sum_{\substack{0<a<(N/2) \\ \gcd(a,N)=1}} a^{m-1} \tag{19}$$

$$\equiv \begin{cases} 2^{n-1} + \dfrac{B_m^{[2N]}(N)}{m} \ (\text{mod} \ 2^n), & \text{if } m = 2 \text{ and } u = 0, \\[2mm] \dfrac{B_m^{[2N]}(N)}{m} \ (\text{mod} \ 2^n), & \text{otherwise.} \end{cases}$$

We first prove the lemma if $m = 1$ or 2. Then by definition we have

$$B_1^{[2N]}(N) = B_0^{[2N]} N = \frac{\phi(N)}{2},$$

$$B_2^{[2N]}(N) = B_0^{[2N]} N^2 + B_2^{[2N]} = \frac{\phi(N)}{2}\left(N - \frac{1}{3}(-1)^{\nu(N)}\right),$$

and so

$$\frac{B_m^{[2N]}(N)}{m} \equiv \begin{cases} 0 \ (\text{mod} \ 2^n), & \text{if } u > 0, \\ 2^{n-1} \ (\text{mod} \ 2^n), & \text{if } u = 0. \end{cases} \tag{20}$$

Therefore the lemma when $m = 1$ or 2 follows from (19).

If $m \geq 3$ the proof is based on the following observation. By (15) (when $X = 1, Y = N - 1$), we have

$$\frac{B_m^{[2N]}(N)}{m} = \frac{\phi(N)}{2mN}(N-1)^m + \frac{\phi(N)}{2N}(N-1)^{m-1}$$

$$+ (m-1)\frac{B_2^{[2N]}(1)}{2}(N-1)^{m-2} + \sum_{k=0}^{m-3} \binom{m-1}{k} \frac{B_{m-k}^{[2N]}(1)}{m-k}(N-1)^k,$$

and in view of Lemma 7 we conclude that (recall that $N - 1$ is even)

$$\frac{B_m^{[2N]}(N)}{m} \equiv 0 \ (\text{mod} \ 2^n).$$

Therefore congruence (18) follows from (19).

In order to prove the congruence

$$2^{2m-1} \sum_{\substack{0<a<(N/4) \\ \gcd(a,N)=1}} a^{m-1} \equiv 0 \ (\text{mod} \ 2^n), \tag{21}$$

we use Lemma 4(b) for the principal character χ modulo N. Then, in view of (13) if $p = 2$, we have

$$2^{2m-1}\frac{B_m^{[N]}}{m} = \frac{2^{2m-1}}{2m}\prod_{\substack{p \mid N \\ p \text{ prime}}}(1 - p^{m-1})(2B_m) \equiv 0\,(\bmod\,2^n).$$

Therefore Lemma 4(b) yields

$$2^{2m-1}\sum_{\substack{0<a<(N/4) \\ \gcd(a,N)=1}} a^{m-1} \equiv \frac{B_m^{[2N]}(N)}{m} - \chi_{-4}(N)\frac{B_{m,\chi_{-4}}^{[4N]}(N)}{m}\,(\bmod\,2^n).$$

Thus congruence (21) follows from Lemma 8 and (20).

In order to prove that

$$2^{3m-1}\sum_{\substack{0<a<(N/8) \\ \gcd(a,N)=1}} a^{m-1} \equiv 0\,(\bmod\,2^n), \tag{22}$$

we apply Lemma 4(c) for the principal character χ modulo N. Then, in view of (13) if $p = 2$, we have

$$2^{3m-1}\frac{B_m^{[N]}}{m} = \frac{2^{3m-1}}{2m}\prod_{\substack{p \mid N \\ p \text{ prime}}}(1 - p^{m-1})(2B_m) \equiv 0\,(\bmod\,2^n).$$

Therefore Lemma 4(c) together with Lemma 8 and congruence (20) gives

$$2^{3m-1}\sum_{\substack{0<a<(N/8) \\ \gcd(a,N)=1}} a^{m-1}$$

$$\equiv -\chi_{-8}(N)\frac{B_{m,\chi_{-8}}^{[8N]}(N)}{m} + \chi_8(N)\frac{B_{m,\chi_8}^{[8N]}(N)}{m}\,(\bmod\,2^n).$$

Now congruence (22) follows immediately from Lemma 8.   ∎

4.3   The Main Theorem

Let m be a natural number. Let $\mathcal{D}$ be the discriminant of a quadratic field such that $\mathcal{D} \neq -4$ and $\chi_{\mathcal{D}}(-1) = (-1)^m$. Then the rational numbers $(B_{m,\chi_{\mathcal{D}}}/m)$ are 2-integral, see for example [Carlitz, 1959] or [Leopoldt, 1958]. More precisely, we have $\mathrm{ord}_2(B_{m,\chi_{\pm 8}}/m) = 0$ and $\mathrm{ord}_2(B_{m,\chi_{\mathcal{D}}}/m) \geq 1$ if $m \geq 2$ and $\mathcal{D} \neq \pm 8$, see [Urbanowicz, 1990/1991a,b] for details.

THEOREM 70 ([Fox, Urbanowicz and Williams, 1999, Theorem 6]) *Let m be a natural number. Let $\mathcal{D}$ be the discriminant of a quadratic field such that $\mathcal{D} \neq -4$ and $\chi_{\mathcal{D}}(-1) = (-1)^m$. Assume that $\chi_{\mathcal{D}}$ is the corresponding character and n is the number of distinct prime factors of $\mathcal{D}$. Then we have*

$$\frac{B_{m,\chi_{\mathcal{D}}}}{m} \equiv 0 \,(\,\mathrm{mod}\, 2^{n-1}\,).$$

PROOF. Let d be an odd fundamental discriminant. The proof falls naturally into three cases

(i) $\mathcal{D} = d$,

(ii) $\mathcal{D} = -4d$,

(iii) $\mathcal{D} = \pm 8d$.

In order to prove the congruence of Theorem 70, we proceed by induction on the pairs (m, n) ordered lexicographically, that is,

$$(k, r) \leq (m, n)$$

if and only if $k < m$, or $k = m$ and $r \leq n$.

If $m = 1$ Theorem 70 follows from Theorem 66. We now make the inductive hypothesis that

$$\frac{B_{k,\chi_e}}{k} \equiv 0 \,(\,\mathrm{mod}\, 2^{n-1}\,),$$

whenever $1 \leq k < m$ $(m \geq 2)$ and $e \neq 1$ is a fundamental discriminant having r distinct prime factors where $1 \leq r \leq n - 1$ $(n \geq 2)$.

Case (i). In this case Theorem 67 together with Lemma 9 gives the congruence

$$\sum_{\substack{e \mid d \\ e \neq 1}} (-1)^{\nu(e)}(\chi_e(2) - 2^m)\frac{B^{[d]}_{m,\chi_e}}{m}$$

$$+ \sum_{\substack{e \mid d \\ e \neq 1}} (-1)^{\nu(e)}\chi_e(2)\left(\frac{B^{[2d]}_{m,\chi_e}(|d|) - B^{[2d]}_{m,\chi_e}}{m}\right) \equiv 0 \,(\,\mathrm{mod}\, 2^n\,). \qquad (23)$$

Let us first consider the first sum on the left hand side of the above congruence. If $e = d$ we have

$$(\chi_d(2) - 2^m)\frac{B^{[d]}_{m,\chi_d}}{m} = (\text{odd integer}) \times \frac{B^{[d]}_{m,\chi_d}}{m},$$

and if $e \mid d$ and $e \neq d$, by the inductive assumption, we have

$$(\chi_e(2) - 2^m)\frac{B_{m,\chi_e}^{[d]}}{m} = \text{(odd integer)} \times \Big(\prod_{\substack{p \mid (d/e) \\ p \text{ prime}}} (1 - \chi_e(p)p^{k-1})\Big)\frac{B_{m,\chi_e}}{m}$$

$$= \text{(integer)} \times 2^{\nu(d/e)} \times 2^{\nu(e)-1} = \text{(integer)} \times 2^{n-1}.$$

We now turn to the second sum of the left hand side of congruence (23). It consists of the numbers

$$\frac{B_{m,\chi_e}^{[2d]}(|d|) - B_{m,\chi_e}^{[2d]}}{m},$$

where $e \mid d$ and $e \neq 1$. Therefore the summands of the second sum are up to sign of the form

$$\binom{m-1}{k-1}\frac{B_{k,\chi_e}}{k}(1 - \chi_e(2)2^{k-1}) \prod_{\substack{p \mid (d/e) \\ p \text{ prime}}} (1 - \chi_e(p)p^{k-1})|d|^{m-k},$$

where $1 \leq k \leq m - 1$. If $k = 0$ the summand disappears since

$$B_{0,\chi_e} = 0$$

if $e \neq 1$. Thus, by the inductive assumption, the summands of the second sum are of the form

$$\text{(integer)} \times 2^{\nu(e)-1} \times 2^{\nu(d/e)} = \text{(integer)} \times 2^{n-1}$$

and all these observations together give Theorem 70 in case (i).

Case (ii). In this case $n-1 = \nu(d)$. We make use of Theorem 68 and Lemma 9, which give the congruence

$$\sum_{\substack{e \mid d \\ e \neq 1}}(-1)^{\nu(e)}(1 - \chi_e(2)2^{m-1} - 2^{2m-1})\frac{B_{m,\chi_e}^{[d]}}{m}$$

$$+ \sum_{\substack{e \mid d \\ e \neq 1}}(-1)^{\nu(e)}\Big(\frac{B_{m,\chi_e}^{[2d]}(|d|) - B_{m,\chi_e}^{[2d]}}{m}\Big) - \chi_{-4}(|d|)\sum_{\substack{e \mid d \\ e \neq 1}}(-1)^{\nu(e)}\frac{B_{m,\chi_{-4e}}^{[4d]}(|d|)}{m}$$

$$\equiv 0 \,(\, \text{mod}\, 2^{\nu(d)} \,).$$

We first consider the first and second sums on the left hand side of the above congruence. The numbers

$$\frac{B_{m,\chi_e}^{[d]}}{m}, \quad \frac{B_{m,\chi_e}^{[2d]}(|d|)}{m}$$

are divisible by $2^{\nu(d)-1}$, this follows from Theorem 70(i). Thus, by virtue of congruence (23), we deduce that

$$\sum_{\substack{e \mid d \\ e \neq 1}} (-1)^{\nu(e)} (1 - \chi_e(2)2^{m-1} - 2^{2m-1}) \frac{B^{[d]}_{m,\chi_e}}{m}$$

$$+ \sum_{\substack{e \mid d \\ e \neq 1}} (-1)^{\nu(e)} \left(\frac{B^{[2d]}_{m,\chi_e}(|d|) - B^{[2d]}_{m,\chi_e}}{m} \right) \equiv 0 \, (\mathrm{mod}\, 2^{\nu(d)}).$$

Thus we obtain the congruence

$$\sum_{\substack{e \mid d \\ e \neq 1}} (-1)^{\nu(e)} \frac{B^{[4d]}_{m,\chi_{-4}\chi_e}(|d|)}{m} \equiv 0 \, (\mathrm{mod}\, 2^{\nu(d)}). \tag{24}$$

The summands of the sum on the left hand side of (24) are up to sign

$$\frac{B^{[4d]}_{m,\chi_{-4}\chi_e}(|d|)}{m} = \sum_{k=1}^{m} \binom{m-1}{k-1} \left(\prod_{\substack{p \mid (d/e) \\ p \text{ prime}}} (1 - \chi_{-4}\chi_e(p)p^{k-1}) \right) \frac{B_{k,\chi_{-4}\chi_e}}{k} |d|^{m-k}.$$

(Recall that

$$B_{0,\chi_{-4}\chi_e} = 0.)$$

If $e \neq d$ then, by the inductive assumption, the summands of the sum on the right hand side of the above equation have the form

$$(\text{integer}) \times 2^{\nu(d/e)} \times 2^{\nu(e)} = (\text{integer}) \times 2^{\nu(d)}.$$

If $e = d$ the summands of the sum on the left hand side of congruence (24) have up to sign the form

$$\frac{B^{[4d]}_{m,\chi_{-4}\chi_d}(|d|)}{m} = \sum_{k=1}^{m} \binom{m-1}{k-1} \frac{B_{k,\chi_{-4}\chi_d}}{k} |d|^{m-k}.$$

Thus, by the inductive assumption, they are

$$(\text{integer}) \times 2^{\nu(d)} \pm \frac{B_{m,\chi_{-4}\chi_d}}{m}$$

and the theorem in case (ii) follows.

Case (iii). In this case $n - 1 = \nu(d)$. We make use of Theorem 69 and Lemma 9, which yield the congruence

$$\sum_{\substack{e \mid d \\ e \neq 1}} (-1)^{\nu(e)} (\chi_e(2) - 2^{m-1} - 2^{3m-1}) \frac{B^{[d]}_{m,\chi_e}}{m}$$

$$+ \sum_{\substack{e \mid d \\ e \neq 1}} (-1)^{\nu(e)} \chi_e(2) \left(\frac{B^{[2d]}_{m,\chi_e}(|d|) - B^{[2d]}_{m,\chi_e}}{m} \right)$$

$$- \chi_{-4}(|d|) \sum_{\substack{e \mid d \\ e \neq 1}} (-1)^{\nu(e)} \chi_e(2) \frac{B^{[4d]}_{m,\chi_{-4}\chi_e}(|d|)}{m}$$

$$- \sum_{\substack{e \mid d \\ e \neq 1}} (-1)^{\nu(e)} \chi_e(2) \left(\frac{\chi_{-8}(|d|) B^{[8d]}_{m,\chi_{-8}\chi_e}(|d|)}{m} - \frac{\chi_8(|d|) B^{[8d]}_{m,\chi_8\chi_e}(|d|)}{m} \right)$$

$$\equiv 0 \, (\bmod \, 2^{\nu(d)}).$$

We first consider the first, second and third sums on the left hand side of the above congruence. The numbers

$$\frac{B^{[d]}_{m,\chi_e}}{m}, \quad \frac{B^{[2d]}_{m,\chi_e}(|d|)}{m}$$

are divisible by $2^{\nu(d)-1}$, this follows from Theorem 70(i), and the numbers

$$\frac{B^{[4d]}_{m,\chi_{-4}\chi_e}(|d|)}{m}$$

are divisible by $2^{\nu(d)}$, this follows from Theorem 70(ii). Thus, by virtue of congruence (23), we conclude that

$$\sum_{\substack{e \mid d \\ e \neq 1}} (-1)^{\nu(e)} \chi_e(2) \left(\frac{\chi_{-8}(|d|) B^{[8d]}_{m,\chi_{-8}\chi_e}(|d|)}{m} - \frac{\chi_8(|d|) B^{[8d]}_{m,\chi_8\chi_e}(|d|)}{m} \right)$$

$$\equiv 0 \, (\bmod \, 2^{\nu(d)}).$$

The summands of the sum on the left hand side of the above congruence have up to sign the form

$$\frac{B^{[8d]}_{m,\chi_{-8}\chi_e}(|d|)}{m} \pm \frac{B^{[8d]}_{m,\chi_8\chi_e}(|d|)}{m} = \pm \sum_{k=1}^{m} \binom{m-1}{k-1} \mathcal{B}(k,e,d)|d|^{m-k},$$

where

$$
\mathcal{B}(k, e, d) = \begin{cases} \displaystyle\prod_{\substack{p \,|\, (d/e) \\ p \,\text{prime}}} (1 - \chi_{-8}\chi_e(p)p^{k-1})\frac{B_{k,\chi_{-8}\chi_e}}{k}, & \text{if } \chi_e(-1) \neq (-1)^k, \\[2em] \displaystyle\prod_{\substack{p \,|\, (d/e) \\ p \,\text{prime}}} (1 - \chi_8\chi_e(p)p^{k-1})\frac{B_{k,\chi_8\chi_e}}{k}, & \text{if } \chi_e(-1) = (-1)^k. \end{cases}
$$

Recall that

$$
B_{0,\chi_{\pm 8}\chi_e} = 0
$$

and

$$
\left(\frac{B_{k,\chi_{-8}\chi_e}}{m}\right) \times \left(\frac{B_{k,\chi_8\chi_e}}{m}\right) = 0. \tag{25}
$$

If $e \neq d$ then, by the inductive assumption, the summands of the above sum have the form

$$
(\text{integer}) \times 2^{\nu(d/e)} \times 2^{\nu(e)} = (\text{integer}) \times 2^{\nu(d)}.
$$

If $e = d$ the corresponding summand has the form

$$
\sum_{k=1}^{m} \binom{m-1}{k-1}\left(\frac{B_{k,\chi_{-8}\chi_e}}{k} \pm \frac{B_{k,\chi_8\chi_e}}{k}\right)|d|^{m-k},
$$

and so, by the inductive assumption, this summand is

$$
(\text{integer}) \times 2^{\nu(d)} + \left(\frac{B_{m,\chi_{-8}\chi_d}}{m} \pm \frac{B_{m,\chi_8\chi_d}}{m}\right).
$$

Now the theorem in case (iii) follows from (25) if $e = d$ and $k = m$.   ∎

4.4   *Final Remarks*

One of the most important properties of the generalized Bernoulli numbers is that they give the values of Dirichlet L-functions at negative integers. Namely, we have $L(1 - m, \chi) = -(B_{m,\chi}/m)$, where $m \geq 1$ (see section 4.5 of Chapter I). Thus, we can rewrite the congruence of Theorem 70 in the form

$$
L(1 - m, \chi_d) \equiv 0 \,(\bmod\, 2^{n-1}),
$$

where $\chi_d(-1) = (-1)^m$ and d has n distinct prime factors. In particular Gauss' congruence considered in Theorem 66 was proved for the numbers $L(0, \chi_d) = h(d)$ (d negative), see section 8.1 of Chapter I. It is surprising that as yet no one has deduced Gauss' congruence for positive d using complex analytic methods, that is, from Dirichlet's class number formula, which in this case has the form $L(1, \chi_d) = 2d^{-1/2}h(d)\log\varepsilon$, where ε is the fundamental unit of the quadratic field with discriminant d.

As usual, the complex and p-adic formulae differ by an Euler factor. The corresponding p-adic formulae for class numbers of quadratic fields are of the form $L_p(1, \chi_d) = 2(1 - \chi_d(p)p^{-1})d^{-1/2}h(d)\log_p \varepsilon$ if d positive and $\log_p$ denotes the p-adic logarithm (this is Leopoldt's formula), and $L_p(0, \chi_d\omega_p) = (1-\chi_d(p))h(d)$ if d negative and ω_p is the Teichmüller character at p. Similarly, we have $L_p(1 - m, \chi\omega_p^m) = -(1 - \chi(p)p^{m-1})(B_{m,\chi}/m)$ if $m \geq 1$, see sections 2.3 and 2.4 of Chapter IV. Uehara [Uehara, 1990, Lemma 3] found an inductive p-adic formula corresponding to that in Lemma 3. Uehara's formula was proved for any quadratic field. It was noticed in [Urbanowicz and Wójcik, 1995/1996] that Uehara's formula is a special case of a more general inductive formula for the so-called p-adic multilogarithms introduced by Coleman [Coleman, 1982]. Let χ be a primitive Dirichlet character modulo $M > 1$ and $\tau(\chi)$ denote the normalized Gauss sum attached to χ. Write $\zeta_M = \exp(2\pi i/M)$. Denoting by $l_{k,p}(s)$ the p-adic multilogarithms, it was proved by Coleman [Coleman, 1982] that

$$L_p(k, \chi\omega_p^{1-k}) = (1 - \chi(p)p^{-k})\frac{\tau(\chi)}{M} \sum_{n=1}^{M} \bar{\chi}(n)l_{k,p}(\zeta_M^{-n}),$$

see section 3.2 of Chapter IV. Denoting by $l_{k,\infty}(s)$ the complex multilogarithms, we have the well known formula

$$L(k, \chi) = \frac{\tau(\chi)}{M} \sum_{n=1}^{M} \bar{\chi}(n)l_{k,\infty}(\zeta_M^{-n}).$$

Let N be a positive multiple of M such that N/M is a rational squarefree integer prime to M. Uehara's inductive formula in its most general form is

$$\sum_{\substack{1 \leq k \leq N \\ \gcd(k,N)=1}} \chi(k)l_{k,p}(\zeta_N^k)$$

$$= (-1)^{\nu(N/M)} \prod_{\substack{p \mid (N/M) \\ p \text{ prime}}} (1 - \chi(p)p^{1-k}) \sum_{k=1}^{M} \chi(k)l_{k,p}(\zeta_M^k),$$

see Lemma 1 in Chapter IV.

The above identity implies many new congruences for the values of $L_p(k, \chi\omega_p^{1-k})$. The most general such congruences were proved in Chapter IV. One may expect that both Gauss' congruence (for any quadratic field) and its generalization for generalized Bernoulli numbers presented in this chapter can be deduced from these congruences. This means that with high probability we should be able to prove Gauss' congruence for any quadratic field using p-adic analytic methods. For further discussion see also [Gras, 1987, 1991/1992] and [Pioui, 1990, 1992]. The related complex problem seems hopeless today.

Bibliography

Akiyama, S. (1994). Refinements of class number formulas for quadratic fields. In: *Proc. Sympos. RIMS Kyoto, 1993*, Sūrikaisekikenkyūsho Kōkyūroku 886, pp. 170–177. Kyoto.

Amice, Y., and Fresnel, J. (1972). Fonctions zêta p-adiques des corps de nombres abéliens réels. *Acta Arith.*, 20:353–384.

Ankeny, N. C., Artin, E., and Chowla, S. (1952). The class-numbers of real quadratic number fields. *Ann. of Math.* (2), 56:479–493.

Apostol, T. M. (1976). *Introduction to Analytic Number Theory*. Springer-Verlag, New York.

Ayoub, R., Chowla, S., and Walum, H. (1967). On sums involving quadratic characters. *J. London Math. Soc.*, 42:152–154.

Barkan, P. (1975). Une propriété de congruence de la longeur de la période d'un développement en fraction continue. *C. R. Acad. Sci. Paris Sér. I Math.*, 281:825–828.

Barner, K. (1969). Über die Werte der Ringklassen-L-Funktionen reell-quadratischer Zahlkörper an natürlichen Argumentstellen. *J. Number Theory*, 1:28–64.

Barrucand, P., and Cohn, H. (1969). Note on primes of type $x^2 + 32y^2$, class-number, and residuacity. *J. Reine Angew. Math.*, 238:67–70.

Bauer, H. (1971). Zur Berechnung der 2-Klassenzahl der quadratischen Zahlkörper mit genau zwei verschiedenen Diskriminantenprimteilern. *J. Reine Angew. Math.*, 248:42–46.

Bauer, H. (1972). Die 2-Klassenzahlen spezieller quadratischer Zahlkörper. *J. Reine Angew. Math.*, 252:79–81.

Belabas, K., and Gangl, H. (1999). Appendix to "Computing the tame kernel of quadratic imaginary fields" by J. Browkin. Preprint. *Math. Comp.*, to appear.

Berger, A. (1890/1891). Recherches sur les nombres et les fonctions de Bernoulli. *Acta Math.*, 14:294–304.

231

Berger, R. I. (1990). Quadratic extensions of number fields with elementary abelian 2-prim $K_2(O_F)$ of smallest rank. *J. Number Theory*, 34:284–292.

Berndt, B. C. (1975a). Character analogues of the Poisson and Euler-Maclaurin summation formulas with applications. *J. Number Theory*, 7:413–445.

Berndt, B. C. (1975b). Periodic Bernoulli numbers, summation formulas and applications. In: Askey, R. A., editor, *Theory and Application of Special Functions*, pp. 143–189. Academic Press, New York.

Berndt, B. C. (1976). Classical theorems on quadratic residues. *Enseign. Math.* (2), 22:261–304.

Berndt, B. C., and Chowla, S. (1974). Zero sums of the Legendre symbol. *Nordisk Mat. Tidskr.*, 22:5–8.

Berndt, B. C., and Schoenfeld, L. (1975, 1980/1981). Periodic analogues of the Euler-Maclaurin and Poisson summation formulas with applications to number theory. *Acta Arith.*, 28:23–68; Corrigendum: *ibid.*, 38:328.

Birch, B. J. (1969). K_2 of global fields. *Proc. Sympos. Pure Math.*, 20:87–95.

Bloch, S. (1977). Applications of the dilogarithm function in algebraic K-theory and algebraic geometry. In: *Proceedings of the International Symposium on Algebraic Geometry (Kyoto, 1977)*, pp. 103–114. Kinokuniya, Tokyo.

Boldy, M. C. (1991). *The 2-primary Component of the Tame Kernel of Quadratic Number Fields*. Ph. D. Thesis, University of Nijmegen.

Bölling, R. (1979). Bemerkungen über Klassenzahlen und Summen von Jacobi-Symbolen. *Math. Nachr.*, 90:159–172.

Borel, A. (1972). Cohomologie réelle stable de groupes S-arithmétiques classiques. *C. R. Acad. Sci. Paris Sér. I Math.*, 274:1700–1703.

Borel, A. (1977, 1980). Cohomologie de SL_n et valeurs de fonctions zêta aux points entiers. *Ann. Sci. Norm. Sup. Pisa* (4), 4:613–636; Errata, *ibid.*, 7:373.

Borel, A. (1995). Values of zeta-functions at integers, cohomology and polylogarithms. In: Adhikari, S. D., editor, *Current Trends in Mathematics and Physics, a Tribute to Harish-Chandra*, pp. 1–44. Narosa, New Delhi.

Borevich, Z. I., and Shafarevich, I. R. (1966). *Number Theory*. Academic Press, New York.

Brauckman, B. (1991). The 2-Sylow-subgroup of the tame kernel of number fields. *Canad. J. Math.*, 43:255–264.

Browkin, J. (1982). The functor K_2 of the ring of integers of a number field. In: Traczyk, T., editor, *Universal Algebra and Applications*, Banach Center Publ. 9, pp. 187–195. PWN, Warsaw.

Browkin, J. (1991). K-theory, cyclotomic equations, and Clausen's function. In: Lewin, L., editor, *Structural Properties of Polylogarithms*, Math. Surveys Monographs 37, pp. 233–273. Amer. Math. Soc., Providence, RI.

Browkin, J. (1999). Computing the tame kernel of quadratic imaginary fields. Preprint. *Math. Comp.*, to appear.

Browkin, J., and Gangl, H. (1999). Tame and wild kernels of quadratic imaginary number fields. *Math. Comp.*, 68:291–305.

Browkin, J., and Schinzel, A. (1982). On Sylow 2-subgroup of K_2O_F for quadratic number fields. *J. Reine Angew. Math.*, 331:104–113.

Brown, E. (1972). The class number of $\mathbb{Q}(\sqrt{-p})$, for $p \equiv 1 \pmod 8$ a prime. *Proc. Amer. Math. Soc.*, 31:381–383.

Brown, E. (1973). The power of 2 dividing the class-number of a binary quadratic discriminant. *J. Number Theory*, 5:413–419.

Brown, E. (1974a). Class numbers of complex quadratic fields. *J. Number Theory*, 6:185–191.

Brown, E. (1974b). Class numbers of real quadratic number fields. *Trans. Amer. Math. Soc.*, 190:99–107.

Brown, E. (1974c). A lemma of Stark. *J. Reine Angew. Math.*, 265:201.

Brown, E. (1975). Class numbers of quadratic fields. *Sympos. Math.*, 15:403–411.

Brown, E. (1981). The class number of $\mathbb{Q}(\sqrt{-pq})$, for $p \equiv -q \equiv 1 \pmod 4$ primes. *Houston J. Math.*, 7:497–505.

Brown, E. (1983). The class number and fundamental unit of $\mathbb{Q}(\sqrt{2p})$, for $p \equiv 1 \pmod{16}$ a prime. *J. Number Theory*, 16:95–99.

Brown, E., and Parry, C. J. (1973). Class numbers of imaginary quadratic fields having exactly three discriminantal divisors. *J. Reine Angew. Math.*, 260:31–34.

Candiotti, A., and Kramer, K. (1989). On the 2-Sylow subgroup of the Hilbert kernel of K_2 of number fields. *Acta Arith.*, 52:49–65.

Carlitz, L. (1953). Some congruences for the Bernoulli numbers. *Amer. J. Math.*, 75:163–172.

Carlitz, L. (1959). Arithmetic properties of generalized Bernoulli numbers. *J. Reine Angew. Math.*, 202:174–182.

Cauchy, A. L. (1882). Note XII. In: *Oeuvres* (1), Tome III, pp. 359–390. Gauthier-Villars, Paris.

Chowla, S. (1961). On the class number of real quadratic fields. *Proc. Nat. Acad. Sci. U.S.A.*, 47:878.

Coates, J. (1973). On Iwasawa's analogue of the Jacobian for totally real number fields. *Proc. Sympos. Pure Math.*, 25:51–61.

Coates, J. (1977). p-adic L-functions and Iwasawa's theory. In: Fröhlich, A., editor, *Algebraic Number Fields, Durham Sympos., 1975*, pp. 269–353. Academic Press, London.

Cohn, H., and Cooke, G. (1976). Parametric form of an eight class field. *Acta Arith.*, 30:367–377.

Coleman, R. F. (1982). Dilogarithms, regulators and p-adic L-functions. *Invent. Math.*, 69:171–208.

Conner, P. E., and Hurrelbrink, J. (1986). A comparison theorem for the 2-rank of $K_2(\mathcal{O})$. In: Bloch, S. J. *et al.*, editors, *Applications of Algebraic K-theory to Algebraic Geometry and Number Theory*, II, Contemp. Math. 55, pp. 411–420. Amer. Math. Soc., Providence, RI.

Conner, P. E., and Hurrelbrink, J. (1988). *Class Number Parity*, Ser. Pure Math. 8. World Scientific, Singapore.

Conner, P. E., and Hurrelbrink, J. (1989a). The 4-rank of $K_2(O)$. *Canad. J. Math.*, 41:932–960.

Conner, P. E., and Hurrelbrink, J. (1989b). *Examples of quadratic number fields with $K_2(O)$ containing no elements of order four*. Preprint.

Conner, P. E., and Hurrelbrink, J. (1995). On elementary abelian 2-Sylow K_2 of rings of integers of certain quadratic number fields. *Acta Arith.*, 73:59–65.

Costa, A. (1989). *Modular Forms and Class Number Congruences*. Ph. D. Thesis, University of Pennsylvania.

Costa, A. (1992). Modular forms and class number congruences. *Acta Arith.*, 61:101–118.

Costa, A. (1993). A generalization of a result of Barrucand and Cohn on class numbers. *J. Number Theory*, 45:254–260.

Currie, J., and Williams, K. S. (1982). Class numbers and biquadratic reciprocity. *Canad. J. Math.*, 34:969–988.

Damey, P., et Payan, J.-J. (1970). Existence et construction des extensions galoisiennes et non-abéliennes de degré 8 d'un corps de caractéristique differente de 2. *J. Reine Angew. Math.*, 244:37–54.

Davenport, H. (1980). *Multiplicative Number Theory*. Second edition, Springer-Verlag, New York.

Dedekind, R. (1863). Bemerkungen zur Abhandlung K. F. Gauss, 'De nexu inter multitudinem classium, in quas formae binariae secundi gradus distribuunter, earumque determinantem'. In: Gauss, C. F., *Werke*, zweiter Band, pp. 292–303. Königliche Gesellschaft der Wissenschaften, Göttingen.

Deligne, P., and Ribet, K. (1980). Values of abelian L-functions at negative integers over totally real fields. *Invent. Math.* 59:227–286.

Desnoux, P.-J. (1987). *Congruences dyadiques entre nombres de classes de corps quadratiques*. Thèse, Université Paris 7.

Desnoux, P.-J. (1988). Congruences dyadiques entre nombres de classes de corps quadratiques. *Manuscripta Math.*, 62:163–179.

Dilcher, K., Skula, L., and Slavutskiĭ, I. Sh. (1991, 1994). *Bernoulli Numbers, Bibliography* (1713–1990). Queen's Papers in Pure and Appl. Math. 87, 175 pp.; Appendix, unpublished manuscript, 30 pp.

Dirichlet, P. G. L. (1828, 1889). Question d'analyse indéterminée. *J. Reine Angew. Math.*, 3:407–408; or in: *Werke*, Bd. 1, pp. 107–108. Georg Reimer, Berlin.

Dirichlet, P. G. L. (1839, 1889). Recherches sur diverses applications de l'analyse infinitésimale à la théorie des nombres, première partie. *J. Reine Angew. Math.*, 19:324–369; or in: *Werke*, Bd. 1, pp. 411–496. Georg Reimer, Berlin.

Dirichlet, P. G. L. (1840, 1889). Recherches sur diverses applications de l'analyse infinitésimale à la théorie des nombres, seconde partie. *J. Reine Angew. Math.*, 21:1–12, 134–155; or in: *Werke*, Bd. 1, pp. 411–496. Georg Reimer, Berlin.

Endô, A. (1973a). On the 2-rank of the ideal class groups of quadratic fields. *Mem. Fac. Sci. Kyushu Univ. Ser. A*, 27:7–12.

Endô, A. (1973b). On the 2-class number of certain quadratic number fields. *Mem. Fac. Sci. Kyushu Univ. Ser. A*, 27:111–120.

Ernvall, R. (1979). Generalized Bernoulli numbers, generalized irregular primes, and class number. *Ann. Univ. Turku, Ser. A1*, 178, 72 pp.

Ernvall, R. (1983). Generalized irregular primes. *Mathematika*, 30:67–73.

Euler, L. (1740, 1924). De summis serierum reciprocarum. *Comment. Acad. Sci. Petropolit.* 7 (1734/5):123–144; or in: *Opera omnia*, Ser. 1, Bd. 14, pp. 73–86. B. G. Teubner, Leipzig-Berlin.

Federer, L. J. (1982). Regulators, Iwasawa modules, and the main conjecture for $p = 2$. In: Koblitz, N., editor, *Number Theory Related to Fermat's Last Theorem*. Progr. Math. 26, pp. 287–296. Birkhäuser, Boston, MA.

Fox, G. J., Urbanowicz, J., and Williams, K. S. (1999). Gauss' congruence from Dirichlet's class number formula and generalizations. In: Győry, K., Iwaniec, H., and Urbanowicz, J., editors, *Number Theory in Progress (Zakopane, 1997)*, vol. II, pp. 813–840. Walter de Gruyter, Berlin-New York.

Fresnel, J. (1967/1968). Nombres de Bernoulli et fonctions L p-adiques. *Ann. Inst. Fourier (Grenoble)*, 17:281–333.

Friesen, C., and Williams, K. S. (1985). Remark on the class number of $\mathbb{Q}(\sqrt{2p})$ modulo 8 for $p \equiv 5 \pmod 8$ a prime. *Proc. Amer. Math. Soc.*, 93:198–200.

Frobenius, F. G. (1910, 1968). Über die Bernoullischen Zahlen und die Euler-schen Polynome. *Sitzungsber. Preuss. Akad. Wiss.*, 2:809–847; or in: *Gesammelte Abhandlungen* III, pp. 440–478. Springer-Verlag, Berlin-Heidelberg-New York.

Garland, H. (1971). A finiteness theorem for K_2 of a number field. *Ann. of Math.* (2), 94:534–548.

Gauss, C. F. (1828, 1897). Letter to P. G. L. Dirichlet dated 30 May 1828. In: Dirichlet, P. G. L., *Mathematische Werke*, vol. II, pp. 378–380. Georg Reimer, Berlin.

Gauss, C. F. (1863). De nexu inter multitudinem classium, in quas formae binariae secundi gradus distribuunter, earumque determinantem. In: *Werke*, zweiter Band, pp. 269–291. Königliche Gesellschaft der Wissenschaften, Göttingen.

Gebhardt, H. M. (1977). Zur Berechnung des Funktors K_2 von einigen euklidischen Ringen. *Schriftenreihe Math. Inst. Univ. Münster*, Ser. 2. Heft 13.

Glaisher, J. W. L. (1901). Formulae derived from Gauss's sums with applications to the series connected with the number of classes of binary forms. *Quart. J. Math.*, 33:289–330.

Glaisher, J. W. L. (1903a). On the distribution of the numbers for which $(s\,|\,p) = 1$ or -1, in the octants, quadrants, etc. of p. *Quart. J. Math.*, 34:1–27.

Glaisher, J. W. L. (1903b). On the expression for the number of classes of a negative determinant, and on the number of positives in the octants of P. *Quart. J. Math.*, 34:178–204.

Goren, E. Z. (1999). Hilbert modular forms modulo p^m. The unramified case. CICMA Reports, Concordia Laval McGill.

Gradshteyn, I. S., and Ryzhik, I. M. (1994). *Table of Integrals, Series and Products*. Fifth edition, Academic Press, New York.

Gras, G. (1973). Sur les l-classes d'idéaux dans les extensions cycliques relatives de degré premier l, II. *Ann. Inst. Fourier (Grenoble)*, 23(4):1–44.

Gras, G. (1987). Pseudo-mesures p-adiques associées aux fonctions L de $\mathbb{Q}$. *Manuscripta Math.*, 57:373–415.

Gras, G. (1989). Relations congruentielles linéaires entre nombres de classes de corps quadratiques. *Acta Arith.*, 52:147–162.

Gras, G. (1991/1992). *Mesures p-adiques*. Publ. Math. Fac. Sci. Besançon, Théorie des Nombres, 107 pp.

Greither, C. (1992). Class groups of abelian fields, and the Main Conjecture. *Ann. Inst. Fourier (Grenoble)*, 42:449–499.

Halter-Koch, F. (1984). Über den 4-Rang der Klassengruppe quadratischer Zahlkörper. *J. Number Theory*, 19:219–227.

Hardy, K., Kaplan, P., and Williams, K. S. (1986). Divisibilité par 16 du nombre des classes au sens strict des corps quadratiques réels dont le deux-groupe des classes est cyclique. *Osaka J. Math.*, 23:479–489.

Hardy, K., and Williams, K. S. (1986). A congruence relating the class numbers of complex quadratic fields. *Acta Arith.*, 47:263–276.

Hardy, K., and Williams, K. S. (1987). Congruences modulo 16 for the class numbers of complex quadratic fields. *J. Number Theory*, 27:178–195.

Hasse, H. (1952, 1985). *Über die Klassenzahl abelscher Zahlkörper*. Akademie-Verlag, Berlin; reedition Springer-Verlag, Berlin-New York.

Hasse, H. (1964). *Vorlesungen über Zahlentheorie*, zweite neubearbeitete Auflage. Springer-Verlag, Berlin-New York.

Hasse, H. (1969a). Über die Klassenzahl des Körpers $P(\sqrt{-2p})$ mit einer Primzahl $p \neq 2$. *J. Number Theory*, 1:231–234.

Hasse, H. (1969b). Über die Klassenzahl des Körpers $P(\sqrt{-p})$ mit einer Primzahl $p \equiv 1 \bmod 2^3$. *Aequationes Math.*, 3:165–169.

Hasse, H. (1970a). Über die Teilbarkeit durch 2^3 der Klassenzahl imaginärquadratischer Zahlkörper mit genau zwei verschiedenen Diskriminantenprimteilern. *J. Reine Angew. Math.*, 241:1–6.

Hasse, H. (1970b). Über die Teilbarkeit durch 2^3 der Klassenzahl der quadratischen Zahlkörper mit genau zwei verschiedenen Diskriminantenprimteilern. *Math. Nachr.*, 46:61–70.

Hecke, E. (1924, 1959). Analytische Funktionen und algebraische Zahlen, zweiter Teil. *Abh. Math. Sem. Univ. Hamburg*, 3:213–236; or in: *Mathematische Werke*, pp. 381–404. Vandenhoeck & Ruprecht, Göttingen.

Hikita, M. (1986a). On the congruences for the class numbers of the quadratic fields whose discriminants are divisible by 8. *J. Number Theory*, 23:86–101.

Hikita, M. (1986b). Quadratic fields and factors of cyclotomic polynomials. *Tokyo J. Math.*, 9:341–355.

Holden, H. (1906a). On various expressions for h, the number of properly primitive classes for a determinant $-p$, where p is a prime of the form $4n + 3$ (first paper). *Messenger Math.*, 35:73–80.

Holden, H. (1906b). On various expressions for h, the number of properly primitive classes for a determinant $-p$, where p is of the form $4n + 3$, and is a prime or the product of different primes (second paper). *Messenger Math.*, 35:102–110.

Holden, H. (1906c). On various expressions for h, the number of properly primitive classes for any negative determinant, not involving a square factor (third paper). *Messenger Math.*, 35:110–117.

Holden, H. (1907a). On various expressions for h, the number of properly primitive classes for a negative determinant (fourth paper). *Messenger Math.*, 36:69–75.

Holden, H. (1907b). On various expressions for h, the number of properly primitive classes for a determinant $-p$, where p is of the form $4n + 3$, and is a prime or the product of different primes (addition to the second paper). *Messenger Math.*, 36:75–77.

Holden, H. (1907c). On various expressions for h, the number of properly primitive classes for a negative determinant not containing a square factor (fifth paper). *Messenger Math.*, 36:126–134.

Holden, H. (1908). On various expressions for h, the number of properly primitive classes for any negative determinant, not containing a square factor (sixth paper). *Messenger Math.*, 37:13–16.

Hudson, R. H., and Williams, K. S. (1982). Class number formulae of Dirichlet type. *Math. Comp.*, 39:725–732.

Hurrelbrink, J. (1989). Class numbers, units and K_2. In: Jardine, J. F., and Snaith, V., editors, *Algebraic K-theory: Connection with Geometry and Topology*, NATO ASI Series C 279, pp. 87–102. Kluwer Acad. Publ., Dordrecht.

Hurrelbrink, J. (1994). Circulant graphs and 4-ranks of ideal class groups. *Canad. J. Math.*, 46:169–183.

Hurrelbrink, J., and Kolster, M. (1998). Tame kernels under relative quadratic extensions and Hilbert symbols. *J. Reine Angew. Math.*, 499:145–188.

Hurwitz, A. (1882, 1932). Einige Eigenschaften der Dirichlet'schen Funktionen $F(s) = \sum(D/n)/n^s$, die bei der Bestimmung der Classenzahlen binärer quadratischer Formen auftreten. *Z. Math. Phys.*, 27:86–101; or in: *Mathematische Werke*, vol. I, pp. 72–88. Birkhäuser, Basel.

Hurwitz, A. (1895, 1933). Über die Anzahl der Klassen binärer quadratischer Formen von negative Determinante. *Acta Math.*, 19:351–384; or in: *Mathematische Werke*, vol. II, 208–235. Birkhäuser, Basel.

Ireland, K., and Rosen, M. (1990). *A Classical Introduction to Modern Number Theory*. Second edition, Springer-Verlag, New York.

Iwasawa, K. (1969). On p-adic L-functions. *Ann. of Math.* (2), 89:198–205.

Iwasawa, K. (1972). *Lectures on p-adic L-functions*, Ann. of Math. Stud. 74. Princeton Univ. Press, Princeton, NJ.

Johnson, W., and Mitchell, K. J. (1977). Symmetries for sums of the Legendre symbol. *Pacific J. Math.*, 69:117–124.

Kaplan, P. (1972). Divisibilité par 8 du nombre des classes des corps quadratiques réels dont le 2-sous-groupe des classes est cyclique. *C. R. Acad. Sci. Paris Sér. I Math.*, 275:887–890.

Kaplan, P. (1973a). Divisibilité par 8 du nombre des classes des corps quadratiques dont le 2-groupe des classes est cyclique, et réciprocité biquadratique. *J. Math. Soc. Japan*, 25:596–608.

Kaplan, P. (1973b). 2-groupe des classes et facteurs principaux de $\mathbb{Q}(\sqrt{pq})$ où $p \equiv -q \equiv 1 \,(\mathrm{mod}\,4)$. *C. R. Acad. Sci. Paris Sér. I Math.*, 276:89–92.

Kaplan, P. (1974). Comparaison des 2-groupes des classes d'idéaux au sens large et au sens étroit d'un corps quadratique réel. *Proc. Japan Acad. Ser. A Math. Sci.*, 50:688–693.

Kaplan, P. (1976). Sur le 2-groupe des classes d'idéaux des corps quadratiques. *J. Reine Angew. Math.*, 283/284:313–363.

Kaplan, P. (1977a). Unités de norme -1 de $\mathbb{Q}(\sqrt{p})$ et corps de classes de degré 8 de $\mathbb{Q}(\sqrt{-p})$ où p est un nombre premier congru à 1 modulo 8. *Acta Arith.*, 32:239–243.

Kaplan, P. (1977b). Cycles d'ordre au moins 16 dans le 2-groupe des classes d'idéaux de certains corps quadratiques. *Bull. Soc. Math. France*, Mém. 49–50:113–124.

Kaplan, P. (1981). Nouvelle démonstration d'une congruence modulo 16 entre les nombres de classes d'ideaux de $\mathbb{Q}(\sqrt{-2p})$ et $\mathbb{Q}(\sqrt{2p})$ pour p premier $\equiv 1 \pmod 4$. *Proc. Japan Acad. Ser. A Math. Sci.*, 57:507–509.

Kaplan, P., and Williams, K. S. (1982a). On the class numbers of $\mathbb{Q}(\sqrt{\pm 2p})$ modulo 16, for $p \equiv 1 \pmod 8$ a prime. *Acta Arith.*, 40:289–296.

Kaplan, P., and Williams, K. S. (1982b). Congruences modulo 16 for the class numbers of the quadratic fields $\mathbb{Q}(\sqrt{\pm p})$ and $\mathbb{Q}(\sqrt{\pm 2p})$ for p a prime congruent to 5 modulo 8. *Acta Arith.*, 40:375–397.

Kaplan, P., and Williams, K. S. (1984). On the strict class number of $\mathbb{Q}(\sqrt{2p})$ modulo 16. *Osaka J. Math.*, 21:23–29.

Kaplan, P. and Williams, K. S. (1986). Cycles d'ordre au moins 16 dans le deux groupe des classes d'ideaux au sens strict de corps quadratiques. In: *Proceedings of the International Conference on Class Numbers and Fundamental Units of Algebraic Number Fields, held in Katata, Japan, June 24–28*, pp. 189–203. Nagoya Univ., Nagoya.

Karpinski, L. C. (1904). Über die Verteilung der quadratischen Reste. *J. Reine Angew. Math.*, 127:1–19.

Kenku, M. A. (1977). Atkin-Lehner involutions and class number residuality. *Acta Arith.*, 33:1–9.

Keune, F. (1989). On the structure of the K_2 of the ring of integers in a number field. *K-theory*, 2:625–645.

Kisilevsky, H. (1982). The Rédei-Reichardt theorem: another proof. In: Taussky, O., editor, *Ternary Quadratic Forms and Norms*, pp. 1–4. Marcel Dekker, New York.

Kleboth, H. (1955). *Untersuchung über Klassenzahl und Reziprozitätsgesetz im Körper der 6l-ten Einheitswurzeln und die Diophantische Gleichung* $x^{2l} + 3^l y^{2l} = z^{2l}$ *für eine Primzahl l grösser als* 3. Dissertation, Universität Zürich, 37 pp.

Klingen, H. (1962). Über die Werte der Dedekindschen Zetafunktion. *Math. Ann.*, 145:265–272.

Koblitz, N. (1984). *p-adic Numbers, p-adic Analysis and Zeta Functions*. Second edition, Springer-Verlag, New York.

Koch, H., and Zink, W. (1972). Über die 2-Komponente der Klassengruppe quadratischer Zahlkörper mit zwei Diskriminantenteilern. *Math. Nachr.*, 54:309–333.

240   *Congruences for L-functions*

Kohno, Y., and Nakahara, T. (1993). Oriented graphs of 2-class group constructions of quadratic fields (Japanese). In: *Combinatorial Structure in Mathematical Models (Kyoto, 1993)*. Sūrikaisekikenkyūsho Kōkyūroku, 853, pp. 133–147. Kyoto.

Kolster, M. (1986). The structure of the 2-Sylow subgroup of $K_2 O_F$, I. *Comment. Math. Helvetici*, 61:376–388.

Kolster, M. (1987). The structure of the 2-Sylow subgroup of $K_2 O_F$, II. *K-theory*, 1:467–479.

Kolster, M. (1989). A relation between the 2-primary parts of the main conjecture and the Birch-Tate conjecture. *Canad. Math. Bull.*, 32:248–251.

Kolster, M. (1992). K_2 of rings of algebraic integers. *J. Number Theory*, 42:103–122.

Kolster, M. (2000). Appendix to "Two-primary algebraic K-theory of rings of integers in number fields" by J. Rognes and C. Weibel. *J. Amer. Math. Soc.*, 13:1–54.

Kolster, M., and Nguyen Quang Do, T. (1998). Syntomic regulators and special values of p-adic L-functions. *Invent. Math.*, 133:417–447.

Kolster, M., Nguyen Quang Do, T., and Fleckinger, V. (1996). Twisted S-units, p-adic class number formulas and the Lichtenbaum Conjectures. *Duke Math. J.*, 84:679–717.

Kronecker, L. (1884). *Über bilineare Formen mit vier Variablen*, 60 pp. Berlin.

Kubota, T., and Leopoldt, H. W. (1964). Eine p-adische Theorie der Zetawerte, Teil I: Einführung der p-adischen Dirichletschen L-Funktionen. *J. Reine Angew. Math.*, 214/215:328–339.

Kudo, A. (1975). On a class number relation of imaginary abelian fields. *J. Math. Soc. Japan*, 27:150–159.

Lagarias, J. C. (1980). On the determining the 4-rank of the ideal class group of a quadratic field. *J. Number Theory*, 12:191–196.

Lang, H. (1968). Über eine Gattung elementar-arithmetischer Klasseninvarianten reell-quadratischer Zahlkörper. *J. Reine Angew. Math.*, 233:123–175.

Lang, H. (1985). Über die Klassenzahl quadratischer Zahlkörpern, deren Diskriminanten nur ungerade Primteiler $p \equiv 1 \bmod 4$ besitzen. *Abh. Math. Sem. Univ. Hamburg*, 55:147–150.

Lang, H. (1988). Über die Restklasse modulo 2^{e+2} des Wertes $2^e n \zeta (1-2^e n, K)$ der Zetafunktion einer Idealklasse aus dem reell-quadratischen Zahlkörper $\mathbb{Q}(\sqrt{D})$ mit $D \equiv 3 \,(\bmod 4)$. *Acta Arith.*, 51:277–292.

Lang, H., and Schertz, R. (1976). Kongruenzen zwischen Klassenzahlen quadratischer Zahlkörper. *J. Number Theory*, 8:352–365.

Lang, S. (1986). *Algebraic Number Theory*. Springer-Verlag, New York.

Lang, S. (1990). *Cyclotomic Fields I and II*, combined second edition. Springer-Verlag, New York.

Lehmer, E. (1971). On the quadratic character of some quadratic surds. *J. Reine Angew. Math.*, 250:42–48.

Leonard, P. A., and Williams, K. S. (1982). On the divisibility of the class numbers of $\mathbb{Q}(\sqrt{p})$ and $\mathbb{Q}(\sqrt{-2p})$ by 16. *Canad. Math. Bull.*, 25:200–206.

Leonard, P. A., and Williams, K. S. (1983). On the divisibility of the class number of $\mathbb{Q}(\sqrt{-pq})$ by 16. *Proc. Edinburgh Math. Soc.*, 26:221–231.

Leopoldt, H. W. (1958). Eine Verallgemeinerung der Bernoullischen Zahlen. *Abh. Math. Sem. Univ. Hamburg*, 22:131–140.

Lerch, M. (1905). Essai sur le calcul du nombre de classes de formes quadratiques binaires aux coefficients entiers. *Acta Math.*, 29:333–424.

Lichtenbaum, S. (1973). Values of zeta-functions, étale cohomology, and algebraic K-theory. In: Bass, H., editor, *Algebraic K-theory, II: "Classical" Algebraic K-theory and Connections with Arithmetic*, Lecture Notes in Math. 342, pp. 489–501. Springer, New York.

Mazur, B., and Wiles, A. (1984). Class fields of abelian extensions of $\mathbb{Q}$. *Invent. Math.*, 76:179–330.

Meyer, C. (1967). Über die Bildung von elementar-arithmetischen Klasseninvarianten in reell-quadratischen Zahlkörpern. In: Hasse, H., Roquette, P., editors, *Algebraische Zahlentheorie (Oberwolfach, 1964)*, pp. 165–215. Bibliographisches Institut, Mannheim.

Milnor, J. W. (1971). *Introduction to K-theory*, Ann. of Math. Stud. 72. Princeton Univ. Press, Princeton, NJ.

Mordell, L. J. (1961). The congruence $((p-1)/2)! \equiv \pm 1 \,(\bmod\, p)$. *Amer. Math. Monthly*, 68:145–146.

Mordell, L. J. (1963). On Lerch's class number for binary quadratic forms. *Ark. Mat.*, 5:97–100.

Morton, P. (1983). The quadratic number fields with cyclic class groups. *Pacific J. Math.*, 108:165–175.

Nagell, T. (1923). Sur la distribution des nombres qui sont premiers avec un nombre entier donné. *Arch. Math. og Naturvidenskab*, 38(2), 34 pp.

Narkiewicz, W. (1990). *Elementary and Analytic Theory of Algebraic Numbers*. PWN-Polish Scientific Publishers, Warszawa, and Springer-Verlag, Berlin.

Neukirch, J. (1988). The Beilinson conjectures for algebraic number fields. In: Rapoport, M., Schappacher, N., and Schneider, P., editors, *Beilinson's Conjectures on Special Values of L-functions*, Perspect. Math. 4, pp. 193–247. Academic Press, New York.

Nizioł, W. (1995). *On the image of p-adic regulators*. Preprint.

Oriat, B. (1977). Relations entre les 2-groupes d'idéaux des extensions quadratiques $k(\sqrt{d})$ et $k(\sqrt{-d})$. *Ann. Inst. Fourier (Grenoble)*, 27:37–59.

Oriat, B. (1978). Sur la divisibilité par 8 et 16 des nombres de classes d'ideaux des corps quadratiques $\mathbb{Q}(\sqrt{2p})$ et $\mathbb{Q}(\sqrt{-2p})$. *J. Math. Soc. Japan*, 30:279–285.

Osborn, G. (1896). Some properties of the quadratic residues of primes. *Messenger Math.*, 25:45–47.

Pepin, P. (1874). Nombre des classes de formes quadratiques pour un déterminant donnée. *Ann. Sci. École Norm. Sup.* (2), 3:165–208.

Pioui, R. (1990). *Mesures de Haar p-adiques et inteprétation arithmétique de* $\frac{1}{2}L_2(\chi, s) - \frac{1}{2}L_2(\chi, t)$, $s, t \in \mathbb{Z}_2$ *(χ quadratique)*. Thèse, Université de Franche-Comté, Besançon.

Pioui, R. (1992). Module de continuité des fonctions L 2-adiques des caractères quadratiques. *Manuscripta Math.*, 75:167–195.

Pizer, A. (1976). On the 2-part of the class number of imaginary quadratic number fields. *J. Number Theory*, 8:184–192.

Plancherel, M. (1908). *Sur les congruences* ($\mathrm{mod}\, 2^m$) *relatives au nombre des classes des formes quadratiques binaires aux coefficients entiers et à discriminant negatif*. Thèse, Paris.

Pumplün, D. (1965). Über die Klassenzahl imaginär-quadratischer Zahlkörper. *J. Reine Angew. Math.*, 218:23–30.

Qin, H. (1994a). 2-Sylow subgroups of $K_2 O_F$ for real quadratic fields F. *Sci. China Ser. A*, 37:1302–1313.

Qin, H. (1994b). Computation of $K_2 \mathbb{Z}[\sqrt{-6}]$. *J. Pure Appl. Algebra*, 96:133–146.

Qin, H. (1995a). The 2-Sylow subgroups of the tame kernel of imaginary quadratic fields. *Acta Arith.*, 69:153–169.

Qin, H. (1995b). The 4-rank of $K_2 O_F$ for real quadratic fields. *Acta Arith.*, 72:323–333.

Qin, H. (1996). Computation of $K_2 \mathbb{Z}[(1 + \sqrt{-35})/2]$. *Chinese Ann. Math. Ser. B*, 17:63–72.

Qin, H. (1998). *Tame kernels and Tate kernels of quadratic number fields*. Preprint.

Quillen, D. (1973). Higher algebraic K-theory, I. In: Bass, H., editor, *Algebraic K-theory*, I: *Higher K-theories*, Lecture Notes in Math. 341, pp. 85–147. Springer-Verlag, New York.

Ramakrishnan, D. (1989). Regulators, algebraic cycles, and values of L-functions. In: Stein, M. R., and Dennis, R. K., *Algebraic K-theory and Algebraic Number Theory*, Contemp. Math. 83, pp. 183–310. Amer. Math. Soc., Providence, RI.

Rédei, L. (1928). Über die Klassenzahl imaginär-quadratischer Zahlkörper. *J. Reine Angew. Math.*, 159:210–219.

Rédei, L. (1932a). On the class-number and the fundamental unit of a real quadratic number field (Hungarian, German summary). *Mat. Természett. Értes.*, 48:648–682.

Rédei, L. (1932b). On the class-number of a quadratic field (Hungarian, German summary). *Mat. Természett. Értes.*, 48:683–707.

Rédei, L. (1934a). Arithmetischer Beweis des Satzes über die Anzahl der durch 4 teilbaren Invarianten der absoluten Klassengruppe im quadratischen Zahlkörper. *J. Reine Angew. Math.*, 171:55–60.

Rédei, L. (1934b). Eine obere Schranke der Anzahl der durch 4 teilbaren Invarianten der absoluten Klassengruppe im quadratischen Zahlkörper. *J. Reine Angew. Math.*, 171:61–64.

Rédei, L. (1934c). Über die Grundeinheit und die durch 8 teilbaren Invarianten der absoluten Klassengruppe im quadratischen Zahlkörper. *J. Reine Angew. Math.*, 171:131–148.

Rédei, L. (1936). Über einige Mittelwertfragen in quadratischen Zahlkörpern. *J. Reine Angew. Math.*, 174:15–55.

Rédei, L. (1939a). Ein neues Zahlentheoretisches Symbol mit Anwendungen auf die Theorie der quadratischen Zahlkörper. *J. Reine Angew. Math.*, 180:1–43.

Rédei, L. (1939b). Die Diophantine Gleichung $mx^2 + ny^2 = z^4$. *Monatsh. Math. Phys.*, 48:43–60.

Rédei, L. (1949/1950). Über die Wertverteilung des Jacobischen Symbols. *Acta Sci. Math. (Szeged)*, 13:242–246.

Rédei, L., and Reichardt, H. (1934). Die Anzahl der durch 4 teilbaren Invarianten der Klassengruppe eines beliebigen quadratischen Zahlkörpers. *J. Reine Angew. Math.*, 170:69–74.

Reichardt, H. (1934). Zur Struktur der absoluten Idealklassengruppe im quadratischen Zahlkörper. *J. Reine Angew. Math.*, 170:75–82.

Reichardt, H. (1970). Über die 2-Klassengruppe gewisser quadratischer Zahlkörper. *Math. Nachr.*, 46:71–80.

Schinzel, A., Urbanowicz, J., and van Wamelen, P. (1999). On short sums of the Kronecker symbols. *J. Number Theory*, 78:62–84.

Schneider, P. (1988). Introduction to the Beilinson Conjectures. In: Rapoport, M., Schappacher, N., and Schneider, P., editors, *Beilinson's Conjectures on Special Values of L-functions*, Perspect. Math. 4, pp. 1–35. Academic Press, New York.

Scholz, A. (1935). Über die Lösbarkeit der Gleichung $t^2 - Du^2 = -4$. *Math. Z.*, 39:95–111.

Serre, J.-P. (1971). *Cohomologie des groupes discrets*, Ann. of Math. Stud. 70. Princeton Univ. Press, Princeton, NJ.

Serre, J.-P. (1973, 1975). Formes modulaires et fonctions zêta p-adiques. In: Kuyk, W., and Serre, J.-P., editors, *Modular Functions of One Variable*, III, Lecture Notes in Math. 350, pp. 191–268. Springer, New York; Corrigendum: Birch, B. J., and Kuyk, W., editors, *Modular Functions* IV, Lecture Notes in Math. 476, pp. 149–150. Springer, New York.

Siegel, C.-L. (1937, 1966). Über die analytische Theorie der quadratischen Formen, III. *Ann. of Math.* (2), 38:212–291; or in: *Gesammelte Abhandlungen*, pp. 469–548. Springer-Verlag, New York.

Siegel, C.-L. (1968). Bernoullische Polynome und quadratische Zahlkörper. *Nachr. Akad. Wiss. Göttingen, Math.-Phys. Kl. II*, 7–38.

Skałba, M. (1987). Master's dissertation, Warsaw University.

Skałba, M. (1994). Generalization of Thue's Theorem and computation of the group $K_2 O_F$. *J. Number Theory*, 46:303–322.

Stevenhagen, P. (1988). *Class Groups and Governing Fields*. Ph. D. Thesis, University of California, Berkeley; Preprint 88-08 of the University of Amsterdam.

Stevenhagen, P. (1993). Divisibility by 2-powers of certain quadratic class numbers. *J. Number Theory*, 43:1–19.

Sueyoshi, Y. (1996). Comparison of the 4-ranks of the narrow ideal class groups of the quadratic fields $\mathbb{Q}(\sqrt{m})$ and $\mathbb{Q}(\sqrt{-m})$ (Japanese). In: *Algebraic Number Theory and Fermat's Problem (Kyoto, 1995)*, Sūrikaisekikenkyūsho Kōkyūroku 971, pp. 134–144. Kyoto.

Sueyoshi, Y. (1997). On a comparison of the 4-ranks of the narrow ideal class groups of $\mathbb{Q}(\sqrt{m})$ and $\mathbb{Q}(\sqrt{-m})$. *Kyushu J. Math.*, 51:261–272.

Suslin, A. A. (1987). Algebraic K-theory of fields. In: Gleason, A. M., editor, *Proceedings of the International Congress of Mathematicians (Berkeley, CA, 1986)*, vol. I, pp. 222–244. Amer. Math. Soc., Providence, RI.

Szmidt, J., and Urbanowicz, J. (1994). *Some new congruences for generalized Bernoulli numbers of higher orders*. Preprint FI94-LF05 of the Fields Institute for Research in Math., Waterloo, Canada.

Szmidt, J., Urbanowicz, J., and Zagier, D. (1995). Congruences among values of Dirichlet L-series at negative integers. *Acta Arith.*, 71:243–248.

Szymiczek, K. (1996). Knebusch-Milnor exact sequence and parity of class numbers. *Acta Math. Inform. Univ. Ostraviensis*, 4:83–95.

Tate, J. (1971). Symbols in arithmetic. In: *Actes du Congrès International des Mathématiciens (Nice, 1970)*, Tome 1, pp. 201–211. Gauthier-Villars, Paris.

Tate, J. (1973). Appendix to "The Milnor ring of a global field" by H. Bass and J. Tate. In: Bass, H., editor, *Algebraic K-theory*, II: *"Classical" Algebraic K-theory and Connections with Arithmetic*, Lecture Notes in Math. 342, pp. 429–446. Springer-Verlag, New York.

Tate, J. (1976). Relations between K_2 and Galois cohomology. *Invent. Math.* 36:257–274.

Uehara, T. (1989). On the 4-rank of the narrow ideal class group of a quadratic field. *J. Number Theory*, 31:167–173.

Uehara, T. (1990). On linear congruences between class numbers of quadratic fields. *J. Number Theory*, 34:362–392.

Urbanowicz, J. (1983, 1993). On the 2-primary part of a conjecture of Birch-Tate. *Acta Arith.*, 43:69–81; Corrigendum: *ibid.*, 64:99.

Urbanowicz, J. (1990a, 1991). Connections between $B_{2,\chi}$ for even quadratic characters χ and class numbers of appropriate imaginary quadratic fields, I. *Compositio Math.*, 75:247–270; Corrigendum: *ibid.*, 77:119–123.

Urbanowicz, J. (1990b, 1991). Connections between $B_{2,\chi}$ for even quadratic characters χ and class numbers of appropriate imaginary quadratic fields, II. *Compositio Math.*, 75:271–285; Corrigendum: *ibid.*, 77:123–125.

Urbanowicz, J. (1990c). A generalization of the Lerch-Mordell formulas for positive discriminants. *Colloq. Math.*, 59:197–202.

Urbanowicz, J. (1990/1991a). *On Some New Congruences for Generalized Bernoulli Numbers*, I. Publ. Math. Fac. Sci. Besançon, Théorie des Nombres.

Urbanowicz, J. (1990/1991b). *On Some New Congruences for Generalized Bernoulli Numbers*, II. Publ. Math. Fac. Sci. Besançon, Théorie des Nombres; Corrigendum: *ibid.*, Années 1992/93-1993/94.

Urbanowicz, J. (1999). On optimal linear congruences for $L_2(k, \chi\omega^{1-k})$. To appear.

Urbanowicz, J., and Wójcik, A. (1995/1996). *On Linear Congruence Relations Related to 2-adic Dilogarithms*. Publ. Math. Fac. Sci. Besançon, Théorie des Nombres.

Vazzana, A. (1997a). On the 2-primary part of K_2 of rings of integers in certain quadratic number fields. *Acta Arith.*, 80:225–235.

Vazzana, A. (1997b). The 4-rank of $K_2(O_F)$ and related graphs in certain quadratic number fields. *Acta Arith.*, 81:253–264.

Vazzana, A. (1999). 8-ranks of K_2-rings of integers in quadratic number fields. *J. Number Theory*, 76:248–264.

Washington, L. C. (1976). A note on p-adic L-functions. *J. Number Theory*, 8:245–250.

Washington, L. C. (1997). *Introduction to Cyclotomic Fields*, second edition. Springer-Verlag, New York.

Waterhouse, W. C. (1973). Pieces of eight in class groups of quadratic fields. *J. Number Theory*, 5:95–97.

Wiles, A. (1990). The Iwasawa conjecture for totally real fields. *Ann. of Math.* (2), 131:493–540.

Williams, K. S. (1976). Note on a result of Barrucand and Cohn. *J. Reine Angew. Math.*, 285:218–220.

Williams, K. S. (1979). The class number of $\mathbb{Q}(\sqrt{-p})$ modulo 4, for $p \equiv 3$ (mod 4) a prime. *Pacific J. Math.*, 83:565–570.

Williams, K. S. (1981a). On the class number of $\mathbb{Q}(\sqrt{-p})$ modulo 16, for $p \equiv 1$ (mod 8) a prime. *Acta Arith.*, 39:381–398.

Williams, K. S. (1981b). The class number of $\mathbb{Q}(\sqrt{-2p})$ modulo 8, for $p \equiv 5$ (mod 8) a prime. *Rocky Mountain J. Math.*, 11:19–26.

Williams, K. S. (1981c). The class number of $\mathbb{Q}(\sqrt{p})$ modulo 4, for $p \equiv 5$ (mod 8) a prime. *Pacific J. Math.*, 92:241–248.

Williams, K. S. (1982). Congruences modulo 8 for the class numbers of $\mathbb{Q}(\sqrt{\pm p})$, $p \equiv 3$ (mod 4) a prime. *J. Number Theory*, 15:182–198.

Wójcik, A. (1998). Linear congruence relations for 2-adic L-series at integers. *Compositio Math.*, 111:289–304.

Wolke, D. (1973). Eine Bemerkung über das Legendre-Symbol. *Monatsh. Math.*, 77:267–275.

Yamamoto, K. (1965). On Gaussian sums with biquadratic residue characters. *J. Reine Angew. Math.*, 219:200–213.

Yamamoto, Y. (1977). Dirichlet series with periodic coefficients. In: Iyanaga, S., editor, *Algebraic Number Theory (Kyoto, 1976)*, pp. 275–289. Japan Soc. Promotion Sci., Tokyo.

Yamamoto, Y. (1988). Class number problems for quadratic fields (concentrating on the 2-part) (Japanese). *Sûgaku*, 40:167–174.

Zagier, D. (1991). Polylogarithms, Dedekind zeta functions, and the algebraic K-theory of fields. In: van der Geer, G., Oort, F., and Steenbrink, J., editors, *Arithmetic Algebraic Geometry (Texel, 1989)*, Progr. Math. 89, pp. 391–430. Birkhäuser, Boston, MA.

Zhang, X. (1989). Congruences modulo 8 for class numbers of general quadratic fields $\mathbb{Q}(\sqrt{m})$ and $\mathbb{Q}(\sqrt{-m})$. *J. Number Theory*, 32:332–338.

Author Index

Akiyama, S., 32, 231
Amice, Y., 119, 231
Ankeny, N. C., 13, 231
Apostol, T. M., 8, 11, 231
Artin, E., 13, 231
Ayoub, R., 39, 231
Barkan, P., 55, 58–59, 231
Barner, K., 27, 231
Barrucand, P., 56, 59, 82, 231
Bauer, H., 75, 231
Belabas, K., 25, 231
Berger, A., 13, 231
Berger, R. I., 25, 232
Berndt, B. C., 6–9, 33–34, 39–42, 45, 60, 62, 64, 72, 84, 232
Bernoulli, J., 10
Birch, B. J., 26, 232
Bloch, S., 20, 26, 232
Boldy, M. C., 25, 232
Bölling, R., 6, 232
Borel, A., 24, 26, 232
Borevich, Z. I., 3, 10–12, 19, 21, 28, 30, 72, 232
Brauckman, B., 25, 232
Browkin, J., 20, 24–26, 77, 81, 95–96, 231–233
Brown, E., 55–56, 58–59, 61–62, 74–75, 82, 84, 233
Candiotti, A., 25, 81, 233
Carlitz, L., 10, 14–15, 124, 233
Cauchy, A. L., 39, 233
Chowla, S., 7–8, 13, 39, 53, 73, 231–233
Clausen, T., 10, 20
Coates, J., 27, 119, 233
Cohn, H., 56–57, 59, 82, 231, 233
Coleman, R. F., xi, 64, 122–123, 125–126, 132, 140–141, 234
Conner, P. E., 25, 27, 30, 76, 81, 95–96, 234
Cooke, G., 57, 233
Costa, A., 63–64, 234
Currie, J., 8, 45, 60, 234
Damey, P., 75, 234

Davenport, H., 2, 184, 234
Dedekind, R., 41, 55, 234
Deligne, P., 119, 234
Desnoux, P.-J., 76, 167, 234
Dilcher, K., 10, 234
Dirichlet, P. G. L., 2–3, 6, 39–41, 53, 55, 235–236
Endô, A., 76, 235
Ernvall, R., 13, 15, 235
Euler, L., 12, 19, 235
Federer, L. J., 27, 235
Fleckinger, V., 26, 122, 240
Fox, G. J., xi, 77, 203, 207–208, 235
Fresnel, J., 119, 231, 235
Friesen, C., 74, 235
Frobenius, F. G., 14, 122, 235
Gangl, H., 25, 231, 233
Garland, H., 24, 235
Gauss, C. F., 6, 41, 55–56, 60, 234, 236
Gebhardt, H. M., 25, 236
Glaisher, J. W. L., 41, 55, 236
Goren, E. Z., 97, 236
Gradshteyn, I. S., 36, 236
Granville, A., 133–134, 181
Gras, G., x–xi, 64, 72, 76, 97, 117, 128, 134, 150, 160–161, 165, 167, 169, 175, 230, 236
Greither, C., 27, 236
Halter-Koch, F., 76, 236
Hardy, K., ix, xi, 6, 49, 51, 61–64, 71, 75, 128, 165, 167, 169, 174, 236
Hasse, H., 2, 56, 58, 60, 75–76, 83–84, 184, 237
Hecke, E., 22, 27, 237
Ilikita, M., 76, 168, 237
Holden, H., 3, 6, 40–41, 45, 237–238
Hudson, R. H., 3, 6, 44, 238
Hurrelbrink, J., 25, 27, 30, 63, 76, 81, 95–96, 234, 238
Hurwitz, A., 13, 41, 54, 238
Ireland, K., 3, 10–12, 16, 238
Iwasawa, K., 12, 119, 238

Johnson, W., 3, 5, 7, 45, 202, 238
Kaplan, P., 56–58, 61–62, 75–76, 168, 172–173, 236, 238–239
Karpinski, L. C., 3, 41, 239
Kenku, M. A., 64, 167, 239
Keune, F., 24–25, 239
Kisilevsky, H., 63, 239
Kleboth, H., 13, 239
Klingen, H., 27, 239
Koblitz, N., 11, 117, 239
Koch, H., 76, 239
Kohno, Y., 76, 240
Kolster, M., 25–27, 81, 122, 238, 240
Kramer, K., 25, 81, 233
Kronecker, L., 53, 240
Kubota, T., 119, 240
Kudo, A., 168, 240
Kummer, E., 11
Lagarias, J. C., 63, 240
Lang, H., 27, 63, 76, 97, 168, 173, 240
Lang, S., 16–19, 21–22, 27, 119, 240
Lehmer, E., 57, 241
Leonard, P. A., 57, 59–61, 241
Leopoldt, H. W., 12–13, 15, 119, 240–241
Lerch, M., 3, 6, 33–34, 39–41, 43, 45, 55, 58, 241
Lichtenbaum, S., 26, 165, 241
Mazur, B., 27, 241
Meyer, C., 27, 241
Milnor, J. W., 24–25, 241
Mitchell, K. J., 3, 5, 7, 45, 202, 238
Mordell, L. J., 6, 43, 45, 52, 241
Morton, P., 76, 241
Nagell, T., 49, 241
Nakahara, T., 76, 240
Narkiewicz, W., 16–17, 21–22, 30, 241
Neukirch, J., 122, 241
Nguyen Quang Do, T., 26, 122, 240
Nizioł, W., 122, 241
Oriat, B., 76, 241–242
Osborn, G., 41, 242
Parry, C. J., 61, 233
Payan, J.-J., 75, 234
Pepin, P., 39, 41, 242
Pioui, R., 76, 97, 167, 230, 242
Pizer, A., 58–62, 64, 82, 84, 167, 242
Plancherel, M., 63, 242
Pumplün, D., 61, 242
Qin, H., 25, 81, 95–96, 242
Quillen, D., 24, 242
Ramakrishnan, D., 26, 242
Rédei, L., 6, 41, 63, 75, 242–243
Reichardt, H., 63, 75, 243
Ribet, K., 119, 234
Rognes, J., 240
Rosen, M., 3, 10–12, 16, 238

Rubin, K., 27
Ryzhik, I. M., 36, 236
Schertz, R., 63, 76, 168, 173, 240
Schinzel, A., xi, 5–7, 24–25, 77, 81, 95–96, 181–182, 184–185, 193, 195, 233, 243
Schneider, P., 122, 243
Schoenfeld, L., 6, 8–9, 232
Scholz, A., 75, 243
Serre, J.-P., 26–27, 119, 243–244
Shafarevich, I. R., 3, 10–12, 19, 21, 28, 30, 72, 232
Siegel, C.-L., 27, 244
Skałba, M., 25, 244
Skula, L., 10, 234
Slavutskiĭ, I. Sh., 10, 234
von Staudt, K. G. C., 10
Stevenhagen, P., 76, 244
Sueyoshi, Y., 76, 244
Suslin, A. A., 26, 244
Szmidt, J., xi, 1, 6, 31, 36, 51, 64, 71–72, 94, 128, 174, 244
Szymiczek, K., 76, 244
Tate, J., 24–26, 244–245
Uehara, T., xi, 64, 72, 76, 117, 127–129, 134, 137, 140, 150, 160–161, 163, 165, 167–173, 175, 180, 245
Urbanowicz, J., x–xi, 1, 5–7, 31, 35–36, 45, 47, 49, 51, 64, 71–72, 77–79, 82, 86–88, 90–94, 97, 99, 101–104, 106–107, 111–112, 114–117, 128–129, 133, 137, 143–144, 156, 160–161, 163, 165–167, 174–177, 179–182, 184–185, 193, 195, 203, 207–208, 230, 235, 243–245
Vazzana, A., 25, 245
Walum, H., 39, 231
van Wamelen, P., xi, 5–7, 181–182, 184–185, 193, 195, 243
Washington, L. C., 2, 10–13, 15–19, 21–23, 27, 117, 119, 151, 245
Waterhouse, W. C., 76, 245
Weibel, C., 240
Wiles, A., 27, 29, 241, 245
Williams, K. S., ix, xi, 3, 6, 8, 44–45, 49, 51, 54, 56–64, 71, 73–77, 128, 165, 167–169, 172–174, 203, 207–208, 234–236, 238–239, 241, 246
Wójcik, A., xi, 64, 72, 117, 123, 128–129, 133–135, 137, 141, 143–144, 151, 156, 160–163, 166–167, 180, 230, 245–246
Wolke, D., 7, 246
Yamamoto, K., 55, 246
Yamamoto, Y., 40, 76, 246
Zagier, D., xi, 1, 6, 9, 26, 31–32, 42, 51, 64, 71–72, 128, 174, 181, 244, 246
Zhang, X., 76, 246
Zink, W., 76, 239

Subject Index

Ambiguous forms and classes, 30, 51, 55–56, 62, 74–75

Ars Conjectandi, 10

Bernoulli
   numbers, 9, 11–12
   polynomials, 10

Binomial coefficient identities, 133, 135, 151

Character analogue of the Poisson formula, 8, 28–29, 31, 34

Class number formulae, 21, 121

Clausen function, 20

CM-fields, 23

Coleman formulae, xi, 122

Congruences for
   a^k ($\mathrm{mod}\, 2^{\mathrm{ord}_2 k+6}$), 98–100
   $B_{k,\chi}^{[T]}(N)/k$, 217, 220
   $B_{k,\chi}/k$, 69, 77, 224
   $B_{k,\chi}/k$ (mod 64), 97, 108, 111–112, 114–115
   $h(-8p)$ (mod 4, 8, 16, 128), 58–60, 65, 75, 92, 170, 172
   $h(-8pq)$ (mod 8, 16), 61–62, 65
   $h(-8pqr)$ (mod 32), 68–69
   $h(-4p)$ (mod 4, 8, 16, 128), 55–59, 65, 92, 170, 172–173
   $h(-4pq)$ (mod 8, 16), 61–62, 65
   $h(-4pqr)$ (mod 32), 68–69
   $h(-p)$ (mod 4, 8, 16, 128), 52, 54, 59, 65, 74, 92, 170, 172
   $h(-pq)$ (mod 4, 8, 16), 60–61, 65
   $h(-pqr)$ (mod 8, 32), 68
   $h(d)$, 51, 76, 168–172
   $h(d)$ $(d < 0)$, ix, 52, 63–64, 78, 80–81, 86, 88, 90–91, 93, 97, 104–107, 111–112, 114–115, 174, 179
   $h(d)$ $(d > 0)$, 53, 176–177
   $h(p)$ (mod 4, 8, 16), 53, 57–58, 73, 170, 172–173
   $h(pq)$ (mod 4, 8), 75

$h(pqr)$, $h(4pq)$ and $h(8pq)$, 75
$h(4p)$ (mod 4, 8, 16), 54, 74, 170, 172
$h(8p)$ (mod 4, 8, 16), 74–75, 170, 172
$k_2(-8p)$ (mod 2, 4, 8), 95–96
$k_2(-4p)$ (mod 2, 4, 8), 95
$k_2(-p)$ (mod 2, 4, 8), 95
$k_2(-pq)$ (mod 4, 8, 16), 96
$k_2(d)$, 77, 175
$k_2(d)$ $(d < 0)$, 176, 179
$k_2(d)$ $(d > 0)$, 78–84, 86, 88, 90–91, 93, 104–107, 111–112, 114–115, 174, 177, 210
$k_2(p)$ (mod 4, 8, 16, 32, 128), 85, 92
$k_2(pq)$ (mod 16, 32), 85
$k_2(4p)$ (mod 8, 32), 85
$k_2(8p)$ (mod 8, 16, 32, 128), 85, 92
$L(k,\chi)$ $(k \leq 0)$, 69
$L_2(k,\chi\omega^{1-k})$, 128, 160, 163
$L_2(k,\chi\omega^{1-k})$ $(k = -1, 0, 1, 2)$, 161, 166, 168, 174–177, 179
$\mathcal{L}_{k,e}(s)$, 156, 158
p-adic numbers, 118
power sums of consecutive natural numbers, 222

Conjecture of
   Birch and Tate, 26–27, 29, 78, 81
   Federer, 27
   Leopoldt, 121
   Lichtenbaum, 26, 30, 165

Cuspidal behaviour of 2-adic modular forms, 64

D-numbers, 13

Dedekind zeta function, 21

Dilogarithm of
   Euler, 19
   Rogers, 20, 125
   Wigner and Bloch, 20, 26

Dirichlet
   characters, 2, 184
   class number formulae, x, 2–4, 16, 21–22, 27–28, 40, 53–54, 58, 72, 198, 203

L-functions, 16, 18, 119–120
  regulator, 21

Euler
  criterion, 1
  factors, 16, 119–120, 123
  formula for $\zeta(2k)$, 12
  numbers, 13, 115
  product, 16
Exact hexagon, 76

Formulae for
  $B_{1,\chi}$, 28, 43
  $B_{2,\chi}$, 29
  $B_{k,\chi}$, 100
  $h(d)$, 30
  $h(d)$ $(d < 0)$, 2–3, 5–6, 27–28, 33–34, 39–41,
       43–44, 79, 101, 103, 183
  $h(d)$ $(d > 0)$, 28
  $k_2(d)$ $(d < 0)$, 30
  $k_2(d)$ $(d > 0)$, 29, 34, 43, 78–79, 97, 101, 104,
       210
  $L_2(k, \chi\omega^{1-k})$, 129
  $l_k(s)$, 125
  $\mathcal{L}_{k,e}(s)$, 141
  $\mathcal{L}_{k,e}(s)$ $(k = -1, 0, 1, 2)$, 137
  $\mathcal{L}_{k,\psi}(s)$, 127–130
Frobenius polynomials, 14, 123–124
Functional equation for
  $L(s, \chi)$, 16
  $\zeta(s)$, 11
  $\zeta_F(s)$, 22
Fundamental discriminants, 2

Gamma function, 11
Gauss
  congruence, 55–56, 58–59, 75, 203, 208
  evaluation of $\tau(\chi)$, 17
  sum, 9, 17
  theory of ambiguous classes, 30, 51, 55, 76
Generalized
  Bernoulli numbers, 12–13, 18, 31, 70, 124, 213
  Bernoulli polynomials, 13–14
  Kummer congruences, 15, 97, 119

Hardy-Williams congruence, 62–64, 68–69, 93
Hasse classical Klassenzahlbericht, 76

Jacobi symbols, 1

K-theoretic background, 23
Kronecker symbols, 2–3
Kummer congruences, 10–11

Legendre symbols, 1
Leopoldt formula, 120, 123
Lerch formulae for short character sums, 5, 33, 43
Linear combinations of
  $L_2(k, \chi\omega^{1-k})$, 156, 160, 163
  $L_2(k, \chi\omega^{1-k})$ $(k = -1, 0, 1, 2)$, 161, 166,
       168, 174–177, 179

$\mathcal{L}_{k,e}(s)$, 156, 158
Linear congruence relations, ix–x, 63, 69, 71, 93,
       160–161, 163, 166, 168, 173–177,
       179–180

Milnor functors, 24
Möbius inversion argument, 64, 70
Modified
  external product, 25
  generalized Bernoulli polynomials, 70–71
Multilogarithms, 122

Nagell formula, 48

Optimal linear congruences, 173

p-adic
  absolute value, 117
  dilogarithm, 125
  formulae for $h(d)$, 166
  formulae for $h(d)$ $(d < 0)$, 121
  formulae for $h(d)$ $(d > 0)$, 121
  formulae for $k_2(d)$, 166
  formulae for $k_2(d)$ $(d < 0)$, 122
  formulae for $k_2(d)$ $(d > 0)$, 122
  function exp, 118
  L-functions, 119–120
  Lichtenbaum conjecture, 122, 165, 167, 175–176
  logarithm, 118
  multilogarithms, 122–123, 125–126
  regulator, 121
  second regulator, 122, 166
  zeta function, 121
Poisson summation formula, 8–9
Power sums
  involving Dirichlet characters, 13
  of consecutive natural numbers, 10

Quadratic characters, 2
Quillen K-functors, 23

Reciprocity biquadratic law, 56
Relative class number for CM-fields, 23
Riemann zeta function, 11–12

Short power sums
  involving Dirichlet characters, 30–31, 36, 192
  involving Kronecker symbols, 34, 101, 103–107
Short sums of
  Dirichlet characters, 6, 40–43, 45, 48, 192
  the Kronecker symbols, x, 7, 33–34, 39, 41, 45,
       47, 181, 193, 198, 201
  the Legendre symbols, 2, 6, 44
Structure of the group K_2O_F, 24–25

Teichmüller character, 118
Theorem of
  Kronecker and Weber, 17
  Matsumoto, 25
  von Staudt and Clausen, 10, 109, 112

Theory of
  elliptic curves, 64
  quaternion algebras, 58
Type numbers of Eichler orders, 60

Uehara functions, 127

Values of
  $L(k, \chi)$, 122
  $L(k, \chi)$ $(k \leq 0)$, 16, 18

$L(1, \chi)$, 18, 28, 31, 40–41
$L(2, \chi)$, 19–20, 31, 42
$L_p(k, \chi)$, 122, 124, 126
$L_p(k, \chi)$ $(k = -1, 0, 1, 2)$, 120–122
$\zeta(k)$ $(k \leq 0)$, 12
$\zeta_F(k)$ $(k \leq 0)$, 22–23
Voronoï congruence, 11

Zagier formula, xi, 31, 42, 45, 64, 181, 192, 203
Zeros of $\zeta_F(s)$, 22

List of symbols

$\mathrm{card}(S)$	number of elements in the finite set S
$\mathbb{N}$	set of natural numbers $\{1, 2, 3, \ldots\}$
$\mathbb{Z}$	domain of rational integers $\{0, \pm 1, \pm 2, \pm 3, \ldots\}$
$\mathbb{Z}_p$	domain of p-adic integers
$\mathbb{Q}$	field of rational numbers
$\mathbb{Q}_p$	field of p-adic numbers
$\mathbb{R}$	field of real numbers
$\mathbb{C}$	field of complex numbers
$\overline{F}$	algebraic closure of the field F
$\mathbb{C}_p$	field of p-adic 'complex' numbers
$G(E/F)$	Galois group of a field extension E/F
$|t|$	absolute value of $t \in \mathbb{C}$
$|t|_p$	p-adic normalized absolute value of $t \in \mathbb{C}_p$
$\mathrm{ord}_p t$	power of p in $t \in \mathbb{Q}$
$\gcd(a, b)$	greatest common divisor of the integers a and b
$\mathrm{lcm}(a, b)$	least common multiple of the integers a and b
$\phi(n)$	Euler phi function
$\mu(n)$	Möbius function
$\mathrm{sgn}\, t$	sign of t ($t \in \mathbb{R}$, $t \neq 0$)
$n!$	n factorial
$n!!$	product of all odd integers $\leq n$ (2-adic factorial)
$s_2(n)$	sum of the digits in the 2-adic expansion of n
$\binom{m}{n}$	binomial coefficient
$\mathrm{Re}\, s = \mathrm{Re}(s)$	real part of the complex number s
$\mathrm{Im}\, s = \mathrm{Im}(s)$	imaginary part of the complex number s
χ	Dirichlet character
$\delta = \delta_\chi$	0, resp. 1 if $\chi(-1) = 1$, resp. -1
$N(\alpha) = N_{F/\mathbb{Q}}(\alpha)$	norm of the element α of F (from F to $\mathbb{Q}$)
$Tr(\alpha) = Tr_{F/\mathbb{Q}}(\alpha)$	trace of the element α of F (from F to $\mathbb{Q}$)
$[t]$	integral part of t ($t \in \mathbb{R}$)
$\delta_{X,Y}$	Kronecker delta function
$\zeta_M; \zeta$	primitive Mth root of unity; $\exp(2\pi i/M)$ (in $\mathbb{C}$ or $\mathbb{C}_p$)
$\left(\dfrac{n}{p}\right)$	Legendre symbol, 1
$\left(\dfrac{n}{m}\right)$	Jacobi symbol, 1

$h(d)$	class number of $\mathbb{Q}(\sqrt{d})$, 2
$\chi_d(n) = \left(\dfrac{d}{n}\right)$	Kronecker symbol, 2
χ_1	trivial primitive character, 2
d	fundamental discriminant, 2
$s(n,k) = s(n,k,p)$	7
$\displaystyle\sum_{a \leq k \leq b}^{*}$	if a or b is an integer, then the associated summands are halved, 8
$\tau(\chi,\zeta)$	Gauss sum, 9
$\tau(\chi)$	normalized Gauss sum, 9
B_n	nth Bernoulli number, 9
$S_n(x)$	10
$B_n(X)$	nth Bernoulli polynomial, 10
$\zeta(s)$	Riemann zeta function, 11
$\Gamma(s)$	Gamma function, 11
$B_{n,\chi}$	nth generalized Bernoulli number, 13
E_n	nth Euler number, 13
D_n	nth D-number, 13
$S_m(N,\chi)$	13
$B_{n,\chi}(X)$	nth generalized Bernoulli polynomial, 14
$R_n(u)$	nth Frobenius polynomial, 14
$L(s,\chi)$	Dirichlet L-function, 16
$\log s = \log(s)$	complex logarithm, 18
$\mathrm{Li}_2(s)$	Euler dilogarithm, 19
$\mathrm{L}(s)$	Rogers dilogarithm, 20
$\mathrm{Cl}_2(t)$	Clausen function, 20
$D(s)$	dilogarithm of Wigner and Bloch, 20
O_F	ring of integers of a number field F, 21
$\zeta_F(s)$	Dedekind zeta function, 21
$N\mathfrak{a} = N(\mathfrak{a})$	21
$R = R(F)$	Dirichlet regulator of F, 21
$w = w(F)$	number of roots of unity in F, 21
r_1, r_2	21
F^+	maximal totally real subfield of F, 23
$h = h(F)$	class number of F, 23
h^+, h^-	23
$W = W(F)$	group of roots of unity in F, 23
$E = E(F)$	group of units in F, 23
Q	unit index, 23
$K_n \ (n \geq 0)$	Quillen K-functors, 23
$\mathrm{rank}(A)$	24
K_0, K_1, K_2	Milnor K-functors, 24
$r_2(A)$	Sylow 2-rank of a finite abelian group A, 24
$g(2)$	24
r	24
$Cl(F)$	group of ideal classes of a number field F, 24
$Cl_2(F)$	24
$F^* \wedge F^*$	modified external product, 25
$A(F), C(F), \mathbb{D}(C(F))$	25
$R_2(F)$	second Borel regulator of F, 26
$w_r = w_r(F)$	largest integer s such that the group $\mathrm{G}(F(\zeta_s)/F)$ is annihilated by r, 26
$w(d)$	$w(F)$ for $F = \mathbb{Q}(\sqrt{d})$, 28
$w_2(d)$	$w_2(F)$ for $F = \mathbb{Q}(\sqrt{d})$, 29

$k_2(d)$	order of the group $K_2 O_F$ for $F = \mathbb{Q}(\sqrt{d})$, 29
$h_0(d)$	number of narrow classes of $\mathbb{Q}(\sqrt{d})$, 30
$\mathcal{L}_\chi(s)$	31
$s_m(r, k, \chi)$	33
$E^*(x), R(x), R^*(x)$	34
$S_m(x, \chi)$	36
$S_m^*(N/8, \chi)$	36
$\varepsilon_{m,\chi}$	36
$\mathcal{A}_{k,\chi}(R)$	37
$S_m(x, r, s, \chi)$	45
$A(t, n)$	number of positive integers $\leq t$ prime to n, 48
t, u, T, U, R, S	53, 57
$F_\pm(X)$	54
$\left(\dfrac{m}{n}\right)_4$	biquadratic symbol, 56, 57, 59
$H(d, \mathcal{D})$	63
$H_1(d), H_2(d)$	63
$\mathcal{T}_r$	set of all fundamental discriminants dividing r, 69
$B_{m,\chi}^{[d]}$	70
$K(d, \mathcal{D})$	93
$K_i(d)$ $(i = 1, 2, 3, 4, 5)$	94
$\mathcal{R}_d, \mathcal{R}_d^+, \mathcal{R}_d^-$	97
ϵ, ρ	97
$\xi(d), \eta(d)$	101
$C_1(d), C_2(d), C_3(d)$	101
$A_i(d, k)$ $(i = 1, 2, 3, 4)$	104, 106, 110
$T_i, (i = 1, 2, 3, 4)$	108
$A_5(d), A_6(d)$	110, 111
$H_k(d)$	113
$\log_p s = \log_p(s)$	p-adic logarithm, 118
$\omega = \omega_p$	Teichmüller character at p, 119
$\langle a \rangle$	119
$L_p(s, \chi)$	p-adic L-function, 119
$\dbinom{X}{n}$	120
$R_p = R_p(F)$	p-adic regulator of F, 121
$\zeta_{F,p}(s)$	p-adic zeta function, 121
$R_{2,p} = R_{2,p}(d)$	second p-adic regulator of $\mathbb{Q}(\sqrt{d})$, 122
$l_k(s) = l_{k,\infty}(s)$	kth complex multilogarithm, 122
$l_k(s) = l_{k,p}(s)$	kth p-adic multilogarithm, 122
$l_k^{(p)}(s)$	126
$\mathcal{L}_{k,\psi}(s)$	127
$\displaystyle\sum_{a=1}^{c}{}'$	sum taken over integers a prime to c, 127
$L^+(s), L^-(s), R_e^{\cdot+}$	127
$S_{k,\chi}(T)$	129
$\Lambda_{k,\psi} = \Lambda_{k,\psi}(N, \chi)$	129
γ_n	133
$\gamma_{n,e}$	134
$W_{k,e}(n)$	134
$\mathcal{L}_{k,e}(\xi)$	137

$w_\alpha = w_\alpha(\xi)$ 137

$G_e(x)$ 139

$\nu_{k,e}$ 139

$x = \{x_{k,e}\}$ 141

$y_n(x)\ (n \geq 0)$ 142

$c = c(y)$ exponent of $y = (y_n)_{n \geq 0}$, 142

$(z_n), (u_n)$ 142, 151

(t_n) 143

$c(L)$ 143

(c_n) 144

$g_{2r+\varrho}, v_{2r+1}$ 159

$L_2^{[m,\Theta]}\left(k, \chi_{ed}\omega^{1-k}\right)$ 160

$\Lambda_1(m), \Lambda(x, m, \Psi), \Lambda(x, m, \Psi, \Theta)$ 161, 164

$H(\mathcal{D}), K_2(\mathcal{D})$ 166

$\Lambda, \Lambda_{-1}, \Lambda_0, \Lambda_1, \Lambda_2, \Lambda'_{-1}, \Lambda'_1$ 166

$\varepsilon_d, \eta_d, t_0, u_0, t, u, T, U, \mu$ 171

$S(\mathcal{D}, q_1, q_2)$ 181

$\mathcal{F}$ 181

$c_\psi = c_\psi(\mathcal{D}, q_1, q_2)$ 181

$C = C(\mathcal{D}, q_1, q_2)$ 182

$\mathcal{P}_0, \mathcal{P}_1$ 182

g_P 184

c_ψ 193

$\mathcal{B}(k, e, d)$ 229